WE ARE NOT
ALONE

Walter Sullivan

BOOKS BY WALTER SULLIVAN

Landprints

Continents in Motion

Black Holes: The Edge of Space, the End of Time

Assault on the Unknown:
The International Geophysical Year

Quest for a Continent

WE ARE NOT ALONE

THE CONTINUING SEARCH FOR EXTRATERRESTRIAL INTELLIGENCE

REVISED EDITION

WALTER SULLIVAN

A DUTTON BOOK

DUTTON
Published by the Penguin Group
Penguin Books USA Inc., 375 Hudson Street,
New York, New York 10014, U.S.A.
Penguin Books Ltd, 27 Wrights Lane,
London W8 5TZ, England
Penguin Books Australia Ltd, Ringwood,
Victoria, Australia
Penguin Books Canada Ltd, 10 Alcorn Avenue,
Toronto, Ontario, Canada M4V 3B2
Penguin Books (N.Z.) Ltd, 182–190 Wairau Road,
Auckland 10, New Zealand

Penguin Books Ltd, Registered Offices:
Harmondsworth, Middlesex, England

First published by Dutton, an imprint of Dutton/Signet, a division of Penguin Books
USA Inc. Distributed in Canada by McClelland & Stewart Inc. Originally published as *We
Are Not Alone: The Search for Intelligent Life on Other Worlds* by McGraw-Hill Book Company.

First Printing (Revised Edition), July, 1993
10 9 8 7 6 5 4 3 2 1

Acknowledgments and permissions for art and photography appear on pp. 337–340.

 REGISTERED TRADEMARK—MARCA REGISTRADA

Library of Congress Cataloging-in-Publication Data
Sullivan, Walter.
    We are not alone : the continuing search for extraterrestrial
intelligence / Walter Sullivan.—Rev. ed.
        p.   cm.
    ISBN 0-525-93674-2
    1. Life on other planets.   I. Title.
QB54.S89   1993
574.999—dc20   92-45744
                                          CIP

Printed in the United States of America
Set in New Baskerville
Designed by Leonard Telesca

To those everywhere
who seek to make
"L"
a large number*

* see page xii

Contents

Preface

Early in the 1960s Philip Morrison told a scientific colloquium at NASA's Institute for Space Studies in upper Manhattan that even as we sat there, signals from civilizations on distant planets might be impinging on the earth. It was the most exhilarating presentation I have ever heard.

Phil was visiting from Cornell University, where he was a physics professor. He makes any subject exciting, but his talk that afternoon, delivered before the customary break for coffee and cookies, inspired me to pursue this subject in a book. What follows is an almost completely revised version of that book, taking into account the many critical related developments of the past thirty years.

Morrison's talk derived from a proposal he had published a few months earlier with Giuseppe Cocconi, a fellow Cornell physicist, in which they identified what they considered a uniquely rational radio frequency by which such a civilization might seek to make contact.

The credentials of both men were impeccable. Morrison had been a group leader at Los Alamos during development of the atomic bomb—and had campaigned later to have it banned. After his Cornell professorship, the Italian-born Cocconi joined CERN, the European center for nuclear research in Geneva.

"Few," they wrote, "will deny the profound importance, practical and philosophical, which the detection of interstellar communications would have. We therefore feel that a discriminating search for signals deserves a considerable effort. The probability of success is difficult to estimate but if we never search the chance of success is zero."

Their proposal, published on September 19, 1959, in the respected British journal *Nature*, created a sensation. In November 1961, the Space Science Board of the National Academy of Sciences called a meeting at the National Radio Astronomy Observatory in Green Bank, West Virginia, to discuss the subject. Two members of that board, Harold C. Urey of the University of California at La Jolla and Joshua Lederberg of Stanford University, were deeply interested in the possibility of extraterrestrial life. Both were Nobel laureates. Because of the subject's sensational nature, no public announcement was made.

The host was Otto Struve, born to a family of famous Russian astronomers and director of the observatory. Some of the others who attended were equally distinguished. J.P.T. Pearman of the Space Science Board, who helped organize the meeting, heard that one of those present, Melvin Calvin of the University of California at Berkeley, might win the Nobel prize in chemistry during the sessions. Since the observatory site, to avoid radio interference, was in a remote region far from wine stores, Pearman secreted three bottles of champagne in his car. They did, indeed, celebrate, for Calvin won the prize.

Participants were specialists in the highly diverse fields that bear on whether life elsewhere may be trying to contact us. Besides Struve, Calvin, Morrison, Cocconi and Pearman the eleven attendees included several who were to devote much of their professional careers to the problem. Frank D. Drake of the observatory had, independent of the Morrison-Cocconi proposal, already attempted an interception and was to attempt many more. Dana W. Atchley, Jr., president of Microwave Associates, Inc., had donated the amplifier that Drake used. Bernard M. Oliver, vice president for research at the Hewlett-Packard Company, maker of advanced electronic equipment, would later propose an extraordinarily ambitious monitoring project. Carl Sagan, aged twenty-seven, had won his doctorate from the University of Chicago the previous year. He had specialized on the planets and agitated for sterilization of spacecraft headed for them and the moon, lest they carry microbes that might destroy fragile life forms already there. Su-Shu Huang was a specialist in the factors that make planets habitable. John C. Lilly, a specialist in dolphins, was there to advise on communication with alien species. The conferees facetiously called themselves the Order of the Dolphin, and Calvin later sent to each a button showing the image of a dolphin on an ancient Greek coin.

On the far side of the world, astronomers in the Soviet Union were aware of the Morrison-Cocconi proposal and the Drake attempt at interception, but as one of them put it, "If I proposed anything of that sort, the government would think that I was out of my mind." Nevertheless, when the Soviet Academy of Sciences asked Iosif Samuilovich Shklovsky to write a book in its popular science series, he chose extraterrestrial life as his subject. It appeared in 1963, and Carl Sagan, after an exchange of letters with Shklovsky, enlarged it into an English-language edition titled *Intelligent Life in the Universe*, published in 1966, two years after the original version of *We Are Not Alone*. In 1967, probably thanks to Shklovsky's initiative, a Soviet edition of my book appeared in Moscow. Shklovsky, one of the USSR's most imaginative and unorthodox thinkers, is now deceased. Sagan became probably the most articulate proponent of the search for life in other worlds.

Since 1964, when the original edition of this book appeared, discoveries concerning the solar system and other planetary systems, as well as the origin of life, have been nothing less than revolutionary. Astronauts, or automated spacecraft, have landed and made observations on three celestial bodies. Spacecraft have flown past all eight planets and their moons. A new generation of far more powerful telescopes has become available to search for other worlds.

In the intervening years some have expressed doubt about the evolution of intelligent life elsewhere, despite the vastness of the universe, but in 1992 the National Academy of Sciences published a 134-page report on the coming "Decade of Astronomy and Astrophysics" by a committee of fifteen leaders in those fields. Concerning the search for extraterrestrial intelligence, the Committee said in part: "Ours is the first generation that can realistically hope to detect signals from another civilization in the galaxy." The effort, it said, "is endorsed by the committee as a scientific enterprise. Indeed, the discovery in the last decade of planetary disks, and the continuing discovery of highly complex organic molecules in the interstellar medium, lend even greater scientific support to this enterprise." The discovery of intelligent life beyond the earth, the committee added, "would have profound effects for all humanity," and it endorsed the space agency's current program for detection.

After NASA had intermittently given financial support to SETI, the Search for Extraterrestrial Intelligence, it approved an effort millions of times more powerful than any in the past. The largest antenna in the world, at Arecibo in Puerto Rico, was assigned to scan the most promising stars, while NASA's global network of deep-space antennas would survey the entire sky. The agency chose October 12, 1992, as the date on which this search would begin, exactly five centuries after Columbus reached the New World.

At the Green Bank conference in 1961 Frank Drake had presented an equation listing seven factors that would determine how many civilizations might be calling us. His formula has become standard for those concerned with SETI. Its factors provide a skeleton for much of this book:

Factor One: The rate at which stars were being formed in our galaxy (the Milky Way) during the period when the solar system was itself born.

Factor Two: The fraction of such stars with planets.

Factor Three: The number of planets in each solar system suitable for life.

Factor Four: The number of such planets on which life appears.

Factor Five: The fraction of life-bearing planets on which intelligence emerges.

Factor Six: The fraction of intelligent societies that develop the ability and desire to communicate with other worlds.

Factor Seven: The longevity of such societies in the communicative state represented by the letter **L**.

The last factor in Drake's equation took into account the rise and fall of past empires on this planet and the potential for mass destruction in an era of nuclear weapons. It was with this in mind that I dedicated the first version of this book to those "everywhere" who seek to make Factor Seven, or **L**, a large number. Today the prospects for survival of technological civilizations seem better than in 1964, but my dedication is the same.

Note to the Reader

Because this book is written for those with a minimum background in science, the arguments are presented in a manner that some specialists will find elementary. For those who wish to pursue the subjects further, references for each chapter are presented at the back of the book. In the interest of brevity, titles such as "Doctor" or "Professor" are omitted. Temperatures, except where otherwise noted, are in degrees Fahrenheit. The expression "billion" is used in the American sense of the word, meaning 1,000 million. Units of measurement are likewise American, but telescope mirrors and wavelengths are, as customary, given in meters and centimeters.

1

Spheres within Spheres

The stage for SETI—the Search for Extraterrestrial Intelligence—was set by more than twenty-five centuries of speculation regarding our world and the panoply of stars above it. Because of the limits that nature places on our vision, our view of the world until we are told otherwise is of a flat land beneath the changing vault of the heavens. So is it in our childhood and so was it for primitive man. Yet long before even our most ancient myths were born, man had begun to speculate about this environment. Such thinking was one of the essential qualities that made him man. He early saw the heavens as a rotating sphere of fixed stars across which moved the seven "wanderers": sun, moon and the five planets visible to the naked eye.

For most of human history it has been thought that the stars were fixed to some sort of sphere or were pinholes in that sphere, permitting glimpses of a universal fire that flamed beyond. Whatever the "model" of the universe, man was at the center—quite natural in view of the fact that the celestial sphere seems equidistant in all directions.

The concept of our central position has been hard a-dying. It is a mark of the climactic period in which we live that after thousands of years of sophisticated observation and speculation, it has only now perished. Some of those who finally laid it to rest (for example, by showing the true, off-center location of the sun within our galaxy) lived into the 1970s. After centuries of resistance, theologians in some of the major religions have begun to grapple with the religious implications of these discoveries—in particular the fact that not only are we not central in the scheme of things, but we may be less equipped, physically, mentally and spiritually, than more highly evolved beings else-where.

As Harold P. Robertson, professor of mathematical physics at the California Institute of Technology, has put it, early man "sat serenely in the middle, the favored onlooker if not the master of all he surveyed." Today he finds himself resident on a planet or-biting an average star in an average galaxy, confronted by the strong possibility that his intelligence—and virtue—are inferior to those in at least some other solar systems.

Almost as soon as men first began to speculate, scientifically, on the nature of the objects that they saw in the sky, some rec-ognized that there may be other worlds like our own. The first flowering of astronomical interest began toward the end of the seventh century before Christ, and the power of objective reason-ing in this area was dramatically illustrated on May 28, 585 B.C. (Julian Calendar), during a battle between the Medes and Lydi-ans. As the two armies fought, the sun was suddenly obscured, even though the sky was clear. The stunned armies ceased fighting and peace was concluded. When the ancient world learned that this eclipse had been predicted shortly beforehand by Thales of Miletus (a Mediterranean port in what is now Turkey), the study of celestial bodies took on new meaning.

Thales is said to have proposed that the stars are other worlds, but it was his pupil, Anaximander, who appears to have been the first to elaborate the idea that the number of worlds is infinite, some of them in the process of birth, some dying. The earth, cylindrical in shape, stands without need for support at the center of this universe, he said, and the sun is as large as the earth (an early hint of the true dimensions of the solar system). Like many of the prophetic concepts that came in subsequent centuries, the idea that there may be many worlds, lacking observational support, was too far ahead of its time to be widely accepted. Even the view of Pythagoras, a young contemporary of Thales and Anaximander, that the earth is spherical, while accepted by many ancient Greeks and by the Arabs, was largely extinguished until the Renaissance.

One of the difficulties was that such concepts were often at variance with the religious myths and dogmas of that time. An-other contemporary of Pythagoras, the philosopher Xenophanes of Colophon, ridiculed the anthropomorphic gods worshiped by his Greek countrymen, noting that cows, if they could make graven images, would worship bovine gods. In seeking to put man in his place he, too, postulated worlds unlimited in number, though not overlapping in time, and said the moon is inhabited. The same views were expressed by Anaximenes of Lampsacus, friend and companion of Alexander the Great, who told that warrior, to his astonishment, that, in his campaigns of the fourth century B.C., he had conquered only one of many worlds.

One of the most radical innovators in the philosophical and scientific thought of ancient Greece was Democritus, inventor of the theory that all matter is made of "atoms"—indivisible particles too small to be seen, indestructible and eternal. The earth was formed, he said, from a whirling mass of these atoms and, since space and time are infinite and the atoms are forever in motion, there must now be, and always have been, an infinite number of other worlds in various stages of growth and decay. His description of natural phenomena, written in the fourth century B.C., avoided any recourse to the supernatural.

His views, as modified by Epicurus, were magnificently set forth 350 years later by the Roman poet Lucretius, in his *De Rerum Natura* (*On the Nature of Things*). Lucretius opened with a ringing tribute to Democritus:

When human life lay groveling in all men's sight, crushed to the earth under the dead weight of superstition whose grim features loured menacingly upon mortals from the four quarters of the sky, a man of Greece was first to raise mortal eyes in defiance, first to stand erect and brave the challenge. Fables of the gods did not crush him, nor the lightning flash and the growling menace of the sky. . . . He ventured far out beyond the flaming ramparts of the world and voyaged in mind throughout infinity. Returning victorious, he proclaimed to us what can be and what cannot . . . superstition in its turn lies crushed beneath his feet, and we by his triumph are lifted level with the skies.

Lucretius' presentation of the ancient atomic theory sounds strikingly familiar to any student of modern physics and quantum mechanics. It also expresses a view that still strongly influences those today who believe life exists elsewhere than on earth, namely that no basic phenomenon in nature, including the emergence of life on a planet, occurs only once.

Granted, then [he wrote], that empty space extends without limit in every direction and that seeds innumerable in number are rushing on countless courses through an unfathomable universe under the impulse of perpetual motion, *it is in the highest degree unlikely that this earth and sky is the only one to have been created* and that all those particles of matter outside are accomplishing nothing. This follows from the fact that our world has been made by nature through the spontaneous and casual collision and the multifarious, accidental, random and purposeless congregation and coalescence of atoms whose suddenly formed combinations could serve on each occasion as the starting-point of substantial fabrics—earth and sea and sky and the races of living creatures. . . . You have the same natural force to congregate them in any place precisely as they have been congregated here. You are bound therefore to acknowledge that in other regions there are other earths and various tribes of men and breeds of beasts.

Nothing in the universe, he said, "is the only one of its kind, unique and solitary in its birth and growth." As there are countless individuals in every species of animal, from "the brutes that prowl the mountains, to the children of men, the voiceless scaly fish and all the forms of flying things," so there must be countless worlds and inhabitants thereof.

During the lifetime of Lucretius and the century that followed, this concept spread far and wide. Plutarch in his *De Facie In Orbe Lunae* (*Regarding the Face of the Moon's Disk*) discussed the habitability of the earth's great satellite—whether its lack of clouds and rain mean it is intolerably dry, and so forth. Elsewhere he wrote of other theories, including one in which the universe is triangular with sixty worlds on each side and one at each apex, making a total of 183 inhabited earths. A picture similar to that of Lucretius was presented in China by Teng Mu, a scholar of the Sung Dynasty, who wrote:

Empty space is like a kingdom, and earth and sky are no more than a single individual person in that kingdom.

Upon one tree are many fruits, and in one kingdom there are many people.

How unreasonable it would be to suppose that, besides the

earth and the sky which we can see, there are no other skies and no other earths.

This belief rested primarily on abstract reasoning and, while it was attractive to philosophers, its general acceptance was prevented until after the Renaissance by the enclosure of our world in a solid sphere, forged by the earliest astronomers. This was the firmament to which the stars seemed to be fixed. It rotated around the earth approximately once a day, traversed by the sun, moon and five visible planets. The chief problem of ancient astronomy was to explain the motions of these wanderers and thus make possible accurate predictions of their movements.

Pythagoras and his followers believed all of these bodies, including the earth, move around a central fire that is forever on the far side of the earth and hence invisible to man. Sunlight, they said, was simply a reflection of this fire. Anaxagoras then identified the sun as a huge, red-hot stone and the moon as a cool body, illuminated by sunlight like the earth, and also inhabited. The Milky Way, he said, was a portion of the celestial sphere shielded from sunlight by the earth. However, despite these advances he clung to the idea that the earth is flat.

He also ran into the bigotry that so often has beset astronomers. This was the age of Pericles, a time of glorious creativity in Athens, but also a time of embitterment by military defeat at the hands of Sparta and its allies and suspicion of "foreign" ideas. The teachings of Anaxagoras, so out of line with classic Greek religious concepts, were deemed impious and subversive. He was tried and, it is said, was to be executed when his friend Pericles intervened. Socrates, reportedly the pupil of Anaxagoras, was not so fortunate.

The efforts of the early astronomers to account for the movements of the wanderers failed because none realized that all the planets, including the one from which we gaze upon the heavens, are traveling in elliptical paths around the sun. Without this knowledge, the motions of the sun, moon and planets appear most peculiar. In fact, at times some planets seem to reverse direction and move backward against the panoply of stars.

In the early days of their astronomical studies the Greeks had only such elementary tools as trigonometry at their disposal, although by the third century B.C. they had advanced to where they could calculate the areas of complex curved surfaces. The simplest and most "perfect" curved surface was, of course, that of a sphere, and this set in motion a train of thought that was not broken for centuries. It plagued all attempts to explain the movements of celestial bodies until the time of Kepler.

Plato, the most noted pupil of Socrates, wrote that God

> made the world in the form of a globe, round as from a lathe, having its extremes in every direction equidistant from the center, the most perfect and the most like itself of all figures . . . and He made the universe a circle moving in a circle, one and solitary, yet by reason of its excellence able to converse with itself, and needing no other friendship or acquaintance.

In the *Timaeus*, a dialogue dating from about 350 B.C., Plato proposed that the Creator assigned a soul to each star and, having created intelligent, moral creatures, he "sowed some of them in the earth and some in the moon . . . ; and when he had sown them he committed to the younger gods the fashioning of their mortal bodies . . . ," but Plato's idea of a perfect universe, with the world at its center, led him to believe there must be only one world. His concept that such perfection was manifested in circles and spheres led his contemporary, Eudoxus of Cnidus, to explain the movements of the celestial bodies in terms of their "attachment" to a series of theoretical spheres. Although complex, his theory was simple compared to those that followed. The stars, as already accepted, were fixed to an outer sphere that rotated uniformly around the earth once in about twenty-four hours. However, to account for the seemingly erratic movements of a planet such as Jupiter, Eudoxus proposed that it is fixed to the equator of a rotating sphere whose poles are, in turn, fixed to the equator of another sphere, rotating in a different direction. That sphere, in turn, is twisted by the rotary motion of a third sphere, and so forth. The combination of these motions was sufficient, in a very approximate way, to account for the observed movements. All told, the theory of Eudoxus called for twenty-seven spheres: one for the stars, three each for the sun and the moon, and four each for the five known planets.

While Eudoxus considered these spheres mathematical devices, his contemporary Aristotle said they were solid, though transparent, objects. In an effort to account for the motions more precisely he proposed fifty-five of them. Another man of that time, Heraclides of Pontus, a pupil of Plato, explained the movement of the stars by saying it is the earth that turns, spinning eastward to make one revolution per day. To account for the changes in brightness of Mercury and Venus, he proposed that they circle the sun, which in turn circles the earth. Because those two planets lie nearer the sun than does the earth and orbit faster, their variations in brilliance are particularly noticeable.

Five years after the death of Heraclides a man was born on the island of Samos, birthplace of Pythagoras, who came to be known as "the ancient Copernicus," for he proposed that the sun—not the earth—is at the center of the universe. The planets and a spinning earth move in circles around the sun, he said. This man of vision, 1,800 years ahead of his time, was Aristarchus of Samos. His idea that the sun must be at the center may have arisen from his calculation that it is some 300 times bigger than the earth and 18 to 20 times more distant than the moon.

We do not know if he made any attempt to measure actual sizes and distances, but his contemporaries were making bold—and remarkably successful—attempts to calculate the size of the earth. This figure could then be applied to the geometry of eclipses to estimate actual, rather than relative, distances and sizes for sun and moon. The best known of these calculations was that made by Eratosthenes, head of the great library at Alexandria, based on observations at the summer solstice in 284 B.C. On that day, every year, he knew that the full image of the sun could be seen at the bottom of a deep well near the edge of the Libyan Desert at Syene, present site of the Aswan Dam on the Nile. Thus the sun at Syene was directly overhead, whereas the shadow cast by a stick at Alexandria showed the angular distance of the sun from the zenith to be one-fiftieth of a full circle. Since Eratosthenes believed Alexandria to be due north of Syene and about 5,000 stades distant, over the curved surface of the earth, he calculated the total circumference of the earth to be 50 × 5,000, or 250,000 stades. He later refined this to 252,000 stades. If one uses Pliny's version of the stade to which Eratosthenes referred, this works out to a circumference of 24,662 miles, compared to an actual distance, through the poles, of 24,860 miles. While the accuracy was largely luck, the calculation gave the ancient world an idea of the size of our planet. When this was applied to the calculations of distance to sun and moon, in terms of the earth's diameter, the vast size of the solar system became apparent.

The universe, as it was envisioned until the Renaissance, is shown in this illustration from Peter Apian's *Cosmographia*, published in 1539. The earth is at the center and the sun moves in a circle between the circles of Venus and Mars. Beyond the star-studded shell lies the Empirean heaven, habitation of God and all the Elect.

As the glory of Greece faded, there were these glimpses of the true scale of our corner of the universe, as well as far-reaching philosophical speculations regarding the existence of other universes or other worlds. But the celestial spheres were confining, and bold inquiry into such matters was on the decline. Furthermore, the truest concept of the solar system, that of Aristarchus, went into eclipse and its author was accused of impiety in downgrading our planet by removing it from the center of things. More serious still, Hipparchus, regarded by many as the greatest astronomer of antiquity, rejected on scientific grounds the idea of Aristarchus that the sun is central. The reason was that Aristarchus had failed to explain the observed movements of the wanderers. We know today that he could not do so because he saw the planets moving in circles instead of ellipses.

Thus this six-hundred-year period of ferment and progress in astronomy came to an end with the concept of spheres within spheres more deeply entrenched than ever. Its final great proponent was Ptolemy, who lived in Alexandria in the second century A.D. His dogma, which remained virtually unchallenged for fourteen hundred years, was an elaboration of the concentric spheres described by Aristotle, but with the addition of "epicycles." These were comparatively small circles described by planetary bodies around points that, in turn, moved in larger circles. The earth sat majestic and motionless at the center of everything. However, some of the circles were off-center, the better to explain various movements. Ptolemy, like some of the early thinkers, regarded the spheres as mathematical concepts, rather than crystalline objects. His epicycles still proved inadequate and, as observing methods improved and records of past observations became more extensive, more and more epicycles were added to overcome the difficulties. By the sixteenth century more than eighty spheres were being called upon to account for the observed motions.

Astronomy became, in fact, a most disheartening activity. It has been said that not one discovery of significance was made from the death of Ptolemy until the time of Copernicus, fourteen centuries later. The fundamental concepts were not to be challenged. A scholarly work was looked upon with suspicion if it did not pay homage, in particular, to the ideas of Aristotle. But this was a state of affairs that could not go on forever. It came to a dramatic end in the Renaissance, when men's minds were suddenly opened to the vastness of the universe and the idea that there must be many inhabited worlds was revived.

2

Science Reborn

When we think of the overthrow of the earth-centered view of the universe that occurred during the Renaissance, the name of Nicolaus Copernicus comes immediately to mind, although the great Polish astronomer who originated what we call the Copernican system clung to many of the old ideas.

Copernicus not only shifted the center of things from the earth to the sun, but, in the minds of men, he set the world spinning. At long last it began to be accepted that the celestial sphere stands still; that it is our earth that turns. To Copernicus, however, it was necessary to retain the epicycles—the movement of planets in small circles around points that travel in large circles—although his proposals halved the number of circles needed to account, in a gross way, for observed motions within the solar system. In the words of Herbert Dingle, professor of history and the philosophy of science at University College, London, "He clung as firmly as the most orthodox medieval philosopher to the machinery of spheres and to the Aristotelian principles of perfect celestial substances and uniform circular motions."

Copernicus' work on the solar system was published as he lay dying, in 1543. Less than a century later, on January 7, 1610, Galileo Galilei, having brought the telescope to bear on the heavens, discovered four of the moons of Jupiter, showing that the orbiting of small bodies around larger ones may be a widespread phenomenon. Yet even Galileo still clung to the epicycles. Meanwhile the Danish astronomer Tycho Brahe demolished the still widely held idea that concentric crystal spheres enclose the earth. He showed that the paths of comets pass unobstructed through these hypothetical shells, but he refused to accept the Copernican model of the solar system, either on religious grounds or because it seemed inadequate to explain the observed motions. He said the sun and moon circle the earth and the planets circle the sun, much as did Heraclides almost two thousand years earlier. However, the painstaking observations and calculations of the movements of Mars, made from Brahe's observatory near Prague, paved the way for his assistant, Johannes Kepler, to recognize the basic laws of orbital movement.

Kepler, like Copernicus, had one foot through the door to modern science and one foot still in the Middle Ages. He was unable to divorce astrology, prophecy and mysticism from his astronomy. His studies were interrupted in 1620 when his mother was put on trial as a witch. His interest, and that of his contemporaries, in Mars was rooted in the strangeness of its orbit, for, among the planets, its movement most stubbornly resisted description in terms of spheres and epicycles. Using Brahe's data, Kepler finally discovered why: its path is an ellipse with the sun at one of the two foci. When Kepler looked at the other planets he found that they, too, flew elliptical paths. Suddenly the need for that absurdly complex system of spheres and epicycles vanished; the solar system became beautifully simple. The fact that satellites travel in ellipses was but one of the laws discovered by Kepler. Another was that an imaginary line joining the center of a planet with the center of the sun would sweep out equal areas in equal periods of time. Thus when a planet is in the portion of its orbit nearest the sun, it must move faster than it does when it is most distant in order to sweep out the same area in the same time.

However, having stated his laws, Kepler could not explain them. That remained for the mathematical genius of Sir Isaac Newton, almost a century later. Newton's laws of gravitation, as demonstrated in the laboratory, accounted for planetary motions. This had profound implications, for it suggested that nature behaves the same everywhere. Those cold pinpoints of light that we call stars were so distant and seemingly unrelated to things on earth that they had been considered residents of another realm, governed by different physical laws.

So completely do we today take for granted the universality of nature's laws that it is difficult to appreciate the extent of this revolution in thought. Modern astronomy is founded on the premise that, for example, the spectrum of hot sodium in the laboratory is identical to the spectrum of hot sodium within the most distant star. Yet until the Renaissance the view was deeply entrenched that there are two realms: the earth with its laws and manifestations, and the heavens with completely unrelated laws and phenomena. Integration of the two into a single structure made life elsewhere seem far more reasonable.

Newton's theory of gravity also suggested to him a mechanism for star formation that was prophetic:

> If the matter was evenly disposed, throughout an infinite space [he wrote] . . . some of it would convene into one mass and some into another, so as to make an infinite number of great masses, scattered at great distances from one to another throughout all that infinite space. And thus might the sun and fixed stars be formed.

The difficulty with Newton's idea that stars are formed by "falling together" was the problem of ignition. What starts them "burning"? It was only with the twentieth-century discovery that great pressure can produce thermonuclear reactions that this question was answered.

The concept of a universe populated by countless suns like our own, as expressed by Newton in the above quotation, was an essential ingredient in restoring the idea that there must be other inhabited worlds, but it was also necessary to shatter, in men's minds, the star-studded sphere that had enclosed the universe since before the time of Plato. The one generally credited with this was Thomas Digges, an English astronomer and mathematician of the sixteenth century who probably did more than any other man to promote acceptance of the Copernican model in Britain. More important, he realized that with the stars stationary instead of circling the earth daily, they no longer had to be wedded to a sphere. From this Digges concluded that the stars were distributed through space, the dimmer ones being farther away. The heavens, he said, are "garnished with perpetvall shininge gloriovs lightes innvmerable, farr excellinge ovr sonne both in qvantitye and qvalitye." Furthermore, he said, the greatest part of the stars "rest by reason of their wonderfull distance inuisable vnto vs."

In spite of the majestic breadth that he attributed to the universe, Digges saw the sun as its center. He said that the sun, like a king, "in the miidest of al raineth and geeveth lawes of motion to ye rest."

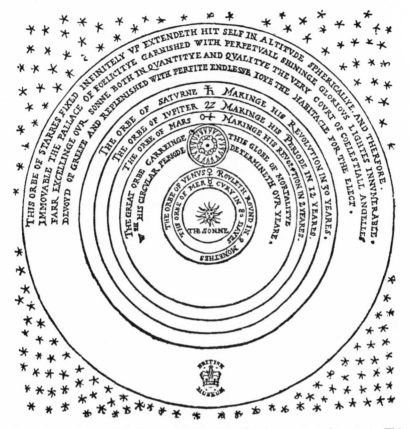

In the universe depicted by Thomas Digges in 1576 the sun is at the center. This was only a generation after Peter Apian produced his earth-centered *Cosmographia*, shown in the previous chapter. In the Digges version the moon circles the earth and the stars are distributed uniformly through space beyond the solar system.

The views expressed by Digges were carried even further, in the time of Newton, by the Dutch mathematician and physicist Christiaan Huygens. He stated that the sun is nothing more than just another star and then asked: "Why may not every one of these stars or suns have as great a retinue as our sun, of planets, with their moons, to wait upon them?" The qualities, he added, that we find on the planets of our own solar system "we must likewise grant to all those planets that surround that prodigious number of suns. They must have their plants and animals, nay and their rational creatures too, and those as great admirers, and as diligent observers of the heavens as ourselves."

Present-day champions of extraterrestrial life could hardly have stated the case better than Huygens, three hundred years ago. His scientific credentials were of the highest. Not only was he the discoverer of Saturn's rings; he was the first to use pendulums in clocks and he developed a wave theory of light that had historic importance.

Huygens anticipated one objection to his proposal: if there are planets orbiting other stars, why can we not see them? His reply was that the stars are too far away. He pointed out that they do not seem to come any nearer, or draw any farther away, in the course of the earth's annual journey around the sun, indicating that they are almost incredibly far away. Furthermore, he said, people are misled by the seeming size of the brighter stars. Thus the brightest star of the heavens, Sirius or the "Dog Star," seems to the naked eye to be as big in the sky as the planet Jupiter. When one looks at Jupiter through a telescope it can be seen as a sphere of considerable size. However, no matter how big the telescope, Sirius is still, in most telescopes, a shapeless blob of light that would appear as a pinpoint were it not for a number of blurring effects, including turbulence in the atmosphere, the response of the eye's retina and even the nature of light itself.

To illustrate his point, Huygens cited "the nature of fire and flame which may be seen at such distances, and at such small angles as all other bodies would actually disappear under." This, he added, was a thing "we need go no further than the lamps set along the streets to prove."

It was this illusion of nearness, said Huygens, that led Brahe to doubt that the earth circles the sun, for he thought that the stars would then vary in brightness. Likewise Kepler was deceived into believing that the stars are closely packed together. If they were as distant from one another as the sun is from the nearest stars, he reasoned, only a few would be visible.

Huygens was not the first man of that period to startle his contemporaries with talk of other worlds and their inhabitants. Roughly a century before he set forth his reasoning in his *Cosmotheros*, a young Dominican monk, Giordano Bruno, was denounced as a heretic to the Inquisition in Naples and fled to become a roving scholar, teaching at the great intellectual centers of Europe. He visited Elizabethan England and, because of resemblances between his poems and certain Shakespearean sonnets and plays, some believe he knew William Shakespeare.

He may also have known Digges, proponent of a boundless universe.

Bruno preached the Copernican model of the solar system with missionary zeal and went on to argue for an infinite universe with infinite worlds, his approach being more philosophical than that of Digges and Huygens. The "fixed" stars are not fixed at all, he argued, "for if we could observe the motion of each one of them, we should find that no two stars ever hold the same course at the same speed; it is but their great distance from us which preventeth us from detecting the variations. . . . There are then innumerable suns, and an infinite number of earths revolve around those suns." Anticipating Huygens by a century, he argued that these planets cannot be seen because the stars are far too distant. He likewise predicted that there may be additional planets in our own solar system, invisible because of their distance, their small size or the poor light-reflecting properties of their surfaces (since then three additional planets and a great number of smaller asteroids have, in fact, been discovered).

If one accepted the view that the universe is infinite, which Bruno believed was unavoidable, then its being peopled by a limited, and therefore "imperfect," population of intelligent beings was to Bruno incompatible with the infinite goodness or perfection attributed to God and His works. Thus, he said, "infinite perfection is far better presented in innumerable individuals than in those which are numbered and finite." He therefore concluded that there must be an infinite number of morally imperfect beings, inhabiting the infinitude of worlds.

Bruno's ideas shocked his contemporaries. The Copernican view was far from being accepted. As early as the eleventh century the concept of many worlds had been declared a heresy, and in 1616, sixteen years after Bruno's death, Rome declared the Copernican system dangerous to the faith, an action which led to Galileo's trial and humiliating recantation before the Inquisition.

Giordano Bruno has been described as a firebrand, as a romantic and as architect of "the greatest philosophical thought structure of the Renaissance." The concept of an infinitude of worlds and rational beings gave flight to his soul. He did not trim his sails to avoid making enemies. Despite the fact that he was sought by the Inquisition, in 1591 he boldly returned to Venice, which had been acting somewhat independently of Rome in such matters. Not long after his arrival he was seized and subjected to prolonged interrogation and trial. Among the charges against him was that he had praised Queen Elizabeth of England, a heretic and enemy of the Church. He replied that his praise was for her fine qualities, not for her opposition to Rome. He was subjected to what today would be called brainwashing and at one point, broken and abject, he confessed to a variety of offenses, only to relapse later to his unorthodox position—much like Savonarola and Joan of Arc.

In 1600, after six months of imprisonment in Rome, he was burned at the stake. He is said to have been defiant to the end. "I have fought, that is much," he said; "victory is in the hands of fate." As he was dying in the flames someone thrust a crucifix into his hands, but he fiercely pushed it away.

The lines of reasoning developed by Bruno, Huygens and the like ignited the minds of thinking men throughout Europe. In Milton's *Paradise Lost*, written in the mid-seventeenth century during Huygens' lifetime, the angel Raphael tantalizes Adam with the thought that the moon and planets in other solar systems may be inhabited. If there be land on the moon, he wrote, might there not be:

> Fields and inhabitants? Her spots thou seest
> As clouds, and clouds may rain, and rain produce
> Fruits in her softened soil, for some to eat
> Alloted there; and other Suns, perhaps,
> With their attendant Moons, thou wilt descry,
> Communicating male and female light—
> Which two great sexes animate the World.
> Stor'd in each Orb perhaps with some that live.

That so vast a realm as the universe should be devoid of life, serving no other purpose than to convey tiny fragments of starlight to the earth, was, the angel said, "obvious to dispute." But he urged Adam to leave such questions to God and, instead, to rejoice in his paradise and his fair Eve:

> Dream not of other worlds, what creatures there
> Live, in what state, condition or degree . . .

Milton's belief that other worlds may exist, but that we can never observe them, was reflected seventy years later in Alexander

Pope's *An Essay on Man*. However, Pope, like those today who have thought about this when direct knowledge of life beyond the earth is conceivable, saw in such elusive knowledge a means of understanding ourselves:

> Of man, what see we but his station here,
> From which to reason, or to which refer?
> Through worlds unnumber'd though the God be known,
> 'Tis ours to trace him only in our own.
> He, who through vast immensity can pierce,
> See worlds on worlds compose one universe,
> Observe how system into system runs,
> What other planets circle other suns,
> What varied Being peoples ev'ry star,
> May tell why Heav'n has made us as we are.

The despair of Milton and Pope at ever observing life on other celestial bodies was not shared by all their contemporaries. In 1638 Bishop John Wilkins, brother-in-law of Oliver Cromwell, published a work (originally anonymous) titled *The Discovery of a World in the Moone, or a Discourse tending to prove that 'tis probable there may be another Habitable World in the Planet*. In a later edition he added a *Discourse concerning the Possibility of a Passage thither*, in which he discussed the journey in terms of air flight.

Wilkins was one of those men of that period who bristled with ideas. He proposed the use of submarines (as yet uninvented) for voyages under the polar ice and helped organize weekly meetings of savants to explore the exciting avenues of scientific speculation opened to them by what they called the New Philosophy. Those who attended the talks, a number of them held in Wilkins's own quarters, included such men as Robert Boyle, famed for his law regarding the compression of gases, Sir Christopher Wren, the architect, and Samuel Pepys, the diarist. The group was finally chartered and became known as the Royal Society, one of the most distinguished associations of scientists ever formed, Newton being an early member.

This was a period of colonial expansion, and Wilkins proudly predicted that the first flag to fly on the moon would be British —in contradiction to the claim of another imaginative scientist, Johannes Kepler, that the first flag there would be German. Nor was Wilkins the only man of that time to propose going to the moon as a way to settle the question of extraterrestrial life. On the Continent a counselor to the French court, Pierre Borel, wrote that "the way whereby one can learn the pure truth concerning the plurality of worlds" is by aerial navigation (what we would call space travel). One of Borel's contemporaries went even further and told of his own voyage to the moon. This pioneer in science fiction was Cyrano de Bergerac, known today chiefly as the hero of Rostand's play about the swaggering, skillful, romantic swordsman who fought all comers to defend the honor of his elongated nose. Cyrano was a real-life figure who, after being wounded in a war with the Spanish, turned to writing satirical novels. In one of them he meets "a little man, entirely naked," who speaks to him in an utterly strange tongue and proves to be a visitor from another planet. In another tale Cyrano himself visits the moon and, for a time, is caged by one of its unscrupulous inhabitants as a freak. He is then charged with impiety for claiming that the "moon" from which he came is a world, whereas the "world" at which he had arrived is a moon. To save himself from death, he goes about among the moon people, assuring them that their moon is a world because "that is what the Council finds it proper that you believe."

Far more effective in stimulating serious thought about the possibility of far-flung life was a book written a generation later by a French scientist-satirist whom Voltaire called "the most universal genius that the age of Louis XIV has produced." His name was Bernard le Bovier de Fontenelle, nephew of the playwright Corneille, and in some respects Voltaire's predecessor as the chief gadfly of the French social scene. He helped write an opera when barely twenty and later became official historian of the French Academy of Sciences. However, the book that brought him fame throughout Europe was *Entretiens sur la pluralité des mondes* (*Conversations on the Plurality of Worlds*), published in 1686 as a series of talks with a fictitious marchioness.

His argument was that other bodies of the solar system are inhabited, although those who live on Mercury, nearest planet to the sun, "are so full of fire, that they are absolutely mad: I fancy they have not any memory at all." On the other hand, he said, the residents of Saturn, outermost of the known planets, "live very miserably . . . the sun seems to them but a little pale star, whose light and heat cannot but be very weak at so great a distance. They

say Greenland is a perfect bagnio, in comparison of this planet."
He proposed that visitors from other worlds travel by comet.

The marchioness replied with a comment applicable to many
of today's speculations: "You know all is very well," she said, "without knowing how it is so; which is a great deal of ignorance,
founded upon a very little knowledge."

Although this fantasy, half-serious in its effort to set people's
minds into new trains of thought, appeared almost a century after
Giordano Bruno died at the stake, such ideas were still considered
dangerous. In particular, Fontenelle's satire on relations between
Rome (Catholicism) and Geneva (Calvinism) led to his denounce-
ment as an atheist by Tellier, confessor to Louis XIV. According
to Voltaire, Fontenelle was saved only through the intervention of
his friend Marc-René de Paulmi, marquis of Argenson, who was
then lieutenant of the police.

Fontenelle's concept of many worlds, expressed in so readable
and popular a manner, excited people far and wide. In the mid-
eighteenth century Mikhail Vasilievich Lomonosov, the so-called
father of Russian science, read the book in its original French and
was deeply impressed. His belief in many inhabited worlds was
reflected in a number of the poems—scientific and satirical—written by this versatile man, and when the first translation of Fontenelle's book was suppressed by the Russian Church, Lomonosov
saw to it that a second translation into Russian was published. It
might even be argued that seeds thus planted in eighteenth-
century Russia bore fruit in the work of the Russian astronomer
Otto Struve after he came to America and presided over mankind's first serious attempt to reach out and make contact with
intelligent beings elsewhere.

One generation after Lomonosov's death, Tom Paine, whose
writings had touched the soul of the American Revolution, argued
eloquently for many inhabited worlds. In his tract, *The Age of Reason*, written while he was in Paris witnessing the French Revolution, he based his argument as much on religious as scientific
grounds. His view of God was derived from the wonders of creation, rather than from dogmas as expressed in the scriptures of
different faiths.

It is only in the CREATION that all our ideas and conceptions
of a *word of God* can unite. The Creation speaketh a universal
language, independently of human speech or human language,

multipled and various as they may be. It is an ever-existing original, which every man can read. It cannot be forged; it cannot be counterfeited; it cannot be lost; it cannot be altered; it cannot be suppressed. . . . It preaches to all nations and to all worlds, and this *word of God* reveals to man all that is necessary for man to know of God. . . . In fine, do we want to know what God is? Search not the book called the Scripture, which any human hand might make, but the Scripture called the Creation.

To manifest adequately the glory of God, the universe and its population of worlds had to be infinite. "If we survey our world," he wrote, "We find every part of it, the earth, the water, and the air that surrounds it, filled, and, as it were, crowded with life, down from the largest animals that we know of to the smallest insects the naked eye can behold, and from thence to others still smaller, and totally invisible without the assistance of a microscope. Since then no part of our earth is left unoccupied. Why is it to be supposed that the immensity of space is a naked void, lying in eternal waste? There is room for millions of worlds as large or larger than ours, and each of them millions of miles apart from each other."

He noted that the stars, as opposed to planets of the solar system, seem fixed in space. "The probability, therefore, is that each of these fixed stars is also a sun, round which another system of worlds or planets, though too remote for us to discover, performs its revolutions, as our system of worlds does around our central sun."

He had apparently sided with the "wrong" faction of the Revolution and hardly had he finished *The Age of Reason* and entrusted it with a friend than he was carted off to jail where, for nearly a year, he lived under threat of the guillotine.

The flights of fancy that flowered in the uninhibited minds of many eighteenth-century thinkers came up against the intellectual discipline of the New Philosophy, which demanded that every hypothesis be subjected to the test of experiment. The curse of scientific thinking before the Renaissance had been the adherence to dogma, which led to perpetual doctoring of inadequate theories—as with the epicycles—rather than to the testing of other concepts in the light of new discoveries. However, there was no means at hand to test the idea that there are intelligent beings on distant planets. In fact the very existence of planets beyond the solar system could not be demonstrated.

Furthermore, a succession of serious difficulties arose in the path of those arguing for such life. The more that was learned of the moon and planets of our own solar system, the less hospitable they appeared. The moon proved to be totally dry and airless. Mercury was far too hot and the outer planets far too cold. Apart from the earth itself, this left only Mars and Venus. Then the argument that other stars must have planets, like our sun, was dealt what seemed a fatal blow with the general acceptance, early in this century, of a new theory of the origin of the solar system. It was proposed that the planets were formed when the gravitational attraction of a passing star sucked out portions of the sun, which then cooled externally to form bodies with solid crusts.

So vast are the distances between the stars that the chances of frequent near collisions, such as the one depicted in this hypothesis, seemed negligible. As noted by the Russian scientist Aleksandr Ivanovich Oparin, if the sun were the size of an apple, on the scale of the universe the nearest star would be as far away as the distance from New York to Moscow. Hence the solar system appeared to be a freak and life on earth came to be regarded as unique. Discussion of life elsewhere was relegated to the authors of science fiction. Only with the step-by-step dissolution of the difficulties has serious discussion of the subject been resumed, much as it was after the Renaissance, and it has even become possible, within limits, to test the hypothesis in the spirit of the New Philosophy in experiments such as Drake's search for extraterrestrial signals.

The accumulation of data that has provided meat for these discussions includes evidence for a boundless, or almost boundless, universe. It also embraces a new theory for formation of the solar system, suggesting that planets are a common phenomenon, along with evidence for the existence of other such systems. And not the least has been the development of biochemistry to the point where some believe life will inevitably arise on any planet whose environment is similar to that of the earth, as well as evidence, from radio observations and meteorites, that at least some steps toward the evolution of life have occurred elsewhere.

3

Is Our Universe Unique?

In recent decades increasingly powerful eyes have been opened on the cosmos. Their mirrors can collect light from a patch of sky smaller than that covered by a pea at arm's length. For more than a generation, beginning in 1948, the giant instrument on California's Mount Palomar, with a 200-inch mirror ground into an extremely precise parabola, dominated astronomy. Then others came into operation, including a larger, but less perfect, Soviet telescope in the Caucasus and instruments not as large but very efficient in Arizona, California, astride the Andes and elsewhere. In 1991 the Keck Telescope atop Hawaii's Mauna Kea "saw first light," using a novel 30-foot-wide array of thirty-six mirrors. Europe's New Technology Telescope, to be built by the end of this century in the Chilean Andes, will be an array of four large mirrors.

Although these instruments weigh many tons, their delicate clockwork can keep them trained for hours on the same spot of the moving heavens while photographic plates soak up extremely dim light from very great distances. Ways have been devised to eliminate the twinkling that, until recently, blurred celestial images, limiting the detail that could be seen. When photographic plates from such telescopes are developed they are peppered with stars—a small sample of the billions in our galaxy. More awesome still: for each of these stars, as one looks beyond the confines of our own galaxy, there can be seen a distant galaxy with its own billions of stars.

Yet information on the core of the great spiral multitude of stars forming our own galaxy is surprisingly meager, for none of the optical telescopes can look through the intervening dust clouds. Only the radio and infrared instruments can "see" the core by analysis of emissions generated there by what is widely thought to be a "black hole"—an unimaginably dense concentration of matter inducing extremely violent manifestations in its vicinity.

The observation and reasoning that have finally relegated our sun to the status of an average star among billions (in an average galaxy among billions) began when Galileo turned his crude telescope on the Milky Way and saw that, at least in part, it was made up of myriad stars. Some proposed that the Milky Way constitutes a ring of stars around the universe, but in 1750 Thomas Wright of Durham, England, private tutor and instrument maker, published *An Original Theory or New Hypothesis of the Universe,* in which he said the stars are scattered more or less uniformly between two parallel planes, as though in a gigantic sandwich of unknown extent. The sun, he said, is midway between the two planes, so that when we look "out" perpendicular to the planes, we see a minimum number of stars, whereas if we look parallel to the planes we see the seemingly impenetrable star clouds of the Milky Way. While the sun is between the planes, it is not necessarily at the center of the whole system, he said. It is, he argued, "not only very possible, but highly probable . . . that there is as great a multiplicity of worlds, variously dispersed in different parts of the universe, as there are variegated objects in this we live upon. Now, as we have no reason to suppose that the nature of our Sun is different from that of the rest of the Stars . . . how can we, with any show of reason, imagine him to be the general center of the whole?"

Despite this reasoning, astronomers continued well into this century to assume that our sun is at or near the center of things, if only because the Milky Way seems to surround the solar system so uniformly.

Wright initiated another controversy resolved only in recent decades. He proposed that our Milky Way (or "galaxy," based on the Greek word for milk) is but one of many such systems. There are, he said, "a Plenum of Creations not unlike the known universe." This, he added, "is in some degree made evident by the many cloudy spots, just perceivable by us, as far without our starry regions in which . . . no one star or particular constituent can possibly be distinguished; those in all likelihood may be external creations, bordering upon the known one, too remote for even our telescopes to reach."

It was the great German metaphysician Immanuel Kant who seized upon this idea and made it famous in his discussion of "island universes" published only five years later, in 1755. He pointed out that, if these universes (galaxies) were disk-shaped, they would, when viewed at an oblique angle, appear elliptical. This was, in fact, the shape of many of the nebulous objects visible in telescopes of that time.

The first systematic study of our galaxy's structure was carried out by William Herschel, astronomer to George III of England, by counting stars in different parts of the sky and then seeing what model of the universe best suited his observations—a technique elaborated upon early in this century by the Dutch astronomer Jacobus Cornelis Kapteyn. In 1785 Herschel announced that the sun is near the center of a flat system of stars roughly five times wider than it is thick. He agreed at first that the nebulae, or cloudy spots, referred to by Wright are distant galaxies of the same sort. However, when his improved telescope failed to show individual stars in many of these luminous objects, he revised his view and said they were strange bodies of "shining fluid."

For more than a hundred years thereafter astronomers wavered back and forth. By the middle of the nineteenth century there were telescopes of sufficient power to resolve stars in some spiral nebulae, but at the same time the emergence of spectroscopy— the science of light analysis—opened a new era in astronomy and raised doubts with regard to the nebulae. Because the stars were too far away to be seen except as points of light, most of what was known of them was based on study of their spectra. From this we learned not only their composition, but their surface temperatures, magnetic properties, rates of rotation, rates of movement toward or away from us and duality (in some binary systems).

Today we know that there are two broad categories of nebulae: those that lie within the galaxy and, in many cases, are clouds of gas made luminous by radiation from nearby stars, and those that lie at great distance and are, in fact, other galaxies. We also know that they seem to be concentrated near the poles of our galaxy simply because dust clouds dim our vision along the plane of the galaxy. It was not until 1924 that it was finally proven, by Edwin P. Hubble of the Mount Wilson Observatory, that there are other galaxies and that our own is no more central in the universe than is our earth in the solar system. The tools with which Hubble demonstrated the true nature of the distant galaxies were the same as those that enabled Harlow Shapley, while at Mount Wilson, to show that the sun is very far from the center of our own galaxy. The tools were a remarkable class of stars known as Cepheid variables.

As early as the sixteenth century it was noticed that a star in the constellation Cepheus appears and vanishes at regular intervals. Its period—that is, the time between each reappearance—is roughly one year. Additional "variable" stars were found and it was discovered that most of them do not vanish, like a light being switched on and off, but rather dim and brighten, often with a characteristic rhythm. With one class of variables it became evident that they are "waltzing" with a dark companion. When the latter comes between the earth and the visible star, the light of the visible star is dimmed or eclipsed. However, another type dims slowly, then rises to maximum brightness rapidly, with accompanying changes in spectrum that show the star to be actually pulsing.

Because one of the most prominent of these stars appears in the constellation Cepheus, they are known as Cepheid variables (a category that in recent years has been further subdivided). The nearest and best known is Polaris, the North Star, whose pulse rate is 3.97 days. The periods of the various kinds of variable stars range from minutes to more than a year.

Toward the end of the nineteenth century the Harvard College Observatory began a systematic search for variables by comparing photographs taken of the same region of the heavens at different times. Various methods were used to speed up the process, such as laying a positive plate over a negative plate of the same star pattern, exposed at a different time, thus canceling out all stars whose brightness was unchanged. By 1910 some four thousand variables were known, three thousand of them discovered at Harvard. Data from the southern heavens was obtained at a field station operated by the university high among the Andes, at Arequipa, Peru. At the observatory in Massachusetts Henrietta S. Leavitt studied the variables appearing in photographs of the two Clouds of Magellan, taken from Arequipa. These two clouds of stars are now classed as irregular galaxies and are our own galaxy's nearest neighbors in the cosmos.

Leavitt and her coworkers listed 1,777 variables in the two clouds, and she picked for study those in the Small Magellanic Cloud for which reliable data on pulse rates and luminosity were available. In 1904 she published preliminary findings on seventeen of them, and eight years later, with firm data on eight more, she felt confident of her discovery. There is, she said, "a remarkable relation between the brightness of these variables and the lengths of their periods."

The slower the pulse rate of the star, the brighter its light, the rule being applicable both at maximum and minimum. Leavitt pointed out that the stars, clustered in a compact and extremely distant cloud, could all be considered the same distance away and therefore the differences in their brightness were intrinsic and not a result of differences in distance.

At first the discovery seemed of interest only in trying to explain why variable stars behave as they do. However, the Danish astronomer Ejnar Hertzsprung almost immediately saw the significance of these stars as a potential yardstick for the universe. Since the dimming of light by distance follows a well-established law, once the distance to one Cepheid variable was known, its intrinsic brightness could be calculated. Then, by simply timing the pulse rate of any other Cepheid, it would be possible to reckon its distance also. The trouble was that none were near enough so that their distance could be triangulated, even using the longest base line available to us: that joining opposite sides of the earth's orbit around the sun. However, a cruder method was available, namely that of analyzing rates of apparent movement against the stellar background. On the average, the faster a star changes position among the other stars the nearer it is presumed to be. In this way Hertzsprung, in 1913, obtained rough distances for the thirteen nearby Cepheids on which he had adequate data. Shapley then attacked the same problem, and the yardstick, though still imperfect, was forged.

The principle of the Cepheid yardstick can be envisioned in terms of a night sky filled with lighthouses flashing at what seem random rates. Some lights are brilliant, some so dim they can only be seen through binoculars, but their brightness cannot be used as a measure of distance, since it is known that some lighthouses contain lights of blinding brilliance, whereas others have bulbs of only a few watts. Then it is found that the rate at which each light flashes is an indication of its intrinsic brightness. All that then has to be done is to find the distance to one lighthouse in order to determine how far it is to all the others.

Having acquired the yardstick, Shapley, then at Mount Wilson, immediately used it to attack a problem that had long perplexed astronomers, namely the peculiar distribution of globular clusters. Each of these clusters resembles a beautifully symmetrical swarm of bees. They consist of several hundred thousand to a million or more stars. The puzzling feature was their concentration near the constellation Sagittarius. One-third of all those that can be seen lie there, within only 2 percent of the celestial sphere, and most of the others are not far away.

Fortunately the clusters were extremely rich in variable stars that seemed identical to those of the Magellanic Clouds. Shapley set to work with his yardstick, charting in three dimensions the true distribution of the globular clusters. He found that instead of being grouped close together, as they seemed, they are spread throughout a spherical region so large it was incredible to most of his contemporaries. Their distribution within this sphere proved to be remarkably symmetrical about the central plane of our galaxy. Of those that he studied, forty-six were above this plane and forty-seven were below it. Most startling of all, the center of this sphere—which he rightly assumed to be the center of the galaxy—lay, according to his calculations, some 50,000 light years away toward Sagittarius, one light year being the distance traveled by light in a year at 186,000 miles a second. By contrast, light takes only eight minutes to reach us from the sun. Shapley estimated that the disklike system of stars forming our galaxy is symmetrically enclosed by the globular clusters and is 300,000 light years wide and 30,000 light years thick.

His findings, published in 1918, created a sensation. His model of the galaxy was ten times bigger than the accepted model and it seemed hardly believable that the sun could be so far from the galactic center. At that time Shapley was, in his own words, "a callow youth," and notable among his critics was Heber D. Curtis of the Lick Observatory, a big name in contemporary astronomy. So heated were the discussions and so widespread the interest that the National Academy of Sciences took the unusual step of staging a public debate between Shapley and Curtis. It took place on April 26, 1920, before a packed house at the academy headquarters in Washington. Subsequent work has shown that, in principle, Shapley was right. Like others of that period, he was unaware of the extent of dust and gas that dim the light from distant stars and hence overestimated his distances. Present figures for the width of the galaxy range from 80,000 to 100,000 light years and place the thickness at some 25,000 light years.

On another question that figured in what some called the "Great Debate" at the academy—the perennial controversy over island universes—Curtis proved to be right and Shapley wrong. Shapley, on the basis of a report (later proven erroneous) that some of these objects (spiral nebulae) are drifting across the stellar background, said they are comparatively near, whereas Curtis said they are other galaxies like our own. Four years later the matter was settled once and for all when Edwin P. Hubble reported that, using the new 100-inch telescope on Mount Wilson, he had identified Cepheid variables in the Great Spiral Nebula of Andromeda, the only spiral star system near enough to be seen with the unaided eye. Cepheids, soon seen as well in other such objects, showed them to lie at distances of 1 million light years or more. Clearly they were sister galaxies to our own, and by looking at their spiral star clouds surrounding a bulbous, bright-glowing core it was possible to form an idea of the nature of our galaxy. It has been said that it is about as difficult for us to chart the Milky Way Galaxy from within it as it would be for someone standing in Central Park to draw a map of New York City.

Shapley was deeply conscious of the significance of his displacement of the sun from near the center of things. "By this move," he wrote in 1958, "we have made a long forward step in cosmic adjustment—a step that is unquestionably irreversible. We must get used to the fact that we are peripheral, that we move along with our star, the sun, in the outer part of a galaxy that is one among billions of star-rich galaxies."

Thus, by the late 1950s, we were coming back to the trains of thought that gripped men in the days of Giordano Bruno. "No field of inquiry," Shapley wrote after his retirement as head of the Harvard College Observatory, "is more fascinating than a search for humanity, or something like humanity, in the mystery-filled happy lands beyond the barriers of interstellar space."

He reasoned that no matter how scarce such lands may be, the almost infinite number of stars demands that they exist. "As far as we can tell," he wrote, "the same physical laws prevail everywhere. The same rules apply at the center of the Milky Way, in the remote galaxies, and among the stars of the solar neighborhood. In view of a common cosmic physics and chemistry, should we not also expect to find animals and plants everywhere? It seems completely reasonable; and soon we shall say that it seems inevitable.''

4

Seeking Other Suns

The quality of science that, above all else, enriches the spirits of those who pursue it is the manner in which it enables us to glimpse the orderliness of nature. One of the most thrilling experiences for the student of chemistry is first exposure to the periodic table of elements and its revelation of the systematic structure of all matter.

A comparable disclosure in astronomy grew, in large measure, out of an ambitious, laborious and seemingly pedantic project at Harvard. In 1886 a classification of stars according to their spectral type was begun. As has already been noted, since even the stars of our own galaxy are too remote to be seen except as diffuse points of light, our knowledge of them depends heavily on their spectra. It had already been found by the Italian astronomer Father Angelo Secchi that the stars fall into certain spectral categories, somewhat as the first walker in a forest might pick out oaks, maples, birches and pines. While the stellar types shade more into one another than tree species, Father Secchi found that they clearly fit into a succession of pigeonholes. In 1890 Harvard published a preliminary catalogue of 10,351 stars, classified according to their spectra, and by 1949 the entries in the catalogue had grown to 359,082, almost all of them processed by a single-minded woman named Annie J. Cannon. Cannon, who had been made extremely deaf by a childhood illness, was equipped with an almost phenomenal memory and unusual eyesight.

Attempts to fathom the significance of the star types were made by astronomers in various lands. Those at Harvard soon guessed that they reflected temperature differences, but it remained for an Indian, Meghnad Saha, to decipher the specific messages contained in the spectra of each group. The stars classed by the letter

"O" displayed the spectral lines of highly ionized helium, silicon and nitrogen and thus were the hottest. Those stars classed as type M proved to be at the cool end of the series. The types are therefore now listed in order of decreasing temperature, as follows:

O, B, A, F, G, K, M

Astronomy students have traditionally kept the sequence in mind with the phrase: "Oh, be a fine girl, kiss me!" but there is a lively debate about a less sexist alternative.

Meanwhile Henry Norris Russell at Princeton and Ejnar Hertzsprung in Europe had come upon another striking pattern in star classification. If the inherent brightness, or luminosity, of the stars is plotted on a graph against their types, arranged in order of temperature, some 90 percent of all stars lie along a narrow band that has come to be called the "main sequence." The hotter a star is, if it lies on the main sequence, the brighter it is, the larger it is and the shorter is its lifetime. The graph that displays this relationship between luminosity and star type is known as the Hertzsprung-Russell diagram (see illustration on page 32). When it was formulated on the eve of World War I, Russell viewed the main sequence as a highway along which stars travel in their long journey from birth to death. Subsequently it was recognized that the true nature of the "main sequence" was different, but it came to be, for astronomers, a testimony of order in the universe as awesome as the evidence of order on the atomic level revealed by the periodic table of elements.

The diagram is not the highway along which stars evolve from one stage to the next. Rather it is their "home," where they remain for a major portion of their lifetimes. An individual star does not change its position on the main sequence. Its location is predetermined by its mass, which in turn determines its luminosity, its temperature, its spectral type and its longevity (as illustrated in the diagram). The main sequence is the home of a star so long as its "fuel" is hydrogen, which is converted by a thermonuclear reaction into helium. The fact that this is by far the longest stage in a star's lifetime explains why 90 percent of the known stars lie on the main sequence.

A star moves onto the main sequence once its internal heat and pressure, from accretion of material, become sufficient to start the hydrogen-burning. Its stay there ends when the supply of

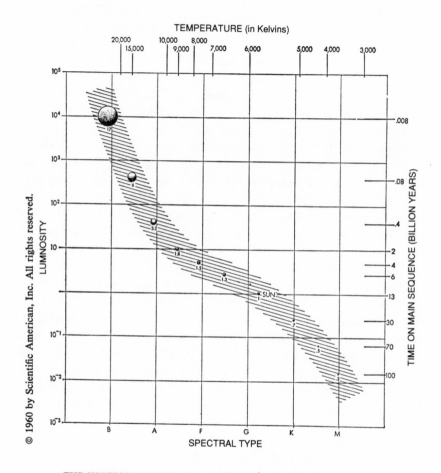

TEMPERATURE (in Kelvins)

SPECTRAL TYPE

THE HERTZSPRUNG-RUSSELL DIAGRAM (SCHEMATIC DRAWING)

When stars are classified by brightness and spectral type (temperature), most lie in a band known as the "main sequence." Their temperature is determined primarily by their mass. The heavier the star, the hotter, brighter and shorter-lived it is. Luminosity, on the left, is shown on a logarithmic scale with the brightness of our sun as one. Temperature, across the top, is in Kelvins. Star sizes are shown schematically, the figures below indicating mass, with the mass of the sun as one.

hydrogen in the core has been exhausted and, with the outflow of heat cut off, the core contracts. Hydrogen in the outer portions of the star then begins to "burn" and increased temperature and pressure in the core make possible the reactions that convert helium to carbon. The star swells into a "giant" or "supergiant," becoming progressively hotter and brighter—and, presumably, destroying any life in its vicinity.

Obviously the only period in a star's history when life could evolve on one or more of its planets would be while it was on the main sequence. Only then would its heat output be stable long enough for such evolution to occur. In orbits around which stars, then, are there likely homes for life? A man who for many years studied this question was Su-Shu Huang, a Chinese who emigrated to the United States in 1947 and obtained his doctorate at the University of Chicago. The astronomy department there was headed by Otto Struve, who was fascinated by the possibility that there might be other worlds like our own. Huang followed Struve to the University of California at Berkeley and then joined NASA.

Huang pointed out that stars with the widest habitable zones around them are the biggest, hottest ones, but their lifetime is too short for life to evolve, some lasting only a million years. Life on earth seems to have originated when the planet was a billion years old (but soon after it became habitable). The present age of the earth and the rest of the solar system is about 4.6 billion years. The smallest stars have a life expectancy of more than 100 billion years, but are so cool that the chances of a planet's orbiting within the zone suitable for life are small.

Huang likened the situation to a bonfire in a field on a cold night. If the fire is small, the zone in which people will be comfortable is narrow. If it is big, the zone will be broad, but, if the "bonfire" is a giant star, it will burn out quickly. The temperature extremes whereby he set the limits of his zones were not those that determine the habitat of human beings, but those necessary for the chemical reactions in any conceivable life process. In the solar system Mars lies within that zone, but near its outer limit, and is very cold. Venus lies near its inner border and is extremely hot for a variety of reasons. Mercury, the innermost planet, has no day-night cycle, keeping the same torrid side continuously facing the sun. The outer planets have hydrogen-dominated, frigid atmospheres.

Huang's argument left the moderately small stars—the smaller

F types, the G's and the larger K's ("fine girl kiss") as the most likely candidates. Ours is a wonderful world, for the sun, a G star, lies squarely in the most propitious section of this series and our earth is at the ideal distance from its star. As will be noted in a later chapter, simulations by George W. Wetherill of the Carnegie Institution of Washington suggest that planets so ideally placed may be common.

Another factor, in seeking suns like our own, is their chemical composition and, hence, that of their planets. It is hard to see how life could arise without an abundance of the heavier elements found on our earth, particularly carbon. Hence the first generation of stars in our galaxy are poor candidates. These "Population II" stars are thought to have been formed when our galaxy was still spherical. Motions within its gas clouds were random, but gradually they canceled one another out until there was a single, residual spin that caused the galaxy to flatten into a spiral disk. Meanwhile the first-generation stars had been making the heavier elements in their cores, some of them ending their lives explosively as novae or supernovae, enriching space around them with heavy elements. From this material the later-generation (or "Population I") stars were formed.

From the spectra of these newer stars we can tell that they are ten times richer in metals than stars of the first generation. The latter remain distributed throughout the original sphere of the galaxy, whereas the later-generation stars, including our sun, lie in the spiral arms that formed from the disk. When one looks up at night, virtually all the visible stars are of that later generation.

A curious feature of older stars is that, compared with the younger ones, few are in multiple systems. This may indicate that they have fewer planets, since double stars and solar systems seem products of roughly the same process. All of which points to the more likely existence of life within the arms of our galaxy—where we are—rather than beyond. In *Interstellar Communication*, a 1963 book he edited while at the NASA Institute for Space Studies, Alastair Cameron argued that, since there is great variety in the densities and spin rates of interstellar clouds, there should also be great variety in the distribution and sizes of planets formed from such clouds, as well as in the size of the central star. Because angular momentum tends to be greater near the center of the galaxy, he proposed that solar systems there will have smaller planets in more widely spaced orbits than in our part of the galaxy.

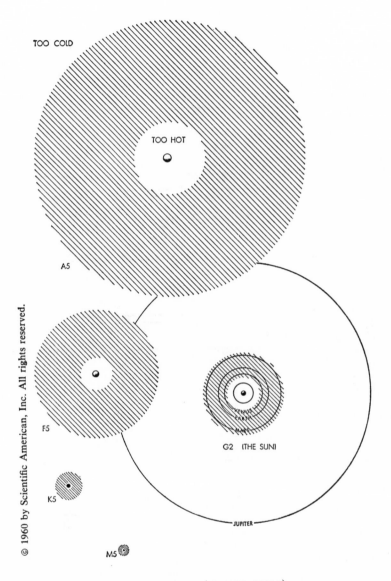

HABITABLE ZONES (SHADED AREAS)

The brightness and, hence, the warmth of a star determines the depth of the habitable zone around it. However, the very large, hot stars such as the type A5 star at the top do not live long enough to give life a chance to develop. The sun, lower right, is a G2 star, which seems "just right" for the emergence of life. The orbit of the earth lies close to the center of its habitable zone. The F5 and K5 stars lie close to the upper and lower limits of star types suitable for life according to Huang's calculations. The tiny M5 star at the bottom has a very long life but its habitable zone is so small that the likelihood of there being a planet within it is slight.

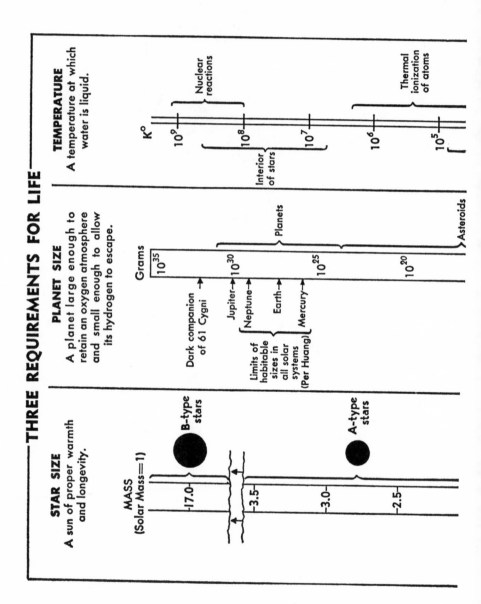

THREE REQUIREMENTS FOR LIFE

STAR SIZE
A sun of proper warmth and longevity.

PLANET SIZE
A planet large enough to retain an oxygen atmosphere and small enough to allow its hydrogen to escape.

TEMPERATURE
A temperature at which water is liquid.

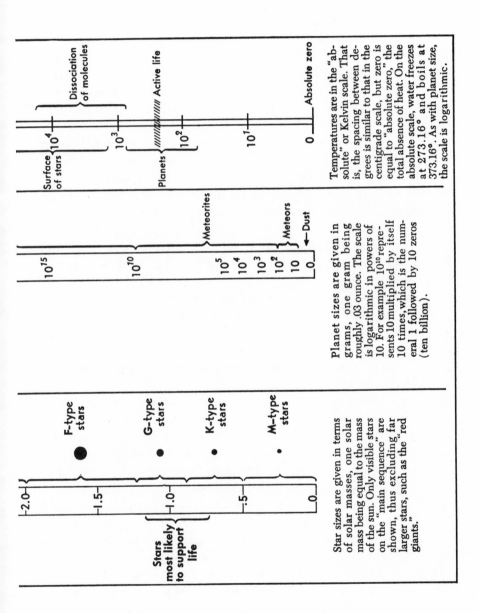

Temperatures are in the "absolute" or Kelvin scale. That is, the spacing between degrees is similar to that in the centigrade scale, but zero is equal to "absolute zero," the total absence of heat. On the absolute scale, water freezes at 273.16° and boils at 373.16°. As with planet size, the scale is logarithmic.

Planet sizes are given in grams, one gram being roughly .03 ounce. The scale is logarithmic in powers of 10. For example 10¹⁰ represents 10 multiplied by itself 10 times, which is the numeral 1 followed by 10 zeros (ten billion).

Star sizes are given in terms of solar masses, one solar mass being equal to the mass of the sun. Only visible stars on the "main sequence" are shown, thus excluding far larger stars, such as the "red giants."

Huang sought to analyze which of our neighbors, within 16 light years, might provide light and warmth to some race of intelligent beings. Our nearest neighbor, Alpha Centauri, which is 4.3 light years away, is a triple system within which habitable orbits seem unlikely. Most of our other neighbors are dwarf-like M stars that produce little heat. Huang did not disqualify them entirely. If, by good fortune, there should happen to be a suitable planet orbiting within the narrow habitable zone around such a star, it would have an extremely long period of stability in which life could emerge. "Consequently," he wrote, "living organisms there can evolve to a very high form." This might come about, in particular, since the M dwarfs are so very numerous, but he thought it unlikely among the stars of this type in our immediate vicinity.

When M stars, as well as members of multiple systems and fast-moving stars that are transients in our part of the galaxy, are all eliminated, he found that only two candidates are left, apart from our sun. Both are some eleven light years away and are dwarfs with an intrinsic brightness about one-third that of our sun. One, Epsilon Eridani, is a K-type star. The other, Tau Ceti, is of the G class, as is the sun.

In acknowledging the role of Otto Struve in stimulating his analysis, Huang said that Struve had several times pointed to Tau Ceti as the possible abode of a life-bearing planet because of its resemblance to the sun. The two astronomers were not the only ones to whom it occurred that life might exist near these two stars, as will be seen later.

Many—probably most—stars are in multiple systems (chiefly in binary, or two-star, combinations), and Huang pointed out that their planets could be intolerable to life if, under the competing gravitational tugs of more than one star, their orbits were far from circular, carrying them close enough to their sun to fry everything on the planet's surface, then swinging out so far that they became frigid. If, however, the stars were far enough apart, each could have its own family of planets unperturbed by the gravity of the other star. In the case of stars similar in size to our sun, the distance between them would have to be at least ten times the earth-sun distance. Or, if the stars were close enough to each other, a planet could orbit the pair in a manner that would keep it within the habitable zone. According to Huang's calculations the stars, in that case, would have to be separated by no more than .05 times the earth-sun distance. That a satellite can, indeed, maintain a stable orbit despite the presence of another large body is demonstrated by the regularity of the moon's monthly flight around the earth in a region where the sun's gravity holds sway.

Binaries display an immense variety in the distances between them. Some of the ones nearest us are far enough apart to be seen as two stars ("visual binaries"). An example is the brightest of all stars, Sirius, accompanied by another star from which it is separated by some forty times the earth-sun distance. The two stars take fifty years to circle one another. Other binaries are so close together that their revolutions take only a few hours. In some such cases they are detected as binaries only because the light of what appears to be a single star is periodically dimmed. This occurs when the two stars are in line, as seen from the earth, with one eclipsing the light of the other ("eclipsing binaries"). Others ("spectroscopic binaries") can be observed only because of periodic splitting of the lines in their spectra. When the two stars (which appear as one in the telescope) are moving transversely, as seen from the earth, none of their orbital motion is toward or away from us and their spectrum is normal. However, later in their orbits, when one star is approaching and the other star receding, the wavelengths of light from each star are shifted in opposite directions.

The probability that single stars or multiple-star systems might have planets seemed closely related to the star's spin rate, and this became a preoccupation of Otto Struve, who, having fought with the Whites in the Russian Revolution, had fled to Turkey and then the United States. By mail he and Grigory Abramovich Shajn, who had remained in the Soviet Union, did an analysis of stellar rotation which was published in 1929 in the *Monthly Notices* of Britain's Royal Astronomical Society, although they did not meet face-to-face until years later. Both astronomers suspected that spectral lines in the light from certain stars were broad because they were spinning very rapidly.

Their suspicions were rooted in the proposal, made in 1842 by the Austrian physicist Christian Johann Doppler, that light from a moving source, such as a star (or part of a star), would be altered in wavelength by relative motion toward or away from the observer. In today's world of fast-moving vehicles this "Doppler effect" is well known, but it was a major discovery in Doppler's time. The principle is applicable to sound waves in that they are compressed, and therefore raised in pitch, if their source is approaching. If it is receding, the waves are lengthened and the pitch lowered. This is vividly evident as a horn-blowing vehicle races past. In the case of light, the shortening of wavelength shifts lines of the spectrum toward the violet, whereas motion away, and consequent lengthening of the waves, causes a shift in the opposite direction, toward the red end of the spectrum.

In 1909 this principle was used by Frank Schlesinger of Yale University to show that certain stars are spinning. If the spin axis of a star is perpendicular to the line of sight from the viewer, then, Schlesinger reasoned, spectral lines in light from the side of the star moving toward him would be shifted in one direction and light from the opposite side would be shifted in the reverse direction. The effect would be to fatten spectral lines in combined light from the whole star.

It also became evident that the orientation, in space, of the spin axes of stars was random. Thus, if there was no broadening of spectral lines, it did not necessarily mean a star was not spinning. It was possible one pole of its axis was pointing directly at the earth so that its spin produced no relative motion toward or away from the observer. A statistical analysis of spectra from a great number of stars was therefore necessary to determine whether there was any relationship between spin rate and the star's other characteristics. It was found that the bigger, hotter stars characteristically rotate fast and thus have considerable angular momentum; the cooler stars are turning so slowly that their rotation speeds are hard to detect. What Struve found "startling" was the abruptness of the cleavage between fast and slow stars. It occurs in the middle of the F-type stars. The cooler ones in that category are slow; the hotter ones are fast. Some of the O and B stars, at the hot end of the catalogue, rotate as fast as 340 miles per second at their equator, completing each revolution in hours or less. It has been deduced that 340 miles per second is the speed limit beyond which centrifugal forces exceed the star's gravity and it begins throwing off material.

Our sun, a G-type star, spins at only one mile per second, a point on its equator taking twenty-five days to make one revolution. This seems typical of stars in this group and among its neighboring types, the K and M stars, as opposed to the big, fast-spinning stars. This was puzzling, for if all the nearby stars, big and small, had condensed from the same cloud of dust and gas in our part of the galaxy, as widely suspected, they should all have similar angular momentum. The galactic clouds themselves spin very slowly—perhaps only once for each orbit around the galaxy, which at our distance from the core takes about 200 million years. However, in the cloud the local spin rate increases as a star condenses.

In an oft-cited analogy, figure skaters can slow their spin by spreading their arms. Conversely, skaters can speed their spin up by bringing their arms close to their sides. To keep angular momentum the same, their spin rates must increase. Likewise, if a dust cloud many light years broad takes millions of years for a single revolution, that part of it concentrated into a star must spin every few hours or months.

A long-standing puzzle was why almost all the angular momentum of the solar system is concentrated in the planets, rather than the sun. If they all fell into the sun, they would add only 0.1 percent to the mass of our star but increase its spin a hundredfold. Struve concluded that at least part of the "missing" angular momentum in the K, G, M and cooler F stars must have been transferred to planets. There may, he said, be some "fundamental difference" between the way in which these slow-spinning stars were formed and that of the other, fast-spinning types. The "slow" stars of our galaxy number in the billions. Struve, in effect, proposed that all are centers of solar systems.

Among other proposals was one by Alastair Cameron, then at NASA's Institute for Space Studies in New York City, that the sun, for its first few thousand years, was more than a hundred times brighter than today and ejected gas at a far greater rate than the "solar wind" now blowing past the earth. This outrushing gas, he said, may have carried off much of the sun's angular momentum. In Sweden Hannes Alfvén, who shared the 1970 Nobel prize for his seminal work in plasma physics, proposed that, over a long period, magnetic interaction between the sun and a cloud of electrically charged particles surrounding it could have produced the necessary braking effect.

That the most massive stars are, in general, the fastest spinners was confirmed by an accumulation of data on several thousand of them by Arne Slettebak of Ohio State University and others. Were the smaller stars rotating slowly because they had shed material that became planets? Was the existence of planets related in some way to multiple-star formation? In hope of answering such questions, Helmut A. Abt of the Kitt Peak National Observatory in

Arizona and his associate Saul Levy aimed the observatory's giant 84-inch reflector at the 123 sunlike stars in the northern sky visible to the naked eye. In a 1977 issue of *Scientific American* Abt termed it reasonable to suppose that, in surface temperature, diameter and life history, "millions of stars are essentially identical with the sun." Abt, who had obtained his doctorate from the California Institute of Technology, was a respected astronomer and managing editor of the *Astrophysical Journal.* He knew Kitt Peak well, having helped choose it as an ideal observing site, overlooking Mexico and high above the Papago Indian Reservation. Abt had worked at the Yerkes Observatory soon after Struve, a champion of the search for other worlds, had been its director.

In his survey each star was allocated twenty observations of its spectrum on randomly chosen nights. He feared that a uniform timetable might conceal cyclic changes in the spectrum caused by its orbit around another star. Of the 123 stars 46 were such "spectroscopic binaries" (only 21 of them previously recognized as such). A total of 88 stars were found to have some sort of companion, of whom 57 had 1, 11 had 2, and 3 had 3, forming assemblages of 4 stars. Each of these quartets were apparently two close binaries orbiting each other. While no evidence was found that anything was orbiting 35 of the 123 stars, Abt assumed 1 or 2 had been missed. He suspected that at least one of the seeming singles might be gravitationally bound to a very distant star, circling it once in as much as a million years. The sun does not appear to be circling another star, Abt wrote, "although the possibility has not been completely eliminated that a faint companion star exists far outside the edge of the solar system." Six years later, after evidence had been found that one or more large objects struck the earth 65 million years ago, perhaps annihilating the dinosaurs and many other species, a group of astronomers in California proposed that such a star may exist. At long intervals, they said, it may come close enough to disrupt the cloud of comets thought to be circling the solar system, sending one or more on a collision course with the earth. The star was appropriately named "Nemesis."

By this time it was widely assumed that large families of stars form in a single, slowly rotating cloud of dust and gas. As denser parts of the cloud condense into stars, their spin becomes intense, finally becoming so extreme that the stars can no longer hold together, splitting into one or more pieces. Where two stars form, wrote Abt, 99 percent of the angular momentum of the parent star is converted into orbital motion of the two stars. Their individual spinning accounts for only about 1 percent. This at least would help explain why sunlike stars are such slow rotators.

It is also believed that many dust-cloud cores never become massive enough to be stars. In 1962 Shiv S. Kumar, an Indian-born astronomer at the NASA institute in New York, calculated that the heat and pressure within the core of a condensed object with less than one-sixteenth the mass of the sun would be insufficient to fuse hydrogen into helium. It could not "ignite" as a star. As gravitational energy from its contraction is released, it may glow, but eventually even that will cease and it will become an invisible "black dwarf." Abt concluded that some of the invisible companions in his survey were either black dwarfs or planets.

Stars like the sun, with no evident companion, he said, could be "escapees" from clusters that have gradually dispersed. He echoed the view that searching for planets should focus on such single stars, and, he added, "one may be allowed to wonder about the conditions on those planets and to speculate about whether radio signals from the earth are likely to raise a listener and perhaps even a speaker at the other end."

The concept of black dwarfs had a long lineage. In 1962 a discussion of the idea by that particularly imaginative astronomer, Harlow Shapley, was published by the National University of Tucumán in Argentina. He pointed out that our theories of planet and star formation allow for the accretion of all sizes of body below a certain ceiling. No star can apparently survive that is more than about seventy times as heavy as our sun because the expansive pressure of radiation from within it would blow the star apart. As noted earlier, big stars are scarce, medium-sized stars like the sun are more numerous, and tiny stars are very plentiful. Hence the bodies too small to shine as stars must be even more numerous. "To me," Shapley wrote, "they seem inevitable."

Where are these objects, now known as "brown dwarfs"? Everywhere, Shapley said. They may be more numerous in the spiral arms of the galaxy or in the great nebulosities than in the "stellar deserts," he added. There is apparently none near enough the solar system for its gravity to have an observable effect on the outer planets. "Myriads" of these bodies, Shapley continued, "are not orbitally obedient to any star—these in addition to the planets of all sizes that are immediately subservient to stars."

But how could that "chemical delicacy" which we call life arise and survive on such bodies in the absence of sunlight? Shapley pointed out that Jupiter, being five times farther from the sun than is the earth, gets little heat from our star, but it is so massive that the heat output of its interior must be considerable. The heat within the earth is generated chiefly by the radioactive decay of materials in its interior, but gravitational contraction also is a contributor. A planet many times the size of Jupiter would generate so much heat that its surface rocks would be molten, but, said Shapley, there must be a critical size whose surface temperature is such that its crust will be solid and water will exist in liquid form. The size must be at least ten times that of Jupiter, he added.

Such a body would not enjoy the solar radiation that seems to have played an important role in stimulating the chemical evolution that led to the emergence of life. However, the lightning discharges that are also thought to have figured in this process would be present, as would an abundant outflow of energy. This internally produced energy is manifest on earth chiefly in volcanoes, hot springs and earthquakes. On so large a body it would be far more evident. The surface would glow brightly in the deep infrared, and this might be exploited by organisms, both for vision and for photosynthesis—the manufacture of carbohydrates by plants. Photosynthesis as we know it would be impossible in the absence of sunlight.

Thus, scattered through the universe, said Shapley, there must be countless bodies of this sort on which the conditions are suitable. "Then something momentous can and undoubtedly does occur," he said. "The imagination boggles at the possibilities of self-heating planets that do not depend, as we do, on the inefficient process of getting our warmth through radiation from a hot source, the sun, millions of miles away. What a strange biology might develop in the absence of the violet-to-red radiation!"

One difficulty, as Shapley recognized, is the enormous gravitational force on the surface of such a body. Presumably few if any earth creatures could endure it. Perhaps only oceanic life could evolve.

Ultimately, he suggested, we may find ways to detect such objects in our corner of the galaxy. Photographic emulsions very sensitive in the deep infrared might provide one method, or the use of electronic image-intensifiers of extraordinary sensitivity, such as are already being used in astronomical work. Also, "in a century or so," he said, star-measuring methods may have advanced to the point where it will be possible to detect the long-range gravitational effect of such bodies on neighboring stars. However, the use of large radio telescopes seemed the best hope, in Shapley's view, since such bodies should emit considerable radiation in the radio part of the spectrum (in wavelengths measured in centimeters and millimeters).

Thus, Shapley, who died in 1962, argued that the nearest life beyond the solar system may not be on a planet orbiting a star, but on one of these lonely bodies. Ultimately its presence may be revealed through its production of heat-generated radio waves or, he said (hinting at the possibility of intelligent life), be "otherwise made manifest."

5

The Solar System: Exception or Rule?

The question of whether or not we are alone in the universe depends critically on the manner in which our planet and the solar system of which it is a part were formed. Running through the long debate on this subject have been two rival theories: either the solar system was a freak, born in a highly unusual, if not improbable, encounter between our sun and another star, or the sun and its planets condensed from a spinning nebula, or cloud, of dust and gas. In the first hypothesis the likelihood of other worlds would be small. In the latter the probability could be very large.

The final decades of the twentieth century have produced so much direct information on the moon, planets, meteorites and comets, as well as newly forming stars, that virtually all astronomers are convinced the answer is nebular. Spacecraft have flown past the eight other planets, photographing from close up their many moons as well as their rings, such as the complex ones surrounding Saturn that form a model of the original nebula. The pictures have shown the dense, overlapping and often huge craters left by massive bombardment during final stages of the formative process. Since the first discovery there by Japanese scientists in 1969, the ice of Antarctica has proved an amazingly efficient collector of meteorites that have fallen on that continent over the millennia. Where the flow of ice is blocked by mountains and becomes stagnant, erosion of the ice uncovers layer upon layer, exposing meteorites that fell on the ice over thousands of years. This has more than tripled the number of specimens in the world's collections, providing chunks apparently knocked off the moon and Mars. The meteorites have also provided what are thought to be samples of the nebula from which the planets were formed. American astronauts and automated Soviet craft have brought back samples from many parts of the moon. Soviet craft have landed on Venus and, despite that planet's thick cloud cover, a radar carried around it by an orbiting spacecraft has mapped virtually all of Venus. In the most virtuoso performance of all, two unmanned American spacecraft have landed on Mars and conducted analyses of its soil.

American, Soviet, European and Japanese probes flew past Halley's comet as it swept through the inner solar system in 1986, obtaining a wealth of data on its composition and that of other such relics of the early solar system. The Infrared Astronomical Satellite (IRAS), scanning the heavens at wavelengths that cannot penetrate the earth's atmosphere, has peered inside the vast dark clouds of our galaxy, recording the formation of new stars and planetary systems.

For the past two and a half centuries the debate on the origin of the sun and planets has engaged many leading scientists, and as one side or the other has predominated, so the belief in the uniqueness of our earth has waxed and waned. In 1745 the French naturalist George-Louis Leclerc, compte de Buffon, proposed that a comet, which he apparently viewed as comparable to a star, struck the sun and knocked off lumps that became the planets. Since the blow was off-center, the sun was set to spinning and the planets not only were cast into orbits around the sun but spun on their own axes. In fact, some spun so fast they threw off additional material, which became moons.

To evaluate the likelihood of such a chance encounter one must be mindful of the distance between our sun and other stars. As pointed out by the Russian A. I. Oparin, if one shrank cosmological distances so that the sun was no bigger than an apple, the distance to the nearest star, on the same scale, would equal that between New York and Moscow. Under such circumstances, the chances of a close encounter by our sun with another star would be so negligible that there might be no other solar systems in the rotating congregation of billions of stars forming our galaxy.

A view close to that of today was set forth only ten years after Buffon's proposal by Immanuel Kant, the great German philosopher, who, inspired by Newton's newly formulated laws of gravity, wrote a book on the formation of the solar system. He envisaged a primordial universe of gas that condensed into blobs of higher density. One mighty blob became the solar system. First, because of its spin, it flattened into a disk whose core, attracted to itself by gravity, "fell" together to form the sun. Other cores formed nearby and became planets. In this manner not one but innumerable solar systems would be formed.

Kant's proposal was, to some extent, anticipated by others. More than a century earlier the French philosopher and mathematician René Descartes suggested that the solar system formed from a cloud of primordial matter, although he made no reference to the role of gravity, Newton's theory not having yet been promulgated. The natural and typical behavior of all matter, Descartes said, was to move in a vortex, like water on the surface of a whirlpool—a theory reminiscent of Democritus' whirling mass of atoms, devised two thousand years earlier. In the beginning, said Descartes, the stuff of the solar system moved in a series of vortices that rubbed against one another, dislodging material that settled to the center of each vortex. The sun thus formed in the center of the large, central vortex and the planets and moon formed within secondary vortices.

In Kant's infancy the Swedish scientist and mystic Emanuel Swedenborg formulated a somewhat similar concept, but like many of Swedenborg's advanced scientific ideas, it received little notice. On the other hand, in 1796, when Kant was in his seventies, Pierre-Simon, marquis de Laplace, published the version of the nebular theory that remained a favorite for many generations and has much in common with today's view. His starting point, like Kant's, was a nebulous cloud circling the sun. As it contracted, he proposed, it gained speed, as would any spinning body, and flattened into a disk that segregated into rings. These ultimately condensed into planets. The same was true of the moons, he said, citing the rings of Saturn as an early stage of such a process. He noted the amazing regularity of the solar system: "The movements of the planets and the satellites are almost circular," he wrote, "and directed in the same direction and in planes that are little different." He pointed out that their rotation was also uniform. Looking beyond the solar system, he said, "Unnumbered suns that may be the seats of as many planetary systems are scattered through the immensity of space."

By then Johann Daniel Tietz (Titius) and Johann Elert Bode of Germany had independently pointed out a striking spacing of planetary orbits that seemed to offer a clue as to how the individual planets were formed. The pattern, known as Bode's law, can be stated thus: the distances between orbits double for each step outward from the sun. Thus, if the distance between the orbits of Mercury and Venus, the innermost planets, is 1, then the Venus-earth spacing is 2 and that between the orbits of earth and Mars is 4. Beyond Mars the pattern is less clear. Where there should be another planet there is, instead, a belt of asteroids whose combined mass is much too small to be the relics of a planet. Beyond the asteroid belt is Jupiter, by far the largest of the planets, and it is now suspected that Jupiter's very powerful gravity threw much of the asteroidal material out of the solar system before it could coagulate into a planet. The space program has extensively used this crack-the-whip ability of a planet's gravity to accelerate a passing body, hurling a spacecraft (with high precision) from one planet to another. While Bode's "law" does not strictly apply to the outer planets, it affects the orbits of many of their moons.

Bode thought his principle applied in other ways. Like many of his contemporaries, he suspected the other planets were inhabited and tried to apply to those worlds a counterpart of his law. It was early recognized that the outer planets are much more lightweight than the inner, "terrestrial," planets (Mercury, Venus, earth and Mars). Just as the planets become lighter at greater distances from the sun, their inhabitants, Bode thought, follow the same pattern, becoming progressively more spiritual as their distances increase. Kant himself regarded the inhabitants of Venus and Mercury as being too underdeveloped morally to be accountable for their own acts, whereas those of Jupiter, he believed, lived in a superior state of happiness and perfection.

Swedenborg said that in a series of dreams that he (and his followers today) regarded as divine revelations, he was visited by spirits from other planets who described to him countless inhabited worlds. He was told that there were two races living on Venus, one gentle and humane, the other savage and cruel. Those inhabiting Mars, on the other hand, were the finest residents of the solar system, resembling in piety the early Christians.

The more solid postulation of Bode and Titius regarding the spacing of planets harked back to the idea of the ancient Greeks that the arrangement of the heavenly bodies was related to the spacing of notes in a musical chord. Even Kepler, discoverer of the laws of orbital motion, had, with his strange admixture of mysticism and mathematics, a somewhat similar view.

A further rule of the solar system that seemed a clue to its formation was recognized in the mid-nineteenth century by Edouard Roche of France, who wondered why the rings of Saturn had never condensed into moons. The answer, he calculated, was the tidal stresses produced by the gravity of a parent body in objects orbiting it at close distances. For example, the gravity of the sun and moon are enough to produce our ocean tides and also to contort slightly the shape of the earth itself. Twice a day, for example, Manhattan Island, in terms of its distance from the center of the earth, rises and falls a few inches, but the gravity of our planet is strong enough to keep such stresses from tearing it apart.

Roche calculated that a satellite with the same density as its parent body would not have strong enough gravity to resist such stresses if the radius of its orbit was less than 2.44 times the radius of its parent. The outside edge of Saturn's rings is at 2.30 times the planet's radius—inside this "Roche limit"—whereas the innermost of its moons visible from earth is at 2.44 radii, outside the limit. Hence, Roche said, the rings consist of material that is too close to Saturn to condense into a moon. Mercury, despite its closeness to the sun, lies well outside the sun's Roche limit.

It was early recognized that such features as the Roche limit and Bode's law should help explain how the sun and planets formed, as should other elements of the solar system, including those cited by Laplace, such as the planetary orbits, which are nearly circular and lie roughly in the plane of the sun's equator. The exception is Pluto, the outermost planet, which may actually have once been a moon of Neptune. They all circle and spin in the direction of the sun's own rotation (apart from Venus, which spins very slowly in the opposite direction, and Uranus, whose spin axis is tilted until it lies almost parallel to the plane of its orbit around the sun).

Thanks to the planetary probes, more than forty moons have been discovered, and those which were presumably by-products of the planet's own formation (as opposed to being captured asteroids or comets) follow the same traffic pattern. When, in 1898, it was found that Phoebe, an outer moon of Saturn, violated the rule, Clarence Day, newly graduated from Yale University (and later to write *Life with Father*), penned an ode:

Phoebe, Phoebe, whirling high
In our neatly plotted sky,
Listen, Phoebe, to my lay:
Won't you whirl the other way?

All the other stars are good
And revolve the way they should.
You alone, of that bright throng,
Will persist in going wrong.

Never mind what God has said
We have made a law instead.
Have you never heard of this
Neb-u-lar Hy-poth-e-sis?

It prescribes, in terms exact
Just how every star should act
Tells each little satellite
Where to go and whirl at night.

Disobedience incurs
Anger of astronomers,
Who—you mustn't think it odd—
Are more finicky than God.

So, my dear, you'd better change.
Really, we can't rearrange
Every chart from Mars to Hebe
Just to fit a chit like Phoebe.

Since then many such wrong-way moons have been discovered.
All are believed to have originated elsewhere and then to have
been captured by a planet's gravity.

In the second half of the nineteenth century the nebular theory
ran into trouble. James Clerk Maxwell of Scotland, who first de-
fined the relationship among light, electricity and magnetism, calculated that if all the material of the planets were spread uniformly
through the solar system as a nebula, there would not—by a long
shot—be enough gravitational attraction between the particles to
draw them together and hold them there in face of the dismembering influence of the sun's gravity. He persuaded the scientific
world that the planets could not have condensed from a nebula,
as proposed by Kant, nor from the rings of Laplace. The problem

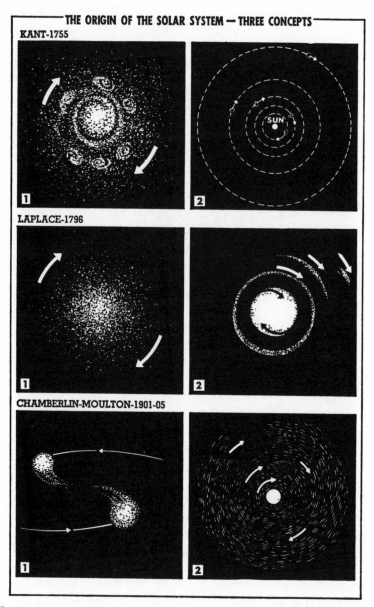

While many proposals have been made regarding formation of the solar system, most of them are related to one of the three basic concepts shown here. In Kant's view the primordial material, drawn to itself by gravity, formed into whirling blobs. The central blob became the sun, surrounded by residual blobs that formed into planets. In the Laplace hypothesis, the remaining material first formed rings, each of which then condensed into a planet. The Chamberlin-Moulton proposal was that gravitational attraction, as another star passed close to the sun, drew forth the material that then formed into planets.

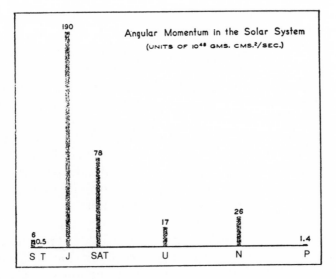

190

Angular Momentum in the Solar System

(UNITS OF 10⁴⁸ GMS. CMS.²/SEC.)

78

26

17

6
⏌0.5

1.4

S T   J   SAT   U   N   P

This diagram shows that within the solar system Jupiter, "J," carries by far the largest proportion of the angular momentum—much more than the sun. The four inner, or "terrestrial" planets, carry so little that their combined angular momentum is shown as "T." The sun, "S," is at left.

was, in fact, not resolved until near the close of the twentieth century, when new ways of peering into space were employed. Older textbooks still propose the collision theory, a comfort to those who would like to think that we and our world are unique.

A persistent difficulty with the nebular hypothesis was the peculiar distribution of angular momentum between the sun and planets. Angular momentum is what keeps a flywheel spinning. Although the major planets together constitute less than 1/700th of the total mass of the solar system, they carry 98 percent of its angular momentum. The sun, with almost all of the mass in the system, spins so slowly (about once a month) that it carries only 2 percent of the system's momentum. It was reasoned that if the sun and planets all condensed out of the same primeval cloud, the sun should spin once in every ten hours or less.

Hence, early in this century, there was a return to the collision idea of Buffon, with attempts to explain how the sun's encounter with another star could have imparted angular momentum to the planets, but not the sun. In the 1920s Thomas C. Chamberlin and Forest R. Moulton at the University of Chicago proposed eruptions of solar material that then assembled into planets. They thought the spiral nebulae seen in telescopes (not yet identified as distant galaxies) were other solar systems forming. Sir James Jeans and Sir Harold Jeffreys in England suggested that the encounter drew out a cigar-shaped filament of solar material whose fat central section produced the largest planets—Jupiter and Saturn—and whose ends became the inner- and outermost planets.

Other freak events were proposed. At Princeton University Henry Norris Russell, whose classification of the stars became a pillar of modern astronomy and who was a specialist in multiple-star combinations, proposed in the 1930s that the solar system was born of a binary, or pairing, of two stars, orbiting each other like waltzers. One of them, he said, was then destroyed in a close encounter with a third star and the debris formed into planets with an excess of angular momentum. In England Sir Fred Hoyle proposed in 1960 that the debris originated when one of the stars in a binary blew up as a supernova—the death throes of a star in which it may briefly become brighter than all the stars of a galaxy.

Nevertheless, objections were raised against these ideas, as they had been against the nebular theories of Kant and Laplace, and new variations of the nebular hypothesis appeared. The Russian mathematician and polar explorer Otto J. Schmidt said in the 1940s that the planets formed from material captured by the sun from the clouds of dust and gas, seen as shadow areas against the backdrop of stars, particularly in the mighty path of the Milky Way. The spacing of planets, he said, was the natural result of competition for material scattered rather uniformly through space. He founded a Soviet school that explained the densities of the planets in terms of the sun's heat. This would have determined what substances condensed at given distances from the sun. Thus the inner planets are heavy, whereas the outer ones contain both light and heavy materials.

A continuing problem was what made the cloud particles stick together in the first place. Among those who attacked it was Gerard P. Kuiper, a Dutch-born astronomer who became director of the Yerkes Observatory in Wisconsin. He said in 1949 that turbulence in the nebula was sufficient to assemble lumps of material big enough for their own gravity to begin consolidation. Indeed, it is now believed that turbulence in the nebula was fierce.

One of the most active theorists on origin of the solar system was Harold C. Urey, whose discovery of heavy water (deuterium oxide) won him a Nobel prize in 1934. Beginning in the 1950s, while at the University of Chicago, he evolved a concept in which objects roughly the size of the moon formed from the dust and gas of the nebula. With the solar system full of such bodies, collisions were bound to occur, and the region became cluttered with debris similar in composition to the meteorites, including those, like meteorites, made chiefly of iron from the cores of the dismembered bodies. The pressure of light and outflowing gas from the sun swept lightweight material from the inner solar system, according to the Urey model, so the debris that fell together to form the inner planets was much heavier than that of the outer planets. The moon, Urey believed, was a lonely remnant of the earlier stage, formed from the dust and gas of the nebula, which would account for its lightweight composition.

Not until astronomers became able to scan the heavens at wavelengths invisible to the eye did elements resolving the long debate on origin of the sun and planets begin to appear.

6

Seeing Infrared

"In the past," Eugene H. Levy of the University of Arizona said in his presentation at the start of a 1985 meeting on the birth of the sun and planets, "conferences on topics such as this one often began with a review of several diverse, competing theories for the origin of the solar system." Today, he said, "there exist no competing theories. . . . There is remarkable agreement about the general character of the events and the physical processes which produced this sun and these planets. Moreover, the events and processes themselves are thought to be unremarkable." Planets suitable for life, he said, "occur widely, in numbers which, while uncertain, are sure to be large." He admitted to "an emotional attachment," but envisaged, on those planets, the evolution of life as "almost inevitable: advanced states of sentient, mobile, and manipulative beings, characterized by opposing thumbs and introspective temperaments, disposed to contemplate the events which led to their own existence."

The observations responsible for this optimism had their birth, in part, with an experiment conducted in 1800 by Sir William Herschel, a musician-turned-astronomer who, among other things, discovered the planet Uranus. To determine the amount of energy in the different wavelengths, or colors, of the solar spectrum he used a device that split sunlight into its rainbow colors and measured the effect of each on a thermometer. When he got past red, at the longest visible wavelength, there was no cutoff. A form of light invisible to the eye was heating his thermometer. He had discovered the infrared.

Later in that century a little-known achievement of Thomas Alva Edison, the great inventor of electrical appliances, was the development of an infrared sensor with which, from a henhouse in Rawlins, Wyoming Territory, he observed an 1878 solar eclipse. He also detected infrared radiation from Arcturus, a bright orange star to the naked eye, and suspected that there are many stars visible only in the infrared. But it was not until the middle of the twentieth century that systematic infrared searching of the heavens began, thanks to those, like Frank J. Low, who devised new ways to observe celestial infrared from the ground, and Gerard P. Kuiper, by then, like Low, at the University of Arizona. In 1965 Kuiper mounted an infrared telescope in an aircraft, NASA's Convair 990, to lift it above most of the atmosphere, since many of the infrared wavelengths from space are absorbed by air. A decade later a C-141 transport, christened the Kuiper Airborne Observatory, was fitted with a 36-inch scope. Even higher observations were made with giant balloons. Meanwhile Robert B. Leighton and Gerry Neugebauer of the California Institute of Technology installed an infrared telescope at the nearby Mount Wilson Observatory and put their graduate students to work tabulating about five thousand stars that were bright in the infrared but otherwise dim, apparently because they were embedded in thick clouds of dust.

Complicating infrared observations is the need radically to cool the observing instrument. Unless cooled by a cryogenic system almost to absolute zero—the absence of all temperature—the warmth of the instrument emits an infrared glow, overwhelming emissions from distant sources. The 1976 edition of the *Encyclopaedia Britannica* noted the huge potential gain if a supercooled infrared telescope were carried in orbit. However, it said, "considering the enormous cost of the cryogenic systems that would be required . . . there is considerable doubt as to whether. infrared space telescopes should be constructed. The maintenance of cryogenic systems on space telescopes would be a formidable task."

Nevertheless, in January 1983, such a 1-ton spacecraft with a 22-inch mirror was launched from Vandenberg Air Force Base in California. It was the coldest man-made object ever sent into space. Special care was needed to make sure its highly sensitive detectors were never fried by being accidentally aimed at the sun, earth, moon, or Jupiter. The cost to the three participating nations—Britain, the Netherlands and the United States—was about $160 million. The following November, as expected, the spacecraft's coolant of superfluid helium gave out, but by then it had more than once surveyed the entire sky, cataloguing some 250,000 sources. It was known as IRAS (the Infrared Astronomical

Satellite), and Neugebauer, codirector of the earlier ground-based survey, was the chief American participant.

IRAS for the first time allowed astronomers to peer inside the vast clouds of dust and gas molecules to see the lumps and disks marking the formation of stars like the sun and, perhaps, their planets. Within the cloud there were often several cores where infrared indicated a density thirty times that of the cloud as a whole. In half those cores primitive stars seemed to be developing. Radio astronomers, observing the radio emissions produced by carbon monoxide and ammonia (which are particularly strong) in dark clouds of the constellations Taurus and Ophiuchus, identified ninety-five cores of intense emission, and the IRAS analysts found that more than half seem to be incipient stars. The IRAS team, in a routine check to verify the function of the scanners, looked at the bright star Vega and discovered in amazement that it is surrounded by a fat ring of infrared radiating particles. Perhaps they will form into planets, but Vega is much larger and shorter-lived than the sun. In the previous decade ground-based astronomers had found that within the great galactic clouds, stars ten times more massive and a thousand times more luminous than the sun were forming, but the earth's atmosphere blocks the longer infrared wavelengths where new stars would be brightest. IRAS was able for the first time to see the weaker emissions where sun-like stars were forming.

A critical step in infrared observations was the development of imaging. The single sensors, like Edison's, could observe only one spot in the sky at a time. A telescope had to scan across the target, like a television raster, to build up an image. An improvement was carried by IRAS, whose sixty-two sensors swept a patch half a degree wide across the sky. This made it possible, after many sweeps, to construct an image. The chief advance, however, came from research on infrared sensing done in California for the Department of Defense by the Santa Barbara Research Center, a subsidiary of the Hughes Aircraft Company, presumably for use on spy satellites. Its astronomical potential was tested by the National Optical Astronomy Observatories in Tucson, using a grid of almost four thousand detectors. It was, said the developers, a revolution comparable to that caused in optical astronomy by the photographic plate.

Another series of observations that were to play a key role in deciphering how the solar system was formed began in 1852 when J. R. Hind, director of a private observatory in Regent's Park, London, discovered a most peculiar star and associated little nebula in the constellation Taurus, the Bull. Both the star, T Tauri, and the nebula varied radically and irregularly in brightness, sometimes within hours. On occasion the nebula disappeared entirely. In 1945 Alfred H. Joy, on Mount Wilson, identified ten more such "T Tauri" stars, which he recognized as a new class of object. All had spectral features reminiscent of the sun and all were among the dark clouds of the Milky Way. Since then many more have been found, frequently occurring in groups, and it has been concluded that they are enveloped in thick clouds of dust that part, at irregular intervals, to permit a glimpse inside, like the moon on a cloudy night. There is ample evidence that they are evolving into stars of the same class as the sun.

One of their peculiarities is that they seem to be contracting, as though forming into a star from infalling dust and gas, yet they are also shooting out material at the incredible speed of 100 miles a second. In retrospect, many astronomers believe the explanation offered in 1980 by Ronald L. Snell of the University of Massachusetts is the answer. Snell realized that such newly forming stars are at a stage when a dense portion of the giant cloud has contracted enough to produce an incipient star, surrounded by remnants of that cloud. The parent cloud, big enough to produce many stars, rotates slowly, but to conserve angular momentum as part of it contracts, the rotation rate increases, shaping the cloud into a disk.

Up and down along the axis of such a rotating disk, said Snell and his colleagues, are jets shooting outward in opposite directions and at amazing speed. Using a radio telescope at the University of Texas's McDonald Observatory and one at the National Radio Astronomy Observatory on Kitt Peak in Arizona, they monitored the powerful emissions from carbon monoxide swept up by jets from a source in L1551, a black cloud in Taurus. Carbon monoxide was chosen because its radio emissions are far more readily observed than those of hydrogen, even though the latter is by far the most plentiful component of the universe and, presumably, of the jets. The recordings showed not only their extreme velocity but their extent, culminating in lobes 1.7 light years above and below the source (one light year being the distance traveled by light in that time). The central object in L1551 from which the jets were radiating was IRS-5, a particularly powerful infrared source numbered 5 in an earlier survey.

Such bipolar jets call to mind those which extend for hundreds of light years from some of the distant quasars and other highly active galaxies, forming the largest and most spectacular single structures in the universe, and theorists wonder whether the mechanisms creating them are related.

The significance of Snell's interpretation had been recognized by 1991, when astronomers from several parts of the world gathered at the University of Arizona to discuss what by then was known about the birth of the solar system. It was the third in a series of such conferences, held at six-year intervals, sponsored by NASA, and partially funded by the National Science Foundation. In an introductory paper Frank Shu, with Joan Najita, Daniele Galli and Eve Ostriker, all at the University of California at Berkeley, and Susana Lizano of the National Autonomous University in Mexico, said almost all aspects of Snell's concept had been confirmed. They outlined five stages in evolution of the sun and planets:

1. Gas and dust in one of the denser parts of a giant cloud slowly contract under their own gravitation until, after a million years, the core becomes dense enough for its infrared glow to be observed from a distance.

2. The core gravitationally collapses, beginning with its innermost region. That is, it collapses from the inside out. Infalling material rains down from the surrounding, slowly rotating cloud, forming a rotating disk that hides a newly forming star from optical, but not from radio and infrared observation. At some point pressure and temperature in the core of the central object become sufficient to fuse hydrogen nuclei into those of helium, and the star "ignites."

3. Powerful winds break out along the rotation poles of the system, locally reversing the infall and sweeping up material over the poles into two outward jets of gas and dust. During this stage there are both inflow around the equator and outflow over both poles. Almost half the sources recorded by IRAS in dense dust cores showed such bipolar jets.

4. The outflow opens "like an umbrella," eventually halting inflow even over the equator. The star at the center becomes vis-

ible to distant observers, as does the disk around it—a situation evident in many of the objects seen by IRAS.

5. The disk itself is cleared, its material condensing into planets. Seminal analysis of this process was done in the 1960s by the Soviet scientist Victor S. Safronov. IRAS observed what seemed to be open areas in some disks, which could be where giant planets were forming.

On a cosmic time scale, stages two and three are very rapid, taking place in a combined period of only 100,000 years, a possible reason they are so rarely observed. The other stages take much longer. The rate at which the formative process is completed is also low. It has been estimated that in our galaxy of many billions of stars, only a single new star is formed each year.

The world's largest array of radio telescopes, the Very Large Array in New Mexico, has recorded jets of ionized wind blowing in opposite directions from IRS-5, but not at the extreme velocities of the carbon monoxide seen by Snell and others. The jet-producing third stage, said Shu's group, was the most surprising —"the one totally unanticipated by prior theoretical developments." Still unresolved is whether the jets originate in the youthful star (perhaps gaining energy from its very fast spin), from the disk, or from the interface between them.

The possibility that Saturn's rings are analogues of solar system formation has been enhanced by the *Voyager* spacecraft, whose photos have shown them composed of thousands of ringlets, like the grooves on an LP record. Spaces between them are maintained by at least eight "shepherding satellites." It may be that, while a planet such as Jupiter was developing, similar satellites in the disk around it formed the nuclei of what are now its inner moons.

Recent observations have made obsolete the idea that the solar nebula was a passive reservoir of material slowly coming together into planets. Rather, attendees at the Tucson meeting were told, it was a dynamic and sometimes violent place. This is shown by the composition of meteorites formed presumably while the solar nebula was evolving. They are larded with small, rounded bodies known as chondrules (from the Greek for grain or seed), only a fraction of an inch in diameter, that clearly have once been molten. It has been said that they were "drops of a fiery rain." They cooled in a few minutes and eventually became embedded in a matrix that has never been heated. In 1966 Fred L. Whipple of the Smithsonian Astrophysical Observatory, dean of comet specialists, proposed that lightning bolts in the nebula could have been responsible for the sudden, transitory heat that melted the chondrules before they became incorporated into the remaining meteorite material. At the Tucson meeting those who argued for lightning pointed out that in thunderstorms on earth, much of the energy in the soaring thunderheads is released as lightning. Violent, heat-driven updrafts in the solar nebula could have similarly produced lightning. Also suggested were magnetic flares. Several theorists have argued that extremely strong magnetic fields were formed in the nebula by such heat-driven currents, producing flares like those that now erupt from the sun.

A remarkable discovery has been that of microscopic diamonds in the more primitive meteorites. At one time it was thought these might have been formed by a nearby supernova, or star explosion, shortly before the solar system was formed. This was considered necessary to shock the solar nebula into coagulating, but that is no longer widely believed, the nebula having itself been a violently turbulent place. Similarly, other explanations are now given for the diamonds and for the occurrence in meteorites of substances once thought to be the decay products of radioactive elements synthesized in such an explosion. Nevertheless, there is still no agreement on how in the disk around the newly forming star, particles began coagulating into "dust balls," which, in turn, grew into objects with strong enough gravity to hold them together. As Alastair Cameron, by then at the Harvard-Smithsonian Astrophysical Observatory, pointed out, on earth, if one rock is thrown at another, it never sticks.

Nevertheless, that dust coalition occurs is shown by the infrared emissions from disks surrounding such relatively mature stars as Vega and Beta Pictoris. Vega, also known as Alpha Lyrae, is the fourth-brightest star in the sky—too big and short-lived to be a likely candidate for the center of an inhabited solar system. The disk around Beta Pictoris, in the southern heavens, was observed by Bradford Smith and Richard J. Terrile of the University of Arizona, following up in 1984 on the 1983 IRAS observations of that star. They used the 100-inch (2.5-meter) telescope operated by the Carnegie Institution of Washington at Las Campanas in Chile. To cut off light from the star itself they inserted a coronagraph, or obscuring device, suspended by silk filaments in front of the star's image. Extending from both sides of the star they could then see, edge-on, an enormous ring extending about twenty times the radius of our own solar system. There seemed hints that the inner part of this disk had been swept clear, as though planets were forming there, but if so they were hidden by the dust.

Most significant was the observation that the infrared wavelengths produced by such disks were typical of those from large grains, rather than the fine dust from which solar systems are formed. The first step in the planet-forming process had begun. It appears that rings of dust surround many mature stars. Hartmut H. Aumann of the Jet Propulsion Laboratory in Pasadena, California, has pointed out that the infrared emissions from the disk around Beta Pictoris are many times stronger than those from the star itself and thousands of times greater than those produced by dust in the solar system. On earth, particularly at low latitudes, that dust can be seen above the horizon where the sun has just set or is about to rise, forming a luminous pyramid known as the zodiacal light. Most, however, believe it is dust shed by orbiting comets, rather than a residue of the solar nebula.

Astronomers from Germany's Max-Planck-Institut für Extraterrestriche Physik, in a survey of eighty-six newly forming stars, found that thirty-seven have disks, leading them to believe that planetary systems are common around solar-type stars. It is thought that the formation of our own planets occurred after the assembly of "planetesimals" 1 to 10 kilometers in radius. These began banging into one another, producing fragments that in a final "Great Bombardment" assembled into the planets of today. Some believe a few planetesimals may remain as asteroids. One of the most dramatic results of the space program has been close-up photographs of the inner planets, showing the overlapping craters of all sizes left by final stages of this bombardment. Only after each of today's planets had swept clear its orbit and nearby space was the battle won. By then so many collisions had occurred that the spin axes of the planets, including the earth, tended to be tilted, rather than upright.

For generations astronomers debated about the origin of earth's moon. It differs from all other moons of the solar system in being relatively large compared to its parent body, like a tennis ball relative to a basketball. The other moons are all tiny in this respect. The moon is far less dense than the earth, having apparently little or no iron core. Hence it was proposed that it originated elsewhere and was then captured into orbit by the earth's gravity. Harold Urey, as noted earlier, believed it was a relic of the planetesimals from which the planets were formed.

By the 1990s, however, the dominant view was that the moon was born, during final stages of the Great Bombardment, when an object as big as Mars struck the earth. In a summary of the theory Cameron, veteran speculator on the origin of the solar system, said that the impact vaporized most of the earth's rocky material, part of which formed a disk that accreted into the moon. The iron core of the impacting body plunged into the earth's interior, forming a very hot layer around the earth's own iron core, which explained why the moon has no substantial core and is made of material much like that of the earth's outer, rocky portion. The earth was so hot, Cameron said, that after the impact it blew off its entire original atmosphere. He noted that a similar impact may account for the lopsidedness of the Uranus spin axis.

Soon after the 1991 Tucson meeting George W. Wetherill of the Carnegie Institution of Washington, a specialist on the solar system, published the results of some two hundred computer simulations of the manner in which the inner, "terrestrial," planets may have formed. Writing in the journal *Science*, he said he used, as a starting point, a disk filled with planetary "embryos," or planetesimals, 1 to 10 kilometers (0.6 to 6 miles) wide, that had formed from the solar nebula. His tests confirmed that if a giant planet formed in an orbit like that of Jupiter, its gravity would accelerate many of the planetesimals that would otherwise have assembled into a planet occupying the next "slot" inside Jupiter's orbit— that is, in the asteroid belt. This would have driven them out of the solar system or inward, past the orbit of Mars, to collide with the sun or inner planets. The effect would also have thinned out planetesimals that might otherwise have coalesced to form Mars, explaining the smallness of that planet.

Wetherill found, however, that whether or not a Jupiter-like planet developed, terrestrial planets would form in approximately the sizes and positions of Earth, Venus and Mercury. For the first 10,000 to 100,000 years, his simulations showed, the planetesimals accreted in "runaway" collisions, forming objects the size of the moon or Mercury, whereupon the growth slowed until, after a billion years, they reached the size of today's inner planets. That is when, based on the dating of impact specimens brought back from the moon by the astronauts, the Great Bombardment ended.

Wetherill said the results did not depend on his detailed starting assumptions. In cases where no giant planet similar to Jupiter formed, one or more large bodies developed in the orbit of the asteroids, but there was always a planet roughly the size of the earth in an orbit like ours.

"These results," he wrote, "suggest the speculation that for stars similar in mass to the sun, surrounded by centrifugally supported gas-dust disks with mass, energy, and angular momentum similar to that of our own primordial solar nebula, an earth-like planet is likely to form near Earth's heliocentric position. . . . This regularity seems to occur despite the extremely chaotic and stochastic nature of the accumulation processes themselves." Characteristics of the earth-like planet, while not exactly matching those of ours, would, he said, be very similar. "These results suggest that occurrence of earth-like planets may be a common feature of planetary systems."

7

Planet-Hunting

lans for a decade-long national effort to search for planets
around other stars were projected at a high official level in
1990 when a panel of twenty-nine experts in many fields met
under the auspices of the Space Science Board of the National
Research Council. The panel, called the Committee on Planetary
and Lunar Exploration, or COMPLEX, acknowledged the lack of
direct evidence that such other planets exist and pointed out that
while existing detection methods are rapidly improving, they are
still only marginally effective.

"Current theories," said the panel's report, "lead us to suspect
that other planetary systems, habitable planets, and perhaps even
life forms are likely, but as yet we have no direct confirmation that
even a single extrasolar planet exists."

The panel therefore proposed a broad search to find out
whether there are other planets in our galaxy. The answer to that
question, it said, would be comparable to such revolutionary discoveries as the rotation of the earth, the earth's flight around the
sun, the sun's far-off-center position in our galaxy, and the limitations of human perception shown by quantum physics.

Each of these revolutions [the report said] had the effect of
displacing humankind from its assumed anthropocentric position. In this context one dramatic question remains. Is it possible that Earth, the habitable conditions on Earth, and indeed
the life that has evolved to fit those conditions constitute rare
accidents? Modern theorists have proposed answers, but we will
never really know, in a scientific sense, until we have surveyed
a statistically valid sample of star systems with enough sensitivity
to determine whether they have Jupiter-size, Uranus-size, and

ultimately even Earth-size planetary bodies near them. Barring interception of signals from intelligent life elsewhere by radio listening searches, only such a survey can address this question.

The strategies discussed were similar to those that had been used in the search for multiple-star systems where one or more members is invisible. In such cases, if the orbit is just right, the unseen star may periodically diminish light from its mate, but the orbit as seen from the earth must permit such an eclipse to occur. In the case of a planet, the star would have to be small and the planet much larger than Jupiter for its passage to dim the star's light to an observable extent.

A tactic more likely to succeed is detection of the wavy motion of a star that has a companion small enough to be considered a planet but large enough for its gravity to produce such motion. The idea that stars move relative to one another dates from 1718, when Edmond Halley, the English astronomer who predicted the return of the comet that bears his name, noticed that some of the bright stars catalogued by Ptolemy sixteen centuries earlier had shifted their relative positions—what is now known as "proper motion."

Even the nearer stars are so far away that such motion, seen against the backdrop of very distant stars, can only be detected by prolonged observations. In 1838 Friedrich Wilhelm Bessel in Germany found that a star in the constellation Cygnus, the Swan, known as 61 Cygni, moved back and forth among the other stars by 0.31 seconds of arc in a twelve-month cycle. He correctly concluded that the apparent motion of the star was, in fact, caused by the earth's flight around the sun. Knowing the diameter of the earth's orbit, it was a matter of simple trigonometry to calculate the distance to 61 Cygni—the first such measurement achieved. A few years later, in 1844, Bessel announced that, on the basis of extended observations, he had found a completely different type of cyclic motion in two of the sky's brightest stars: Sirius and Procyon. In this case the movement was that of the stars themselves, but what struck Bessel was that their paths, instead of being smooth, showed a slight wave motion or wiggle. They must each, he said, be waltzing through space with an invisible companion. As telescopes improved, both of these companions were sighted as dim stars.

The effect that Bessel had identified is like one in which a fat

man waltzes with a skinny little girl. As they spin around one another, his motion across the dance floor is slightly perturbed by
her dainty weight. The same is true of the earth-moon system. We
tend to think that the moon simply circles the earth, but this is
not so, for the moon is large. Its radius is one quarter that of the
earth and its gravity pulls on us just as that of our planet tugs at
the moon. The two bodies circle their common center of gravity
(the "barycenter"), which lies inside the earth, but 2,880 miles
from its center. Hence, it is the barycenter of the earth-moon
system that circles the sun, and if, from a great distance, one recorded the path of the earth, it would be wavelike. For a long time
looking for such motion has been the favorite way to search for
evidence of planets, although it can only apply to stars near
enough to have substantial "proper motion."

Another way to detect the effect of a planet on its parent star
is to see whether the star's motion, toward or away from the
earth, is constant. If it varies in a cyclic manner, this could mean
that it is being affected by the gravity of a circling planet. Such
speed changes would alternately lengthen and shorten the wavelengths of spectral lines in light from the star—the effect first
recognized in 1842 by Doppler. To do so the planet would have
to be in an orbit that produced substantial to-and-from motion,
relative to the line of sight from the earth. An advantage of this
method, however, is that in contrast to the optical tracking
method, it is not limited to nearby stars.

A pioneer in the early search for planets was Peter van de
Kamp, who in the 1930s thought of looking for stars that showed
wavy motions, similar to those observed by Bessel a century earlier,
but with the subtler pattern indicative of planets. The advent of
photography had greatly aided determinations of relative motions
among the stars, since pictures of one patch of sky could be repeatedly monitored over long periods. The task, however, was
challenging. It has been pointed out that if one observed the sun
from one of the nearest stars, the sun's cyclic displacement from
a straight path, under the influence of Jupiter, would equal only
the width of a dime seen from 300 miles. When, in 1937, van de
Kamp came to Swarthmore College's Sproul Observatory near
Philadelphia, he initiated a long-term watch on a number of stars,
most of them within thirty-three light years and hence moving at
observable rates. The path of each was typically measured with
reference to from three to six background stars. Van de Kamp

and his associate Sarah Lee Lippincott thought they saw evidence for the invisible companions of several stars, as did other astronomers. In all of these cases the invisible object seemed larger than Jupiter, and some were probably brown dwarfs—too big to be planets, but too small to generate the fusion reactions that illuminate a star. In 1985 astronomers in Tucson at the National Optical Astronomy Observatories and the University of Arizona reported evidence for what seemed a large planet or brown dwarf orbiting the star Van Biesbroeck 8, but a later effort at confirmation failed.

The greatest excitement followed the observations at Sproul of Barnard's star, discovered in 1916 by the American astronomer Edward E. Barnard. It is six light years away; only the triple system of Alpha Centauri is closer. Barnard's star moves faster, relative to the backdrop of stars, than any other known star, traversing a path equal to the observed width of the moon every 180 years. In 1963 van de Kamp told the American Astronomical Society that Barnard's star has a planet 1.5 times larger than Jupiter, circling the star in a period intermediate between those of Jupiter and Saturn. He later reported that during 1,000 nights between 1916 and 1975, some four thousand photographs had been taken, many of them with Sproul's 61-centimeter refractor, and he told of "additional, marginal evidence" for a second planet with a mass half that of Jupiter. Since Barnard's star is only 15 percent as massive as the sun, it is far more easily influenced by a planet's gravity than would be a larger star. Astronomers at the University of Pittsburgh's Allegheny Observatory and Wesleyan University's Van Vleck Observatory in Middletown, Connecticut, reported seeing supporting evidence. Since then, however, no observatory has been able to confirm this, and the evidence for one or more planets circling Barnard's star remains equivocal.

More promising have been the spectroscopic observations, seeking to detect wavy motion by a star through cyclic changes in its spectrum, caused by variations in its motion toward or away from the earth—the much-used Doppler effect. A method employed by Bruce T. E. Campbell of the University of Victoria in British Columbia and his coworkers has proved exceptionally sensitive in detecting very slight changes in a star's toward or away motion. Light from the star is passed through a cell containing hydrogen fluoride, whose spectral lines are superimposed on similar ones from the star. The extent to which lines from the star are displaced is an index of relative motion. Six times a year Campbell's group has been measuring light from fifteen stars with the Canada-France-Hawaii telescope on Mauna Kea in Hawaii. In almost half of them they have found evidence of speed changes, including the sunlike star Epsilon Eridani, which had been one of the first two stars scanned for radio signals. Its behavior was said to indicate a companion two to five times the size of Jupiter. A clearer case was the star Gamma Cephei, which in observations that began in 1981 seemed to show a cycle of 2.6 years. Its behavior was taken as evidence of a planet 1.5 times the size of Jupiter orbiting at about twice the earth's distance from the sun. In a review of such evidence published in a 1991 issue of *Scientific American,* David C. Black, head of the Lunar and Planetary Institute near Houston, pointed out that, for so large a planet, this was surprisingly close to the parent star. He said, however, that "surprises are the brood stock of scientific inquiry." Some unexplained irregularities in the star's motion have been observed, and, Black said, the star will have to be observed for several 2.6-year orbits before the evidence becomes convincing.

A group from the Harvard-Smithsonian Astrophysical Observatory in Cambridge, Massachusetts, led by David W. Latham, has observed the star HD 114762 for more than twelve years, using the Smithsonian's 61-inch reflector near the village of Harvard, and has detected an eighty-four-day cycle of variations in its light. They believe it may be circled by a small brown dwarf, rather than a giant planet. Its orbit is as close to the star as that of Mercury to the sun. Observations by the University of Texas have seemed to confirm the orbit and to have shown the object to be even larger.

In December 1986, Tom Gehrels and Robert S. McMillan of the University of Arizona began monitoring sixteen solar-type stars from nearby Kitt Peak. By January 1991, they had made 1,576 observations. Some of the stars, they reported, "are showing slow, tantalizing variations in velocity that should become clearer after additional years of observing." Since some planets presumably take many years for each orbit, prolonged observations are needed.

An announcement that threw theories of planet formation into turmoil was made in July 1991 by three astronomers from Britain's Jodrell Bank Observatory. Since 1985, they had been keeping highly precise timing records on forty newly discovered pulsars.

Such objects are the extremely dense remnants of stars that have exhausted their fuel and blown up as a supernova, leaving a rapidly rotating cinder. They are known as pulsars because with each rotation they emit a radio pulse. While their pulse rates slow down very slightly over the years, their rhythm is extremely precise. What the Jodrell Bank astronomers reported was that the pulse rate of one, pulsar 1829–10, was changing in the cyclic manner to be expected if it had a planet. The period of the oscillations was about six months—half the time it takes for the earth to circle its own star—and the mass of the planet was estimated to be ten times that of the earth.

For a pulsar to have a planet was totally unexpected. It was assumed that if a star had any planets when it blew up, they would be blasted out of existence. In subsequent months theorists had a field day devising seemingly unlikely explanations. For example, at Cambridge University in England the leading cosmologist Martin Rees and his colleagues proposed that the pulsar may have collided with the central star of a sun-like planetary system, "cannibalizing" the star but sparing the planets. Then, on January 9, 1992, Alexander Wolszczan of Cornell University and Dale A. Frail of the National Radio Astronomy Observatory in Socorro, New Mexico, reported in the British journal *Nature* that precise timing of the pulsar 1257+10 with Cornell University's giant radio dish at Arecibo, Puerto Rico, had shown that it is being orbited by two or more planets. The pulsar is of the "millisecond" variety—one rotating so fast that its pulses occur more than a thousand times a second. The discovery of a second such strange object, Wolszczan said, may be opening up "a whole new branch of astronomy."

A week later, however, *Nature* carried a letter by the Jodrell Bank group confessing that its report had been in error. In correcting for the effect of the earth's motion around the sun on the timing of the pulsar's signals the group had not taken into account a very subtle factor: the elliptical nature of the earth's orbit. The Arecibo observations seemed free of any such error, and Black singled them out as the only planetary reports so far that seemed persuasive. When the COMPLEX report was written in 1990 none of the reports of evidence for distant planets, widely publicized in previous years, was considered convincing. Such reports "have either not been confirmed by subsequent observations or are beset with such uncertainties that discovery cannot yet be claimed." This does not mean that astronomers, with observations such as those of IRAS, are not seeing evidence of planets being formed, but that is not the same as actually "seeing" them. It seems likely that this will ultimately occur, but the obstacles to direct observation are formidable. Since planets shine by light reflected from their parent star, they are typically more than a billion times dimmer than the star, and their detection is further made difficult by their closeness to the star.

Nevertheless, the prospects for direct imaging of planets are far from hopeless. A revolution is taking place in observational astronomy, both on the ground and in space. The Hubble Space Telescope, now in orbit, is equipped with a "coronagraphic finger," or thin rod, which, on command from the earth, can be projected into the telescope's field of view, cutting off all light from a star, but not its neighborhood. In this way it is hoped that a Jupiter-sized planet could be seen around any of the hundred closest stars. "Are we alone in the universe?" asked George Field, professor of applied astronomy at Harvard, and Donald Goldsmith, coauthors of the definitive book on the Hubble Space Telescope. For the first time in history, they wrote, the space telescope may allow astronomers to observe directly a true planet. Unfortunately the now-famous oversight in testing of the telescope before it was launched prevented the necessary precision, but this may be corrected when astronauts repair the telescope in the mid-1990s.

Meanwhile, on the earth's surface, a new generation of giant telescopes is being built in Arizona, Chile, Hawaii and elsewhere. The one completed in 1948 on California's Mount Palomar, with a massive mirror 5 meters (200 inches) wide, for more than a generation dominated world astronomy. Creating its primary mirror was an engineering triumph. It was ground from a many-ton "blank" of special glass into the highly precise concave (parabolic) shape needed to collect faint starlight and focus it onto a secondary mirror. The primary mirror had to retain its shape despite radical day-night temperature differences and not sag as the 500-ton instrument swung to different targets.

Recently, however, a variety of ways have been found to build bigger ones. A major breakthrough, in the optical shops of the University of Arizona and at the Schott Glaswerke in Germany, has been development of rotating ovens whose motion shapes molten glass (of a type not greatly deformed by temperature changes) into the proper parabola. After the glass has cooled,

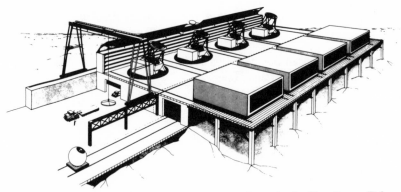

Artist's impression of the linear array of four mirrors in the Very Large Telescope in Chile. For scale, note flat-bed truck at left. *D. Ehard.*

there is minimal need for grinding. A way has also been found to couple as many as six telescopes so that they can observe the same target, greatly increasing the amount of light collected without a mirror of impossible dimensions. This is the design of Europe's Very Large Telescope (VLT) in Chile, whose four mirrors will be linked to serve as a single instrument of 32.8 meters (107 feet).

The biggest single mirror is that of the Keck Telescope on Mauna Kea, the Hawaiian volcano that has become the world's favorite site for new observatories. The Keck mirror, formed of thirty-six hexagonal segments, is 10 meters (33 feet) wide. A twin to sit beside it is already being prepared, financed largely by a $74.6-million commitment from the estate of William Myron Keck, founder of the Superior Oil Company, which donated a similar amount for the first telescope. According to Edward C. Stone, director of the Jet Propulsion Laboratory in California and chairman of the project's board of directors, the two telescopes, in tandem, should be able to distinguish the two headlights of a car at 16,000 miles. At infrared wavelengths, he added, their planet-detection ability (if the planet is warm) should match that envisioned for the Hubble Space Telescope. Nine telescopes larger than the 5-meter on Mount Palomar are built or projected, as listed on the next page.

Building bigger mirrors is not the only way to achieve better seeing. Another is the charge-coupled device, or CCD, developed during the 1970s. It is a grid of thousands, even millions, of tiny

**Optical Telescopes, Built or Projected,
Larger Than the 5-Meter Reflector on Mount Palomar**

Name	Organization & Site	Mirror Area in Square Meters	Mirror Type
Very Large Telescope (VLT)	European Southern Observatory on Cerro Paranal, Chile	210	Four 8.2-meter telescopes
Columbus	Italy, University of Arizona, on Mount Graham, Arizona	110	Two 8.4-meter mirrors
Keck	California Institute of Technology, University of California, on Mauna Kea (to be twinned, with a second telescope next to the first)	76	Thirty-six 1.8-meter hexagonal mirrors
Magellan	Carnegie Institution of Washington, Johns Hopkins University, University of Arizona at Las Campanas, Chile	50	One 8-meter mirror
Gemini North	National Optical Observatories, Britain, Canada, on Mauna Kea	50	One 8-meter mirror
Gemini South	National Optical Observatories, Britain, Canada, on Cerro Tololo, Chile	50	One 8-meter mirror
Japanese National Telescope	Japan, on Mauna Kea	44	One 7.5-meter mirror
Multi-Mirror Conversion	Smithsonian Institution, University of Arizona, Mount Hopkins, Arizona	33	One 6.5-meter mirror
Russian Large Alt-azimuth Telescope	Russia, on Mount Pastoukhow, Caucasus	32	One 6.05-meter mirror

sensors each of which, when activated by a single particle of light, or "photon," releases an electron that, through a succession of stages, can be amplified to thousands of electrons. The output of each sensor produces one point, or "pixel," of the image, much as a published news photograph is formed from thousands of dots of varying intensity. This has improved a telescope's ability to record faint objects a thousandfold and is used to send images of the planets from passing spacecraft.

Another major advance in telescope technology has been the discovery of ways to eliminate the blurring of images from light that has come through the atmosphere. Because of turbulence in the air, the light is constantly distorted. More than two and a half centuries ago this was eloquently described by Sir Isaac Newton in his *Opticks*. He pointed out that the air through which we see stars is in perpetual tremor:

> For the Rays of Light which pass through diverse parts of the aperture, tremble each of them apart, and by means of their various and sometimes contrary Tremors, fall at the same time upon different points in the bottom of the Eye, and their trembling Motions are too quick and confused to be perceived severally. And all these illuminated Points constitute one broad lucid Point, composed of those many trembling Points confused and insensibly mixed with one another by very short and swift Tremors, and thereby cause the Star to appear broader than it is, and without any trembling of the whole. Long Telescopes may cause Objects to appear brighter and larger than short ones can do, but they cannot be so formed as to take away that confusion of the Rays which arises from the Tremors of the Atmosphere.

A star, being a pointlike source of light, twinkles, because while tremors of the atmosphere have deflected its light, the tremors of so narrow a beam are in unison. A bright planet like Jupiter does not twinkle because it is a broad light source. Rays from different parts of its surface travel separate paths, and, as described by Newton, their wiggles cancel one another, creating simply a large, fuzzy image. Very large telescopic mirrors collect much light from a dim source, but the images, until now, have been blurred, presenting an insurmountable problem to those trying to see planets amid this blur.

Only in the last decades of the twentieth century have ways been found to overcome this, thanks to teams of theorists, telescope designers and missile defense specialists. In 1970 a French astronomer, Antoine Labeyrie, devised what he called "speckle interferometry," now in worldwide use. It depends on many extremely short exposures of the target—a few hundred per second—each being so short that it "catches" the twinkles of the incoming rays. After many such "speckles" have been recorded, a computer can rectify and combine them into a single image. In this way, for example, the two stars of binary systems that had previously been visible only as single stars have been shown as separate objects.

Even more revolutionary is the technique known as "adaptive optics," which depends on "rubber mirrors," special electronics and computer assistance, as well (in its latest version) as a trick developed for missile defense. In 1953 Horace W. Babcock, later director of the Mount Wilson and Palomar observatories, proposed a way to correct the constant variation of light waves after they have descended through the atmosphere. Its key element was to be introduction into the telescope's optics of an additional mirror whose shape could be rapidly adjusted to correct for the constant changes. The mirror, deformed in a manner equal but opposite to the distortion of the incoming wave, would restore the wave's uniformity of "phase."

If it were not for the atmosphere, waves of starlight reaching a telescope's mirror would be in line, like the crests of water waves simultaneously striking the entire length of a beach. They would be "in phase." But if they came in at constantly changing angles, the crests would reach different parts of the beach at varying moments. Under such circumstances, if the beach were a mirror, it would reflect a blurred image. Because of the atmosphere, the waves of starlight are constantly changing in phase, and therefore the image of the star becomes blurred. If, said Babcock, a detector measured these changes, it could send electronic commands to what came to be known as a "rubber mirror." These would adjust its shape very slightly, but enough to restore phase uniformity— make the wave crests parallel to the beach—and thus produce a reflected image free of distortion. The process depends on an electronic camera that can measure the tilt of the wave hundreds of times a second. This information then adjusts the rubber mirror (which in no case is actually made of rubber) to reflect the wave in its original shape.

At the time Babcock made his suggestion, the construction of such a system did not seem feasible. The starlight wave changes occur hundreds of times a second and the wave analysis, calculations and mirror distortions had to be even more rapid. Not until the 1970s did advances in electronics and high-speed computers make this seem feasible.

Military as well as astronomical applications were obvious, for such a system, based on the ground, could not only look for distant planets but inspect (and perhaps destroy) satellites and missiles. There began intense efforts to develop some practical form of adaptive optics. Participants included Bell Laboratories in Murray Hill, New Jersey, Itek Corporation in Lexington, Massachusetts, MIT's Lincoln Laboratory near Boston, the University of Chicago, the Lawrence Berkeley and Space Sciences Laboratory of the University of California, and the Phillips Laboratory at Kirtland Air Force Base in New Mexico (most of them working in secret for the Defense Department). Proposals applicable to astronomy came from such leading scientists as Freeman Dyson of the Institute for Advanced Study in Princeton, New Jersey, Robert H. Dicke of Princeton University, and Antoine Labeyrie in France.

The challenge was first to detect how the wave had been altered, then to calculate how the rubber mirror could be figured to correct the distortion. In some designs the mirror is distorted by hundreds of tiny pistons, or "actuators." In other versions it is controlled by electric currents that can subtly bend it in a multitude of ways. An early version produced by Itek in 1976 had twenty-one such "correction zones."

Two methods are used to determine the distortion. One is to split the incoming light between scores, or even thousands, of little "lenslets," each of which is adjusted until its image is in focus, showing how the incoming wave has been distorted. The other method, with a wavefront sensor, records the extent to which the wave has been "tilted" by passage through the air. Either system provides the information needed to properly distort the mirror.

In some early systems a fragment of light extracted from the optics was subjected to the necessary wave analysis, but this robbed so much light that only bright stars could be observed. Another approach is to observe a second star close enough to the target for its light to have traversed the same path and been similarly distorted. Only a tiny percentage of target stars are alongside companions bright enough to serve as such beacons.

In 1985 Labeyrie and Renaud Foy proposed that with a powerful laser, an artificial star could be created in the same direction as the heavenly target. They were unaware that in 1981 such a system had already been secretly proposed in the United States by Julius Feinlab of Adaptive Optics Associates. Feinlab's proposal was seized upon by DARPA, the Defense Advanced Research Projects Agency, and then taken over by SDIO (the Strategic Defense Initiative Organization, or "Star Wars" project), which sought to improve the effectiveness of lasers against incoming missiles.

In 1983 Robert Q. Fugate and his associates at the Phillips Laboratory at Kirtland Air Force Base aimed a laser into the sky from its Starfire Optical Range to see whether the resulting pencil beam could be used as a guide star. Tests were also conducted by MIT's Lincoln Laboratory from the summit of Haleakala, the massive volcano on the Hawaiian island of Maui. The system's effectiveness was limited because the illuminated molecules of air, while high in the sky, were still inside the atmosphere; unlike light from a distant star, the light had not traversed the earth's entire blanket of air.

Meanwhile, Will Happer, a researcher associated with Jason, the highly secret project in which new ideas for defense are discussed, suggested in 1988 how an artificial star could be created above virtually the entire atmosphere. A properly tuned laser would be used to create a fluorescent spot in the sodium particles that envelop the earth at a height of 60 miles—they are believed to be a residue of infalling meteors. The possibility of such a sodium guide star was tested in 1984 by Lincoln Laboratory's Ronald A. Humphreys and Charles A. Primmerman at the White Sands Missile Range in New Mexico. Unaware of these tests, astronomers from the universities of Hawaii and Illinois in 1987 conducted similar experiments high above Mauna Kea in Hawaii.

Because the laser beam is as much distorted on its upward journey as on its return, a natural guide star has to be observed to correct for this effect, but the star need not be bright. Also essential when viewing the sodium "star" is avoiding the pencil beam of light produced by the laser within the atmosphere. This, as pointed out by Claire Max of the Livermore National Laboratory, a leader in laser development, can be achieved in two ways. One is by observing slightly to the side of the laser. The other is by pulsing the beam and recording the flash of sodium light during the fraction of a second after the beam within the atmosphere has vanished.

If a very large telescope is used, the waves of starlight falling on various parts of its giant mirror have traveled such different paths through the atmosphere that more than one artificial star is needed. By 1990 such multiple laser-induced guide stars were being tried secretly by the Lincoln Lab group. The Europeans, who had been using natural stars, were talking of multiple laser beacons for their 8-meter telescopes on Cerro Paranal in Chile.

Then, with the Cold War winding down and "Star Wars" use of lasers abandoned, it seemed absurd to continue keeping the Defense Department's program secret. It was made public in 1991, and those who had been working on it for years finally received public credit. MIT's Lincoln Laboratory was allowed, with financial aid from the Defense Department, to install its sixty-nine-channel adaptive optics on the 60-inch telescope atop California's Mount Wilson.

Like so many other new research devices, adaptive optics is expensive, but will almost certainly be installed at many of the larger telescopes. As stated by Fugate in a 1992 interview with Graham P. Collins of *Physics Today*, "People who have an eight-meter telescope without adaptive optics have no better resolution than a guy in the backyard with an amateur telescope."

It is most effective in the infrared, since the shorter wavelengths of visible light present more of a challenge. Nevertheless, said MIT, such systems in the coming century "should become standard equipment for all ground-based telescopes, and the next generation of 8–10 meter telescopes equipped with adaptive optics will provide unprecedented views of the cosmos."

The COMPLEX report to the National Research Council envisioned ambitious programs for planet detection during the next decade and beyond. One under design study, called Precision Optical Interferometry in Space (POINTS), was viewed in the report as an appropriate follow-up to projected development of such observations from the ground. An array of orbiting antennas, working in combination, could look for stellar wobbles, or perhaps even observe planets directly. Using such optical interferometry in space would be an extension of the highly successful multiple radio telescopes on earth, such as the thirty-six dishes of the Very Large Array in New Mexico. Even on the earth's surface, said the report, arrays sensitive to infrared emissions between the wavelengths of radio and visible light might also be able to detect other solar systems. It noted that experiments already under way, using a baseline of 330 feet, were a step in that direction.

Similarly in a 1991 issue of *Physics Today*, Charles A. Beichman and Stephen Ridgway, who had played leading roles in the most recent survey of American astronomy and astrophysics, said that despite the great advantage of adaptive optics, arrays of telescopes using interferometry should do a hundred times better and "revolutionize stellar astronomy." Once in orbit, they said, such arrays should be able to detect star wobbles out to 500 light years. Such observations should either "find planets or show them to be far less common than is now supposed." The radio astronomer Bernard F. Burke of MIT proposed that such an array would be able not only to detect planets, but also to analyze their reflected light, showing whether wavelengths indicative of oxygen were being absorbed. The presence of oxygen would imply life.

Other new devices are standing in the wings. The original plan for the Space Station called for a long boom on which could be mounted an astrometric telescope, but to cut costs the boom was shortened and such a payload eliminated. Another project, nearer fruition, is SIRTF, the Space Infrared Telescope Facility, which was being prepared for launching near the end of the 1990s but because of budget constraints is on hold.

The abundance of new technology and equipment that has emerged, or is on the horizon, gives promise that the decade of national effort called for by COMPLEX will at least bring us closer to the discovery of planets elsewhere in the galaxy. Considering all the hints of such planets and the properties of our own solar system that make it seem typical, it is hard to believe ours is special. That others will be discovered, sooner or later, seems almost inevitable.

8

Creation or Evolution?

"**W**here did we come from?"

People have asked themselves this question since primitive times but, as in so many other fields, science did not come forth with a plausible answer until the closing years of the twentieth century. New discoveries in biochemistry have suggested the steps whereby life could have arisen from inanimate matter. They also imply that the same thing has occurred elsewhere.

Thus science, having relegated to absurdity the idea that we are in any way central to the universe, and having made it appear unlikely that our planet is unique, now proposes that our proudest quality, life itself, is not unique to the earth.

Not that this idea is new. In ancient times Anaximander of Miletus was prophetic both in his argument that the number of worlds is infinite and in his view on evolution. He suggested that life originated in sea ooze and adapted itself to a multitude of environments. Even man himself, he said, was descended from sea creatures. However, such ideas conflicted with the view that so marvelous and mysterious a phenomenon as life could only come about through some form of divine action. Even at the time of Anaximander, some 600 years before Christ, this concept of the miraculous, simultaneous creation of all species was deeply rooted in many (though not all) of the world's religions.

Indeed, it persists today in the minds of some fundamentalists and led, in 1925, to the famous trial of John T. Scopes, a biology teacher, for violating a Tennessee law that forbade the teaching in public schools of any concept of man's origin that contradicted the Bible.

Yet observant people, from earliest times until a century ago, believed that life could arise from inanimate matter. They noted that maggots and worms appear, as though from nowhere, in rotting material and regarded this as evidence that life can be formed spontaneously. This idea of "spontaneous generation" was accepted by such pillars of seventeenth- and eighteenth-century science as Newton and Harvey, discoverer of the circulation of the blood.

By the seventeenth century, nevertheless, at least one man attacked the problem of spontaneous generation scientifically. He was Francesco Redi, court physician to Ferdinand II de' Medici of the famous Florentine family. Redi showed that the white worms in rotting meat hatch from eggs laid there by flies. Using "controls" like a modern researcher, he placed identical pieces of meat in two jars. One jar he covered with gauze; the other he left open. The flies, plentiful in seventeenth-century Florence, swarmed about the jars, but could not land on the covered meat. The worms, of course, appeared only in the meat accessible to the flies.

Yet Redi's experiment did not destroy the idea of spontaneous generation. The seventeenth-century Flemish scientist Jan Baptista van Helmont, who did some of the earliest experiments in physiology, believed that the key to life lay in fermentation and proposed methods for the generation of scorpions and other creatures that today seem preposterous. His most famous was his recipe for mice:

> If one stuffs a dirty shirt into the orifice of a vessel containing some grains of wheat, the fermentation exuded by the dirty shirt, modified by the odor of the grain, after approximately twenty-one days brings about the transformation of the wheat into mice.

During Redi's lifetime, the debate took a new turn when a draper's apprentice in the Lowlands, Antoni van Leeuwenhoek of Amsterdam, turned the recently discovered microscope to biological use and gazed upon the wonderful world of bacteria and protozoa. To many, such organisms seemed to arise spontaneously, although Leeuwenhoek himself suspected that they drifted through the air. The latter view was strengthened in 1765 when Lazzaro Spallanzani, an Italian naturalist, put soup, of a kind in which bacteria thrive, in a flask whose neck was drawn out to a fine tip so that it could be quickly sealed by melting. While the soup was boiling, Spallanzani sealed the flask and, no matter how long it was kept, no clouding appeared to indicate the activity of bacteria.

When Nicolas-François Appert, a French businessman, heard of this experiment, he realized its significance in food preservation, and the canning industry was born. Still the idea of spontaneous generation refused to die. In England a Jesuit priest, John Turberville Needham, rejected the significance of Spallanzani's experiment. In fact he argued from the Bible that all life, in a sense, arose spontaneously. He pointed out that, according to Genesis, God did not create life directly but commanded the earth and sea to bring it forth:

And God said, Let the earth bring forth grass, the herb yielding seed, and the fruit tree yielding fruit after his kind . . .

And God said, Let the waters bring forth abundantly the moving creature that hath life, and the fowl that may fly above the earth in the open firmament of the heaven . . .

And God said, Let the earth bring forth the living creature after his kind, cattle and creeping thing, and beast of the earth after his kind: and it was so.

Only man, in this account, was directly created by God. Father Needham argued that the earth and sea are still carrying out God's command.

Early in the nineteenth century experiments with electricity allegedly created organisms in the laboratory. Andrew Crosse reported that when he soaked porous stone in a mixture of hydrochloric acid and potassium silicate, then ran an electric current through the stone, fearsome-looking creatures of microscopic size emerged. It is now assumed that the creatures, though unobserved, were there to begin with.

Meanwhile Buffon, the French naturalist who invented the collision theory to explain the origin of the solar system, came forth with an ingenious explanation for the seemingly spontaneous appearance of life. Living matter, he said, consists of "organic molecules" that, during the process of decay, can rearrange themselves to form new organisms from recently deceased material.

Buffon's countryman Laplace, in addition to suggesting a dust-cloud origin for the solar system, proposed that the plants and animals of the earth were brought into existence by the beneficent

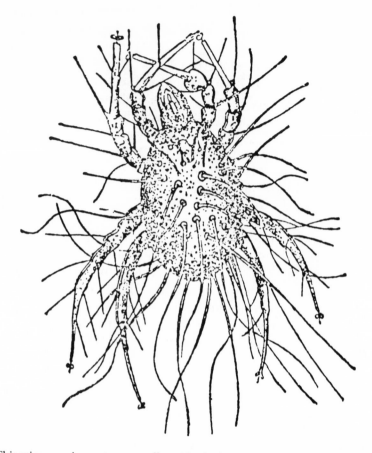

This microscopic creature was alleged by Andrew Crosse to have been synthesized by the action of electricity on a porous stone soaked with hydrochloric acid and potassium silicate. The organism, named *Arcarus electricus*, was shown in H. M. Noad's *Lectures on Electricity* (London, 1849).

action of sunlight. He reasoned that the same thing had occurred on other planets despite the different temperatures existing there. The life that would arise on such planets, he said, would be suitable for that environment.

However, the question as to whether or not lowly forms of life can arise spontaneously did not provoke passionate debate until the theory of evolution proposed that all species, including our own, are descended from some primitive creature. This brought into conflict the ideas of two scientific giants of the nineteenth century: Louis Pasteur and Charles R. Darwin.

Darwin's theory of evolution was rooted in the classification of species that had been going on since the days of the great Swedish botanist Linnaeus, a century earlier. Naturalists, traveling to distant lands and oceans, had vastly enlarged the catalogues of known plants and animals. More significant, when the internal, as well as external, characteristics of these organisms were studied, it became clear that there was some form of "system" in the plant and animal kingdoms. The situation was somewhat analogous to that confronting those who classified the stars, the basic elements, or, in our time, the subatomic particles. In each case there was clearly some inner meaning to the orderliness of nature, the problem being to fathom what it was.

Some of Darwin's predecessors, in particular Jean-Baptiste de Monet, chevalier de Lamarck, had guessed that the grouping of species into family trees was the result of evolution from one species to another. Lamarck thought evolution had taken place through the adaptation of plants and animals to their environment and the passing on, to the next generation, of these acquired characteristics. Darwin then came forward with the idea that new types emerge through occasional, random variations, or errors, in heredity and are then weeded out through survival of the fittest.

In looking back down the long avenue of evolution, Darwin concluded that in the dim past there must have been some single, primitive form of life from which everything else arose. He then logically asked: Whence came that original species? In an 1882 letter thought to have been the last that he dictated and signed before his death, Darwin said contemporary knowledge was so meager that any serious attempt to explain life's origin would be premature. He conceded that he knew of no reliable experiment demonstrating the spontaneous generation of life, but he reaffirmed his belief that life itself must have evolved, as did the species that came later. The "principle of continuity," he wrote, "renders it probable that the principle of life will hereafter be shown to be a part, or a consequence, of some general law."

Darwin recognized the difficulty confronting any theory that depends on the coming together of the key chemicals of the life process. These chemicals cannot exist in the oxygen-rich world of today except inside of living things. In the open, so to speak, they are quickly gobbled up, either in chemical reactions or by living organisms. However, Darwin pointed out that conditions in the past may have been different:

It is often said that all the conditions for the first production of a living organism are now present, which could ever have been present. But if (and oh! what a big if!) we could conceive in some warm little pond, with all sorts of ammonia and phosphoric acid salts, light, heat, electricity, etc., present, that a proteine compound was chemically formed ready to undergo still more complex changes, at the present day such matter would be instantly devoured or absorbed, which would not have been the case before living creatures were formed.

Whereas Darwin's genius lay in his flashes of insight and imagination, Pasteur was passionately devoted to the experimental method. While the French scientist was still a young man, Rudolph Virchow in Germany established the doctrine that the cells constituting all life, from one-celled organisms to man, cannot arise except as the offspring of other cells. The stream of life, both in terms of an individual's growth and the flow from generation to generation, was thus a succession of cell divisions. This seemed incompatible with spontaneous generation, but did not change the beliefs of many highly respected scientists.

Among these was Félix A. Pouchet, director of the Museum of Natural History in Rouen, France. He believed that some constituent of the air, such as oxygen, was the critical factor in stimulating the spontaneous appearance of bacteria in material ripe for decay.

Pasteur said that such spontaneous generation was impossible. Life, even as lowly as bacteria, could only arise from other life. The organisms responsible for fermentation, he argued, were suspended in the air. In 1860, to prove his point, he set forth from Paris with seventy-three sealed flasks, each containing a broth that was fermentable but had been sterilized by heat. His first stop was at Arbois, near his father's tannery in the vicinity of the Jura Mountains, where he opened twenty of the flasks, allowing fresh air to rush in, then sealed them again. Eight of these containers later showed signs of fermentation.

Pasteur and his helpers then climbed to the summit of Mont Poupet in the Jura, some 2,800 feet above sea level, where another twenty flasks were opened in the same manner. Of these, five produced evidence of organic activity. He then went to Chamonix, at the foot of Mont Blanc, highest peak of the Alps, where he hired a guide and mule to carry his equipment up to the Mer de Glace, the great glacier on the side of that mountain. To make sure no bacteria were swept into the flasks except those in the air itself, he broke the sealed neck of each flask with sterile nippers, holding both high over his head, then sealed the neck again by melting the glass with a flame. Of the twenty flasks opened on the Mer de Glace, only one displayed signs of bacterial contamination.

The results showed, Pasteur said, that the decay was produced by organisms that were scarcer at high elevations than at low levels. In no case would one expect all the samples to be infected, any more than one would expect all those encountering someone with a cold to catch it.

Meanwhile Pouchet was doing some collecting of his own, exposing samples to the air on the plains of Sicily, on Mount Etna and at sea. His results were quite different from those of Pasteur. He said they showed that all air is "equally favorable to organic genesis," whether it be in the midst of a crowded city, out at sea or on a mountaintop. When Pasteur's results became known, Pouchet decided to outdo him at his own game. He and his associates organized an expedition into the Maladeta Mountains, highest of the Pyrenees, and opened a series of flasks on the edge of a glacier higher than Pasteur's loftiest experimentation site. The solution in each flask was a decoction of hay that was ripe for decay, and the contents of all the flasks fermented.

Pasteur charged that Pouchet had contaminated each flask by breaking its neck with nonsterile nippers. At first, sentiment in the Paris Academy of Sciences was on Pouchet's side, and a commission was formed to resolve the dispute, which by now had shaken much of French society. The commission wished to settle the matter through a single, joint experiment, but Pouchet refused. Meanwhile, however, the tide had turned and the scientific community had begun to accept Pasteur's results as irrefutable. The climax came on the evening of April 7, 1864, when Pasteur was invited to give one of the newly inaugurated evening lectures on science at the Sorbonne.

When Pasteur, with his stubbly beard, stepped to the rostrum he saw before him the leading lights of Paris—novelists like George Sand and the elder Dumas, scions of Parisian society, and a multitude of scientists. In a deep, firm voice he noted that they were living in a time of great debates: Was man created several thousand years, or several thousand centuries, ago? Are the species of plants and animals fixed in form, or do they evolve from one type to another? (Darwin's *Origin of Species* had been published only five years earlier.) Pasteur said he would direct his remarks to one such question: "Can matter organize itself on its own? In other words, can beings come into the world without parents, without ancestors?"

The controversy on this subject, which had spilled over into the public domain, he said, was essentially between "two great currents of thought as old as the world, and which, in our time, are known as materialism and spiritualism.

"What a victory, gentlemen, for materialism," he continued, if it could be shown that matter can organize itself and come to life. "Ah! if we could give [to matter] that other force which is called life . . . what need to resort to the idea of a primordial creation, before whose mystery one must indeed bow down? What need for the idea of a God-creator?"

Pasteur, a devout Catholic, then ridiculed a book published a few years earlier by Jules Michelet, a rather fanciful writer, who pictured life originating in a nitrogen-rich drop of seawater—a bit of mucus or "fecund jelly" that, in perhaps 10,000 years, evolved "to the dignity of insects" and in 100,000 years to monkeys and man. What most irked Pasteur was Michelet's statement that this concept had taken on new strength, with "much éclat," as a consequence of Pouchet's experiments.

Pasteur showed the audience a flask, part of which had been drawn out to a point. Such a flask, he said, containing fermentable material, had been opened at the tip of its point four years ago, but because it was shaped so that no dust could enter, there had been no fermentation of the material inside. Spontaneous generation, whether by contact with pure air or by any other means, was out of the question, he said:

> No, there is no circumstance known today whereby one can affirm that microscopic beings have come into the world without germs, without parents resembling themselves. Those who claim it are the playthings of illusions, of badly done experiments, tainted with errors that they did not know how to recognize or that they did not know how to avoid.

From the perspective of a century René J. Dubos, French-born bacteriologist at the Rockefeller Institute in New York, who studied the work of both Pasteur and Pouchet, came to the conclusion that both, in a sense, were right and both were wrong. He believed it was, in fact, pure air that activated the bacteria in Pouchet's flasks, not by generating them spontaneously, but by bringing to life bacterial spores already in the flasks. What neither Pasteur nor Pouchet realized, he said, was that bacteria can often withstand high temperatures and did not die when Pouchet originally heated the flasks.

A decade after his lecture at the Sorbonne, Pasteur dealt another blow to the idea that life can spring from nonliving matter. He said there is a peculiar quality to the chemical substances in living things that sets them fundamentally apart from nonliving substances. He referred to the work that first brought him to the attention of the scientific world—his study of the manner in which tartaric acids (derived from fermented grapes) twist light waves (that is, change the plane of polarization either to the right or left). The effect is produced by dissymmetry in the structure of molecules related to life, he said. It is not a matter of chemistry, he added, but rather the manifestation of a "force" that has its roots in the dissymmetry of the universe itself.

So effectively did Pasteur discredit the adherents of spontaneous generation that only a few brave souls dared speculate about how life might have originated. The rest clung to the view that some miraculous "spark" was needed to "breathe life" into the first living thing (or they followed the Oriental philosophy that life has always existed).

Among the few who dared say otherwise was Alexander Winchell, professor of geology at the University of Michigan. When he moved to a similar post in Tennessee, according to an account by Melvin Calvin, his teachings on the origin of life led to such virulent attacks that he was asked to resign. He replied that he would do so only if the trustees announced the reason, which they declined to do. Instead his job was abolished and he returned to Michigan, where his *Sketches of Creation* appeared in 1870.

Later in the nineteenth century there were scattered attempts to discuss how life might have originated on the primitive earth, but not enough was known of the chemistry of life (biochemistry) or of the probable history of the earth to make possible the development of a persuasive theory. At the turn of the century, how-ever, a remarkable proposal was advanced: that life came to earth from elsewhere. The father of this idea called it "panspermia," and it channeled the debate over the origin of life into new directions.

9

Panspermia

The idea that life is, and always has been, everywhere can be said to have originated twenty-five centuries ago with the Greek philosopher Anaxagoras, discoverer of the cause of eclipses. Its first modern proponent, however, was Svante August Arrhenius of Sweden, one of the first winners of the Nobel prize in chemistry. At the start of this century he proposed that spores of life are adrift throughout the universe and named the idea "panspermia." James Clerk Maxwell had shown that light exerts pressure, though extremely weakly, and Arrhenius said that after air currents or a volcanic eruption had carried life spores aloft from their native planet and electric forces had moved them free of the atmosphere, the pressure of sunlight sent them far into space. They were then pushed everywhere by the light of other stars.

The proposal was elaborated in 1954 by J.B.S. Haldane of Britain, who called the life spores "astroplankton," a celestial equivalent of plankton, the drifting, microscopic life of the sea. Three years before the first *Sputnik* he wrote: "One of the earliest parties to land on the moon should be able to look for astroplankton, that is to say, spores and the like, in the dust from an area of the moon which is never exposed to sunlight." If the material were forever shaded, he felt, it would not be altered by prolonged exposure to solar radiation.

Haldane was one of the most innovative speculators on how life might have originated. He was also one of the most unorthodox scientists of his time. His career had begun in a manner that would have destined most men to become members of the British establishment. He went to Eton and Oxford and served with distinction in the Black Watch regiment during the First World War. His interest in biochemistry was, in a sense, thrust upon him, for he was gassed during the fighting and then became a specialist in the effects of poison gas. Soon after becoming a reader (instructor) in biochemistry at Cambridge University, he figured in a controversy strikingly reminiscent of a fictional one placed at that seat of learning in C. P. Snow's novel *The Affair.* At the time, Snow was a student at the University of Leicester, but the incident was still reverberating in the halls of Cambridge when Snow came there three years later.

Haldane, because he had been named correspondent in a divorce case, was asked in 1925 to resign from the university. He refused to do so on the grounds that his actions had not been immoral and that his personal affairs had nothing to do with his qualifications as a teacher. The case was brought before the Sex Viri (or Six Men), a university tribunal which consisted of the masters of three colleges and three professors, with the vice chancellor as chairman. They reaffirmed his discharge, but again Haldane appealed, this time to a court that, it was said, had not been convened in a century. Its members included a distinguished jurist, the member of Parliament for the University, the provost of Eton and two well-known scientists, all of them Cambridge graduates. Meanwhile, men like Bertrand Russell and G. K. Chesterton, a leading Catholic layman as well as a novelist, spoke out for Haldane on the grounds of academic freedom. The higher tribunal promptly reinstated him.

For the rest of his life Haldane was a thorn in the side of the establishment—a witty expositor of science (some called him the George Bernard Shaw of science journalism). A brilliant scientist in his own right, he was for a time chairman of the editorial board of the *Daily Worker* in London. He later ceased his activities as a Communist, apparently because of the suppression, under Stalin, of free scientific inquiry, particularly in his own field of genetics.

Haldane discussed the astroplankton idea as a second-choice explanation for the origin of life. In a reference to the steady-state theory of the universe, promulgated two years earlier by Hermann Bondi, Fred Hoyle and Thomas Gold, he said life itself may have had no origin:

> The Universe may have had no beginning. I do not think it had. Further, if it had no beginning, some parts of it may at all times have been in the condition of the parts which we know,

and have included some niches where life was possible. . . . On such a view life is co-eternal with matter.

Not only could life spores be carried from one part of the universe to another by light pressure, he said, it was even possible that they were "launched into space by intelligent beings." This may seem farfetched, he admitted, but if chemical evolution finally proves unlikely, he thought such ideas would have to be taken seriously.

Others argued, however, that even bacterial spores of the type that survived boiling in Pouchet's flasks would be killed soon after they left the protecting envelope of the atmosphere of this or any other planet. Carl Sagan, while still at the University of California at Berkeley, calculated that such spores could not even survive the journey from the earth to Mars. The chief hazard was lethal wavelengths of ultraviolet from the sun and other stars. This hazard would be much less in the vast regions remote from stars, but there would be other hazards—for example, from cosmic rays.

Nevertheless, the idea of panspermia has remained alive, and in 1973 Francis Crick and Leslie Orgel revived Haldane's proposal that life on earth had developed from spores dispatched here by a distant civilization. The latter had sent such seeds of life to many planetary systems—"directed panspermia," Crick and Orgel called it. This might have seemed preposterous had it not been suggested by two highly regarded scientists. F.H.C. Crick, who had demonstrated a lifelong affinity for unconventional ideas, was then at the Laboratory of Molecular Biology in Cambridge, England, having in 1972 shared with James D. Watson the Nobel prize for physiology and medicine for his role in discovering the structure of DNA (deoxyribonucleic acid), the key molecule of genetics. The other author, Leslie E. Orgel, at the Salk Institute for Biological Studies in San Diego, was a leading experimenter on the origin of life. Their proposal was published in the journal *Icarus* and, in 1981, was elaborated by Crick in his book *Life Itself,* after he had joined Orgel at the Salk Institute.

They proposed that to avoid damage from lethal ultraviolet or other sources, the spores had been sent in the nose of an unmanned spacecraft. A motivation for the hypothetical senders may have been realization that their own planet was doomed when their parent star reached its final stage, becoming a red giant so big that it destroyed any planets near it. As regards our own fate, the logician and philosopher Bertrand Russell once wrote: "All the labors of the ages, all the devotion, all the inspirations, all the noonday brightness of human genius are destined to extinction in the vast death of the solar system." If passing on our achievements by direct communication with other worlds proves hopeless, directed panspermia would at least perpetuate life.

Fossil evidence on earth shows that life here, for its first few billion years, was single-celled. Hence, if it came from elsewhere, it did not arrive in an evolved form, comparable to mice or insects. The payload of the spaceship would have contained millions or billions of microorganisms, representing a variety of species capable of coping with whatever environments might exist at their destination. The survivors could then evolve into higher life forms. In the deep cold of interstellar space, approaching absolute zero, they would have been freeze-dried for a journey that could have lasted millions of years.

Such missions could be sent to any number of stars likely to have earth-like planets with oceans suitable for life. Even one surviving bacterium dropped into such an ocean would be enough to "infect" the ocean and set in motion the long, long evolution culminating in intelligent beings. It is generally agreed that all life on earth has a common ancestry, as shown by the universal similarity of its basic chemistry. Crick was particularly familiar with the manner in which DNA specifies the structure and function of organisms, using the same genetic code. "It is a little surprising," he and Orgel wrote, "that organisms with somewhat different codes do not coexist." There are also other clues to the universality of life's chemistry, such as the use of ATP (adenosine triphosphate) to transport energy within all cells, from the most primitive to those constituting a human being. Striking, too, has been nature's choice of the same twenty-three amino acids, from the many that are known to chemists, for the construction of our proteins. Furthermore, although the structure of each amino acid can occur in two forms, each a mirror image of the other, those in living things are all of the left-handed variety.

Crick and Orgel pointed out that since the galaxy is many billions of years older than the solar system, there would have been enough time for an earlier civilization to evolve before the earth was formed, allowing it then to send a mission to our newly emerged solar system. If a distant civilization concentrated on trying to infect nearby solar systems, they said, "this might suggest that we have cousins on planets which are not too distant." If so, they added, "perhaps the galaxy is lifeless except for the local village, of which we are a member." They noted the proposal of Carl Sagan that we might send organisms to such lifeless planets as Venus. Conversely, it would be essential to ensure that the spores we dispatched did not land on a planet already inhabited, unleashing a catastrophic epidemic.

Crick and Orgel realized that arguing for panspermia simply removed the problem of life's origin from the earth to some other planet, or planets. Perhaps it occurred there, rather than here, they said, because such a planet was more propitious. They could not prove life's extraterrestrial origin, but they could find no evidence to deny it. "If the probability that life evolves in a suitable environment is low," they said, "we may be able to prove that we are likely to be alone in the galaxy (Universe). If it is high the galaxy may be pullulating with life of many different forms. At the moment we have no means at all of knowing which of these alternatives is correct."

The proposal evoked a somewhat indignant response from Oparin, the grand old inquirer into life's origin. When I visited him in Moscow six years before the appearance of the Crick-Orgel proposal, he was, despite his age, very much up to date on experiments relating to his favorite subject. By the time he heard of directed panspermia, he was some eighty years old. Countering the argument of Crick and Orgel, he said that although the genetic code was the same in all living creatures on earth, that was no evidence of its extraterrestrial origin, since over the millennia natural selection would have eliminated all other less efficient versions.

The most radical proposals regarding panspermia have been those of Sir Fred Hoyle and his associate Chandra Wickramasinghe. They argue that microorganisms are continuously reaching the earth in debris shed by comets, causing many, if not all, of the world's great epidemics. Besides being a theorist, Hoyle combined his imagination with his knowledge of astronomy and physics to become a successful science fiction writer. Soon after World War II he began marveling at the great diversity of organic molecules being identified in the dust clouds of the galaxy. In his fictional account, *The Black Cloud*, published in 1957, he proposed that such molecules became organized into a living entity—a black cloud—that headed directly for the sun, seeking to replenish its energy from our star. As it approached the sun it entirely cut off the earth's sunlight, freezing to death a quarter of the world's population. Communication with the cloud was established by astronomers, who warned it that hydrogen bombs had been sent toward it by trigger-happy governments. The cloud reversed the courses of the bomb-bearing missiles, causing further devastation on earth, but then departed without further vengeance.

As shown by this and other fanciful works of science fiction, Hoyle has not shunned far-out ideas, although much about his past has been more mainstream. From 1967 to 1973 he headed the Institute of Theoretical Astronomy at Cambridge University after his historic success in the 1950s, with William Fowler, Margaret Burbidge and her husband, Geoffrey Burbidge, in explaining how stars and stellar explosions synthesized the heavier elements. In 1983 this work won a Nobel prize for Willi Fowler, but not for the other three, as often happens when there are so many participants.

Meanwhile, radio astronomers, scanning the dust clouds at microwave and infrared wavelengths, had by 1987 identified more than sixty organic molecules. While some, like carbon monoxide, produce a strong signal, others do not, particularly the more complex molecules associated with life. The catalogue included not only water but alcohol, ether, ammonia, acetylene and formaldehyde (which, mixed with water, is used for embalming). Most were combinations of two, three, four and five atoms, but the largest, cyanodecapentyne, is a thirteen-atom molecule (compared to thousands in a protein).

By the 1970s Hoyle and Wickramasinghe suspected that much of the infrared emissions from space were from cellulose, the building material of all plant cell walls. Laboratory tests then persuaded them that they were seeing not only cellulose but a variety of other carbohydrates. Meanwhile, in comets radio astronomers were also finding a variety of substances thought to be precursors of life, such as methyl cyanide and hydrogen cyanide. Likewise the analysis of some carbon-rich meteorites showed that they contained what Hoyle and Wickramasinghe called "cellulose coal." A number of astronomers besides Hoyle and his colleague had proposed that the delivery of such material by comets when the earth was young and the Great Bombardment was barely ended had been much greater than today and could have provided the material needed for the emergence of life. In their 1978 book *Life-*

cloud Hoyle and Wickramasinghe argued that since space is full of prebiotic molecules, "it is almost self-evident that the origin of life on earth merely involved a piecing together of interstellar pre-biotics," much as the units of a child's construction toy can be assembled. They then went an important step further: some 4 billion years ago life reached the earth in microscopic form, not inside a spaceship, as proposed by Crick and Orgel, but in the protective shielding of a comet.

In their 1979 book, *Diseases from Space*, they further proposed that viruses and bacteria causing the diseases of plants and animals have come from beyond the earth, as did the original seeds of life, and are still falling through the atmosphere. They argued that this explains the sudden appearances of new strains of influenza, instead of the conventional view that they developed unobserved, for example, in South China pigs, accounting for the 1968 epidemic of "Hong Kong flu." They took as evidence for an extra-terrestrial source the simultaneous appearances of a new strain in places extremely remote from one another. "The virus," they wrote sarcastically, "apparently has the property of being able to make leaps in days, over many thousand miles." They described cases in which person-to-person contagion seemed impossible. They admitted that the analysis of meteorites and particles from space had not shown any pathogens, but, they wrote, if an infectious virus was contained in "as little as one part in a hundred million of this material," it would be enough to start a worldwide pandemic. They found what they took to be corroborating evidence in a survey of flu outbreaks in British schools. They visited some and examined the health records of many others, including a half dozen near their base of operations at University College in Cardiff, Wales.

Not all types of disease are recent arrivals, they said, having come to earth long ago, and some may be occasionally replenished. As examples they cited polio, mumps, cholera, and Legionnaires' disease (AIDS not having yet appeared). Although global vaccination has eradicated smallpox, they raised the possibility that there may be a new invasion from space.

In the 1981 book *Evolution from Space*, the same authors invoked divine intervention to explain the original appearance of life somewhere (or everywhere) in the universe, as well as the most perplexing steps in its evolution on earth. Those steps, they argued, could not be explained by Darwinian selection. They noted the many breaks in the fossil record of evolution—gaps, they proposed, that were filled by the intervention of some supreme intelligence. They knew that such an argument would not be welcomed by many scientists. "We have received hints and even warnings from friends and colleagues that our views on these matters are generally repugnant to the scientific world," they said, but they continued to defend their hypothesis of divine intervention.

The chief test of their life-in-comets hypothesis came in 1986 when an international fleet of five spacecraft—one European, two Japanese and two Soviet—greeted Halley's comet on its return trip to the sun. In the most ambitious astronomical effort in history, those probes, as well as virtually every observatory on earth, joined in scanning the comet as it swept past the earth and around the sun. Hoyle saw a chance to show that living organisms are carried by such objects. Three of the craft, passing close in front of the comet, were to penetrate the coma, analyzing its dust grains and trying to photograph the solid comet head inside it.

One of the craft was *Giotto*, sent by the European Space Agency and named for the great Florentine painter Giotto di Bondone, whose fresco *Adoration of the Magi* recorded the appearance of Halley's comet in 1301. The other two, *Vega 1* and *Vega 2*, were sent by the Soviet Union with a variety of instruments provided by other nations as well as the USSR. *Vega 1* flew into the coma ten days ahead of *Giotto* and, in the cooperative spirit that marked this effort, the Soviets provided the Europeans with data from *Vega 1* so that *Giotto* could be guided close enough to photograph the comet head and yet not be destroyed by dust bombardment. When *Vega 2* flew through the coma three days after *Vega 1*, the hurricane of dust on its power-generating solar panels cut their efficiency 80 percent. Observations were made at greater distances by the two Japanese spacecraft and from even farther away by the International Cometary Explorer (ICE), which had previously been sent by NASA through the tail of another comet, Giacobini-Zinner.

Astronomers had predicted that while Fred Whipple's traditional concept of comets as "dirty snowballs" would remain valid for the interior of the comet, volatile components of the "snow" on the surface would have been driven off by solar radiation, and, some believed, the remaining crust would also have been darkened by prolonged cosmic radiation. Hoyle predicted that if the comet head came into view, it would be as dark as chocolate. This was confirmed by all whose instruments were able to look inside the coma, revealing a peanut-shaped comet head 10 miles long and 5 wide. Jets of gas and dust were shooting out from several areas on its surface, presumably activated by sunlight. It was estimated from the *Giotto* data that the comet head was losing 6 tons per second. Because the comet was rotating, the jets did not strike all three spacecraft equally. It was calculated that if the comet shed 6 tons per second while near the sun, every seventy-five years, its surface each time would be lowered by 5 feet, averaged over the whole comet, limiting its ultimate lifetime. *Giotto* was so heavily battered while inside the coma that its wobbles affected the aim of its earth-oriented radio antenna.

Hoyle and Wickramasinghe had asked that those testing the dust look for viruslike grains, and the answer came in April 1987. Two scientists from Germany, J. Kissel of the Max-Planck-Institut für Kernphysik in Heidelberg and F. R. Krueger from the Arheiliger Apotheke in Darmstadt, reported in *Nature* the results from *Vega 1*. All three spacecraft carried "dust particle impact spectrometers" for dust analysis, but, they wrote, *Vega 1* "flew through the most dusty environment" and had the clearest results. The devices determined the mass (and, by inference, the composition) of each impacting particle. The interpretation of the results was in part confirmed by laboratory simulations. The impact velocities, as the spacecraft swept past, were, of course, extremely high (48 miles per second).

The two scientists identified more than thirty "organic" molecules—those containing carbon—including many with biological affiliations, such as those (pyrimidines and purines) that spell out the "message" in the genetic code. Some had been inferred from ground-based observations of comets and dust clouds, but in those cases the particles were observed after they had flown away from the comet and sometimes become altered. *Vega 1* showed that a typical particle had a fluffy mineral (silicate) core enveloped in a mantle of even fluffier organic material similar to that in very primitive meteorites (the carbonaceous chondrites in which, as will be seen later, amino acids and other prelife molecules have been found).

Particles from the coma had features in common with those collected from space near the earth by D. E. Brownlee of the University of Washington, flying as high as possible in a U-2 reconnaissance plane, as well as those extracted from sea floor sediments. Nearly a thousand such particles had been analyzed, most of them assumed to be from comets, Brownlee told a 1980 colloquium on "Comets and the Origin of Life" at the University of Maryland. He estimated that 10,000 tons of such dust falls to earth each year. For any of it to carry a panspermia organism without destruction by heat, he said, the particle would have to be microscopic, but he did not rule this out. If the particles enter the atmosphere at a shallow angle, they are often slowed down so gradually that they are not greatly heated. The ones we see as shooting stars typically come in at steeper angles. Analysis of those collected showed that 80 percent of the ones less than a millimeter in size are of the same composition as the primitive meteorites. In 1961 Joan Oro, a Spanish-born biochemist at the University of Texas, had proposed that before the origin of life on earth, comets had brought to the planet a rich variety of life's component molecules.

The German scientists said that at least from the *Vega 1* analysis of dust from Halley, the prediction of "spermiae" by Hoyle and Wickramasinghe had been disproved. Freeze-dried cells, such as those predicted, would contain at least some fraction of sodium and potassium, but less than one part in a million was found. Nor was there any significant number of oxygen-rich compounds, such as sugars and peptides. Nucleic acids should contain phosphorus, but it too was missing in any appreciable quantity.

In a letter to *Nature* Hoyle and Wickramasinghe replied that they did not claim all organic material in comets is biological. And, they said, the dust grains "could surely not be observed in anything like their pristine state" after such a high-velocity impact. They argued that the infrared spectrum of the comet obtained from the ground by David Allen and Dayal Wickramasinghe, Chandra's brother, using the Anglo-Australian Telescope on Siding Spring Mountain in New South Wales, was strikingly like that from living material. The Germans, in their response, stuck to their guns: "We are, indeed, very astonished that one may draw conclusions of living matter being present merely from infrared spectra."

Despite widespread doubts that living spores are adrift in space, experiments have purported to show that some of them might be able to withstand the space environment for a considerable period. The tests, conducted by J. Mayo Greenberg at the University of Leiden in the Netherlands, were reported to the Royal Astronomical Society in London in 1983. In an effort to mimic conditions in space the tests were done in a deep vacuum under ultraviolet light. The latter broke up some biological molecules and stimulated the synthesis of others. Most, but not quite all, bacteria were killed.

Hoyle and Wickramasinghe, in support of their thesis, quoted Hermann Ludwig Ferdinand von Helmholtz, the great nineteenth-century scientist who first established the conservation of energy as a fundamental principle:

> It appears to me to be fully correct scientific procedure, if all our attempts fail to cause the production of organisms from non-living matter, to raise the question whether life has ever arisen, whether it is not just as old as matter itself, and whether the seeds have not been carried from one planet to another.

Most of those seeking the answer now believe it will be found on earth, and not in space. The Germans who analyzed the *Vega 1* results said in their original report that although they had disproved the Hoyle-Wickramasinghe hypothesis, "we do think that the question of the origin of life in the content of primordial matter as found in comets has become even more exciting. We stress that the substance classes, which we believe are present in cometary dust are highly reactive especially in warm water." Should some mechanism bring such substances into contact with liquid water, they said, the ensuing reactions could produce sugars and components of the genetic molecules. They noted the suggestion that high concentrations of such substances could "trigger the self-organization of nucleic acids." Is this what happened in a previously lifeless ocean billions of years ago? While this discussion continues, there are other scientists who claim that some meteorites contain the fossil remains of life elsewhere.

10

Fossils from Space

At approximately 5:30 P.M. on March 15, 1806, a man named Reboul and his son Mazel, employees of a local landowner, were working in the fields near the village of Valence, in the south of France, when they heard the unlikely sound of cannon shots. Napoleon's armies were far to the east, for the battle of Austerlitz had been fought only three months earlier. Furthermore, the sound seemed to come from the wintry sky and was followed by prolonged thundering of a peculiar sort. This, in turn, was succeeded by a sound that, the two men said later, resembled the scream of a spinning well hoist that has been released, allowing the bucket to plunge downward.

A moment later they were transfixed by the sight of an object flying through the air toward them. It plunged into the ground fifteen paces from Reboul, who cautiously drew near. He found a lump of black material, about the size of a child's head, that had broken into three pieces.

Meanwhile, a few miles away, another father-son team in the fields had a similar experience. They told the scientifically inclined burghers who came to investigate from the nearby town of Alais that they saw a dark body come flying out of the clouds and land near them, bursting into fragments and digging a shallow pit.

Reboul and his son were persuaded to part with one of their three chunks, and another was obtained from the second batch of fragments. The rest have been lost to science—a pity, for this proved to be the first known fall of a carbonaceous chondrite, an extremely interesting visitor from heaven. It is known, because of its proximity to that town, as the Alais meteorite.

Some twenty-eight years later a fragment of Alais reached J. Jakob Berzelius, the great Swedish chemist, who took one look at it and decided there must have been some mistake. This could not be a meteorite. It was too crumbly, and the material disintegrated in water. There were then thought to be three kinds of meteorite, all of them hard and insoluble: those of iron (with an admixture of nickel), those of stone, and those of both iron and stone. Most of the stones were filled, like a raisin cake, with small spherical objects (now known as chondrules).

Berzelius was about to throw the specimen away when he reread the description of the object as it was found, fresh and warm after the fall. This, and the evidence that a paper-thin layer of the original surface had been fused by heat, convinced him that the object on his desk had really fallen from the sky. He set to work, with the best methods of analysis available in 1834, to determine its composition.

To his amazement, it was rich in carbon compounds and responded to his tests much as would humus, the mixture of decayed vegetable and animal matter that enriches our soil. Might it, in fact, be genuine humus or some other assemblage of organic compounds? he asked in his report. "Does it possibly give an indication of the presence of organisms on extraterrestrial bodies?"

His question 130 years later became one of the most hotly debated problems of contemporary science. Berzelius himself had decided in the negative. Despite the resemblance of carbon compounds in the meteorite to those in the soil of our own planet, he said, this "does not appear to justify the presence of organisms in its original source." He believed that rock on the parent body whence came this "chunk of soil" must have been converted into earth by some inorganic process.

Four years after Berzelius did his analysis another of these strange objects was seen to fall, this time near the Cold Bokkeveld Mountains in South Africa. In 1857 a third was recovered at Kaba, near Debrecen, Hungary, and samples of Kaba and Cold Bokkeveld were sent to Berzelius' most famous student, Friedrich Wöhler, at Göttingen, Germany. It was Wöhler who, some years earlier, had achieved the first laboratory synthesis of an organic compound (urea). From Cold Bokkeveld he extracted an oil "with a strong bituminous odor." Although the proportion of organic material was very small, it was definitely there, he said, and after his analysis of both specimens he concluded that "according to our present knowledge" such substances could only have been produced by living organisms.

That was in 1860. Four years later, on the night of May 14, 1864, the peasantry in the south of France had another terrifying experience. An object appearing as large as the full moon, but more like the sun in brilliance and slightly teardrop-shaped, raced across the sky with the sound of an express train, punctuated by peals of thunder. It was visible throughout Aquitaine as it broke into pieces that quickly darkened. Behind it a broad, luminous trail turned to white smoke and slowly dissipated. When dawn came, it was found that parts of the meteorite had rained on the region around the village of Orgueil. Scientists recovered twenty pieces, some head-sized but most smaller than a fist. They found that the specimens could be cut with a knife and, when sharpened, could be used like pencils.

A specimen of the Orgueil meteorite was immediately examined by a French scientist, S. Cloëz, who found it was held together by a water-soluble salt. When moistened, the material immediately turned to dust. In fact, one chunk of this meteorite, preserved in the humid environment of a museum in Calcutta, has disintegrated into dust spontaneously. Cloëz was much impressed by the resemblance of the carbon, hydrogen and oxygen content of his sample with that found in peat or lignite (a woody variety of coal). Such substances in this and other similar meteorites, like Alais, he said, "would seem to indicate the existence of organized substances in celestial bodies."

By "organized substances" he meant life. His report was communicated to the Academy of Sciences in Paris by Gabriel-Auguste Daubrée, one of the six members of its Mineralogical Section, another of whom was Louis Pasteur. It was, in fact, the year of Pasteur's great lecture on spontaneous generation. Pasteur said such generation could not occur on earth, yet here was possible evidence that it had taken place elsewhere. Orgueil and Alais were a new class of meteorite which came to be called carbonaceous chondrites because of their carbon content and the raisin-like chondrules that permeate them.

Although the report of Cloëz came at the height of the dispute between Pasteur and Pouchet regarding the origin of life on earth, the controversy regarding the carbonaceous chondrites did not reach its climax until the 1960s. By then, because they typically disintegrate soon after landing, only 20 had been identified out of a total of more than 1,500 well-authenticated "falls" and "finds." The arrival of a "fall" is witnessed, whereas a "find" is discovered on the ground and may have fallen many years earlier. A fall can produce a great number of fragments. One that struck Pultusk, Poland, in 1868 is said to have broken into at least 100,000 stones, and the Sikhote-Alin meteorite, which fell on the mountains of that name flanking the east coast of Siberia, showered the region with iron on February 12, 1947. At least 60,000 pounds of it have been collected.

Meteorites have been known—and even worshiped—for thousands of years, but not until 1803, when Jean-Baptiste Biot, a French physicist, witnessed a fall of some two thousand stones, did the scientific world fully accept the idea that they come from the sky. For example, the Black Stone in the Kaaba at Mecca is the holiest of holies in the Muslim world, having been kissed by countless pilgrims since before the days of Mohammed. To free his followers from idolatry, Mohammed destroyed various sacred objects in the Kaaba, but he spared the Black Stone.

Even before the Iron Age, meteorites provided material from which knives were made—weapons valued at many times their weight in gold. The largest meteorite on exhibit is a 31-ton chunk of nickel-iron brought back from Cape York, Greenland, by Robert E. Peary, discoverer of the North Pole. It is on display at the American Museum of Natural History in New York. Meteorites, as they exist in space, seem to come in all sizes, from chunks miles in diameter down to dust grains.

It is calculated that any meteorite heavier than 100 tons hits the earth with such force that a major portion of it is vaporized by the resulting heat. In other words, it blows up, producing a characteristic crater with raised rim. The classic example is Meteor Crater in Arizona, whose width is three-quarters of a mile. It was produced some 25,000 years ago by an object thought to have been 140 feet in diameter, hitting the earth at 45,000 miles an hour. Most of it presumably vaporized in the resulting explosion. The overlapping big and little craters of the moon, as well as those photographed by spacecraft on Mercury and Mars, show that they have been struck by an enormous number of objects. The view of Harold Urey that most lunar craters were produced billions of years ago when space was still cluttered with planet-forming objects is now generally accepted. Specimens brought back by the astronauts show that most of these craters and "seas" were formed when the moon (and Earth) were young. Nevertheless, it is obvious that such impacts have been continuing. On Earth weather, through erosion and sedimentation, has erased the most ancient features, as has repeated plowing under of the surface by the processes that are constantly reshaping the crust.

In spite of these influences there are hints of giant craters in the earth's landscape, although until recently there was no way to tell whether or not they were produced by meteorites. The story of how this question was settled goes back to 1891, when a tiny diamond was found in one of the meteorite fragments picked up near Meteor Crater. The fragment is now at the Smithsonian Institution's Museum of Natural History in Washington, D.C. Diamond is a form of carbon created at depths of 200 miles or more within the earth, the reason it is so rarely found near the surface. Whereas graphite is an arrangement of carbon atoms that can be produced under normal pressure, it takes the weight of hundreds of miles of rock to squeeze carbon atoms into the diamond configuration. This, in part, led Urey to presume that diamond-bearing meteorites must have been formed inside bodies at least as large as the moon.

In November 1959, NASA called a conference in Washington on the origin of the moon and meteorites. Among those invited were two members of the faculty of the University of Chicago: Edward Anders of the Enrico Fermi Institute for Nuclear Studies and John C. Jamieson of the Geology Department. Neither knew the other, but they recognized names as they registered at the Willard Hotel, and they went to the bar for a drink. Anders had just been given some diamonds from the debris around Meteor Crater, and the conversation turned to their origin. Anders had been wondering whether it was possible that these crystals had been produced by the shock of impact, rather than by prolonged pressure. This was not a new idea. People long before had wondered whether they could achieve sudden wealth by pounding carbon, but it never worked. However, it occurred to the two men that there is one kind of graphite—the rhombohedral form—whose atoms are arranged in the same manner as in diamond, except they are slightly farther apart. Might it be possible, by hitting this graphite hard enough, to produce synthetic diamonds? Jamieson, by a fortunate coincidence, had been doing some experiments on the effects of shock with Paul De Carli at the Stanford Research Institute in California. The idea was put to the test in January 1960, and sure enough, tiny diamonds were produced.

Soon after this was reported in the journal *Science*, letters began arriving from the big industrial diamond concerns—de Beers, Krupp and others—asking for reprints. Obviously they were tantalized by the thought of "instant diamonds." Meanwhile a number of concerns had been moving toward the high-pressure synthesis of diamonds, an important step being the development of equipment for such work by Loring Coes, Jr., of the Norton Company in Worcester, Massachusetts. In 1953 Coes was able to squeeze quartz into a dense state never before observed. The new substance, like quartz, was still silicon dioxide, the prime constituent of glass. Its chemical composition was unchanged, but its atoms had been compressed into a tighter configuration.

In honor of Coes the new substance was named coesite. In their experiments in transformation by shock, De Carli and Jamieson tried to pound quartz into coesite without success. Eugene M. Shoemaker of the United States Geological Survey looked in vain for coesite in quartz-bearing material subjected to a nuclear explosion.

Then Shoemaker and two colleagues, Edward C. T. Chao and B. M. Madsen, decided to look at the fragments of Coconino sandstone from in and around Meteor Crater. Shattered pieces of this white, quartz-bearing sandstone constitute a large part of the debris that lies, to a depth of 600 feet, under the crater floor. Samples from the crater rim and from a hole drilled 650 feet into the crater floor were analyzed by X-ray diffraction, optical properties and spectrograph. All three methods disclosed the presence of coesite—the first time it had been seen outside the laboratory.

When Shoemaker, Chao and Madsen reported their discovery, in the July 22, 1960, issue of *Science*, they noted that coesite "may afford a criterion for the recognition of other impact craters on the earth and perhaps ultimately on the moon and other planets."

There was still another substance in the Meteor Crater debris that Chao and his coworkers could not identify. When X rays were shined through the tiny grains of this material they were diffracted into a pattern unlike that of any known material. The following year S. M. Stishov, who was doing high-pressure research at Moscow State University, sent to his American colleagues a report on his experiments in the further compression of silicon dioxide. He said that with a pressure of 2,350,000 pounds per square inch and a temperature above 2,200 degrees he had squeezed this substance into an even more compressed phase than coesite. In the lattice structure of its crystals six oxygen atoms, instead of the usual four, were equidistant from each silicon atom. The new material was named stishovite, for its synthesizer.

On a computer Chao calculated the diffraction pattern to be expected when X rays passed through stishovite crystals, and it proved identical with the unidentified pattern from the Meteor Crater debris. It probably takes a shock equivalent to the weight of 300 miles of rock to make stishovite, compared to only 200 miles for diamonds. The Meteor Crater explosion must have been a terrible event indeed.

With such tags for the identification of impact craters, it was now possible to find out whether or not even larger depressions were produced in this way. Coesite was soon found at several of them, including the Ries Kessel, a bowl 15 miles wide in western Bavaria. It has thus become evident that, from time to time, the earth is struck by huge meteorites that explode with violence greater than that of the most terrible nuclear weapon. While meteorites large enough to leave a crater are rare, more than a thousand sizable ones hit every year, most of them unobserved because they land in the sea or in remote areas. They tend to fall on the afternoon side of the earth, overtaking our planet in a manner that suggests they were in orbit farther out in the solar system.

Only one person is known to have been hit by a meteorite. On November 30, 1954, Mrs. E. H. Hodges of Sylacauga, Alabama, was sitting in her house after lunch when a 9-pound stone crashed through the roof and hit her in the upper thigh. She was not seriously hurt. According to a tabulation by Christopher Spratt and Sally Stephens in a 1992 issue of *Mercury*, published by the Astronomical Society of the Pacific, since the eighteenth century more than fifty buildings have been hit in various parts of the world. During the evening of November 8, 1982, Robert and Wanda Donahue of Wethersfield, Connecticut, were watching an episode of *M*A*S*H* on their television when they heard a sound "like a truck coming through the front door." A 6-pound meteorite had plunged through the roof, had broken through the living room ceiling, had bounced back through the ceiling into the attic, and then had fallen again through another ceiling into the dining room. Spratt and Stephens documented four instances, three of them in the United States, in which a meteorite hit within 1 meter (3 feet) of a person.

No rocks on earth resemble those meteorites containing beadlike chondrules. The latter seem to be frozen droplets of a fiery rain, yet are embedded in material that has never been greatly heated. Such relics from the formation of the solar system testify to a time of sound and fury whose nature is still uncertain. Sometimes a meteorite of one type is embedded within one of a quite different kind. In some cases material within a single specimen represents five "generations" of breakup and reassembly. Those meteorites lacking in chondrules, both stones and irons, are assumed to be fragments from collisions involving planets and asteroids. Some are even thought to have been knocked off Mars and the moon. The irons are assumed to be fragments blasted by a catastrophic collision from the nickel-iron core of a body large enough, like the earth, to have differentiated itself and formed a core. When in 1808 Count Alois Josep von Widmanstätten of Vienna sliced, polished and etched the surface of an iron, he found a geometric pattern of exquisite beauty. Such "Widmanstätten patterns" are produced when two mixtures of nickel and iron arrange themselves into eight-sided patterns as they cool. Laboratory tests show they do so only if, after being hotter than 900 degrees Celsius, they cool very slowly. This would have to mean they were inside a large object and were not formed by some sudden event.

In a 1987 summary of worldwide collections of meteorites Harry H. McSween, Jr., of the University of Tennessee reported that 84 percent were stones containing chondrules (including carbonaceous chondrites), 8 percent were stones free of chondrules, 7 percent were irons, and 1 percent were stony-irons (presumably from the edge of the core of some large body).

The total bulk of the asteroids, of which many meteorites are fragments, is not great. The largest, Ceres, is 480 miles in diameter. Only about 2,000 have been catalogued, although it is estimated that 55,000 come within observable range at one time or another. As they rotate, the bigger ones reflect sunlight with constant intensity, indicating they are spherical, rather than jagged fragments. The old idea that they are the debris of two planets that collided (or one that blew up) has gone out of fashion because, if all the asteroids were assembled, they would constitute only a few hundredths of the mass of the moon. While the falls of extremely large meteorites, including those big enough to be classed as asteroids, are rare, the rate seems to increase steadily with smaller sizes until, at the microscopic level, there is a constant rain of material, including the fluffy grains shed by comets. Some of the latter, as noted earlier, have been collected by high-flying U-2s. Others plunge into the atmosphere as shooting stars. Such meteors are, at most, pebble-sized and are destroyed in the atmosphere, as opposed to meteorites, which are substantial enough to reach the earth's surface.

Most meteors, being cometary debris, are still orbiting the sun along the highly elliptical paths of their parent comets. They are seen chiefly on those days each year when the earth crosses one of those paths. The better known of these "meteor showers" are named for the constellations from whose direction they seem to be coming, such as the Lyrids (from the direction of Lyra), which fall on or about April 20, the Perseids of around August 10, the Leonids of mid-November and the Geminids of December 10 to 13. The fact that meteorites do not fall any more often during these showers than at other times seems to show that the two phenomena are unrelated. The icy comets and their offspring are thus an entirely different breed from the planets, asteroids and meteorites, although there is a suspicion that some "asteroids" in highly elliptical (and therefore cometlike) orbits may be the gravelly remains of comets that over millions of years and many close passages of the sun have shed their ice.

It is obvious that comets sometimes strike the earth. On June 30, 1908, a fireball streaked across the Yenisei basin in central Siberia, and from Kirensk, 250 miles away, a "pillar of fire" was seen over the horizon, followed by several detonations. Gusts of wind swept across the steppes, so powerful that horses were knocked down near Kansk, more than 400 miles away. Earthquake recorders from Washington to Java detected the explosion, and shock waves in the atmosphere circled the world in all directions, thus being recorded twice at Potsdam, Germany. For several nights thereafter the sky was strangely bright, from Spain to the Arctic Ocean, because of particles high in the atmosphere.

It was not until 1921 that the first scientific expedition reached the remote site, yet it was possible to locate the area of the explosion because trees had been felled outward in all directions for 20 or more miles. On hilltops they were leveled more than 35 miles away, their bark and branches often stripped off. At closer ranges they were charred. Yet no large crater or fragments were found. This, the Tunguska "meteorite," is thought to have weighed several thousand tons, but it exploded in the atmosphere, like a hydrogen bomb. Some fanciful Russians have proposed that this was

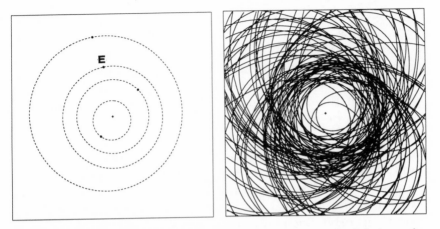

The earth resides in a swarm of asteroids. The diagram on the left shows the orbits of the inner planets, including the position of each on January 1, 1993. On the right, to the same scale, are the orbits of the 100 largest asteroids known to pass near the earth. These diagrams, prepared by Richard P. Binzel of M.I.T., were displayed at the NASA International Near-Earth-Object Detection Workshop held in 1992 at three sites in California.

a nuclear device or space ship from another world. In 1964, however, James R. Arnold of the University of California at San Diego suggested at a meeting of the National Academy of Sciences that Tunguska was a burned-out comet (its approach was never observed). Many specialists believe the earth gains more material from comets and their powdery debris than from meteorites, and that comets are therefore the most likely source of the carbon compounds that figured in the emergence of life.

Arnold warned that other objects whose orbits come close to the earth may follow Tunguska's example and cause another giant explosion. Let us hope, he said, that it does not occur during a period of international tension, since it could easily be mistaken for a hydrogen bomb. More sobering, to say the least, would be the impact of an object as large as the one that struck the earth 65 million years ago, causing, many believe, the global extinctions that wiped out the dinosaurs. Another such impact could have fatal consequences for humanity.

In 1990 NASA began sponsoring a series of workshops on how to detect an object on a collision course with earth and how to divert or destroy it. One, held in 1992 at the Los Alamos National

Laboratory, a center of "Star Wars" research, was attended by some seventy specialists, including astronomers and weapons experts, among them Edward Teller, "father" of the hydrogen bomb. For effective countermeasures a threatening object would presumably have to be detected one or more orbits in advance, allowing enough time for preparations. To blow it up would create a shower of fragments some of which, still headed toward the earth, might do as much damage as the original object. More effective might be use of a nuclear device to change its orbit, applying the appropriate boost at its maximum distance from the sun. Military participants in the 1992 meeting argued that if one spot on the surface of the threatening object were heated by a neutron bomb, material would fly off the object like a jet blast, altering its orbit. Such an interception would be extremely costly and not lightly undertaken. Meanwhile, there is greater support for the effort to find and keep track of such objects.

Beginning in the 1980s several astronomers began devoting all or most of their observing time to the hunt for objects in orbits that cross that of the earth. They report their discoveries and further observations for orbital analysis and prediction at an international center at the Smithsonian Astrophysical Observatory in Cambridge, Massachusetts. By 1992 more than a hundred had been found, but that was only 5 percent of the estimated total. Victor Clube, an astrophysicist at Oxford University, has proposed that during a century or two, a comet may shed a number of fragments 50 to 500 meters (164 to 1,640 feet) in diameter, the Tunguska explosion having been caused by an object at the smaller end of this range.

The study of meteorites was given a great boost in 1969 when a Japanese expedition found nine of them lying on the ice inland from Antarctica's Yamato Mountains. This demonstrated that where the seaward flow of ice from the interior is stalled by mountains, the great ice sheet covering Antarctica is an extraordinary collector of meteorites. Those that have fallen and been buried in the ice during its long, slow journey from deep in the interior are exposed near the mountains as the stalled ice is worn away by wind and weather. On the opposite side of Antarctica, Americans found a similar situation west of McMurdo Sound. As a result the number of specimens in the world's collections has roughly tripled, Antarctica contributing more than 13,000, including at least six of the remarkable carbonaceous chondrites.

Meanwhile, beginning in 1959, the investigation of such meteorites took a sensational turn. In that year Melvin Calvin and his University of California coworkers, exploring ways in which life might have originated, obtained a chunk of the carbonaceous chondrite that had fallen near Murray, Kentucky, in 1950. They subjected it to the most precise techniques of analysis at their disposal, and in January 1960 Calvin and Susan K. Vaughn reported their results. They noted that earlier analyses of two carbonaceous chondrites—Cold Bokkeveld and Orgueil—had shown signs of hydrocarbons, the family of compounds that includes petroleum. Their own analysis, they said, hinted at the presence of material resembling cytosine, one of the four "words" in the "code-of-life" DNA molecule. They did not, however, claim that cytosine had been identified. The California scientists also found hydrocarbons similar to those in waxes and petroleum, although they could not identify any of them. They assumed that all these substances had been formed by nonliving processes and saw this as exhilarating evidence that reactions postulated by Calvin for the young earth had taken place in space.

The following year the debate over carbonaceous chondrites moved into high gear when, on March 16, 1961, a paper was presented at the New York Academy of Sciences describing a new analysis of the Orgueil meteorite. The three authors were Bartholomew Nagy and Douglas J. Hennessy, both of the Department of Chemistry at Fordham University in the Bronx, and Warren G. Meinschein, a petroleum chemist at the Esso Research and Engineering Company in Linden, New Jersey. Nagy, aged thirty-four, had received his early education in his native Hungary and had more recently done research for the American oil company.

They had used techniques typical in petroleum research. Hydrocarbons extracted from the sample proved to be built on chains of as many as twenty-nine carbon atoms and displayed striking similarities to the paraffins and other hydrocarbons in living matter. Such biological hydrocarbons tend to have carbon spines of odd-numbered atoms (twenty-one-atom chains, twenty-three-atom chains, etc.). The experimenters found that the assortment of paraffins in the Orgueil meteorite, when classed by weight, resembled those in butter and in sediments containing recent life remains. One constituent, they said, seemed related to the cholesterol found in blood.

They concluded that they were not analyzing bits of dirt that had entered the sample after its fall to earth, for the specimen's saturated hydrocarbons were far more abundant than in typical soils. Saturated molecules are those whose carbon atoms lack free sites where additional hydrogen atoms could hook on.

The three scientists described their efforts to avoid contamination: glass equipment acid-cleaned and baked in a vacuum; mortars and pestles kept red-hot for at least a half hour; no grease used on stopcocks, and so forth.

The next day Robert K. Plumb, reporting in *The New York Times*, quoted one of them as saying: "We believe that wherever this meteorite originated something lived." It was clear that they regarded this as the first direct evidence of life beyond the earth. Plumb got in touch with two men experienced in the study of meteorites: Harold Urey in California and Brian Mason at the American Museum of Natural History in New York. Both urged caution in the interpretation of this evidence. Nevertheless, the fats—or rather, the saturated hydrocarbons—were in the fire, and the conflagration was not to simmer down for a number of years.

With the surge of public interest that followed the New York meeting a reporter from the news agency Science Service went to see the keeper of the meteorites at the Smithsonian Institution in Washington and discovered that, besides giving Calvin a chunk of the Murray fall, he had presented another piece to a microbiologist at the U.S. Geological Survey named Frederick D. Sisler. The latter was cultivating living organisms that he had ostensibly extracted from the heart of the specimen.

Needless to say Sisler's phone began ringing, and the press quoted him as saying that while he could not guarantee the extraterrestrial origin of his organisms, it was a real possibility. He pointed out to the reporters, however, that water seepage could have carried bacteria into the heart of the porous specimen. *Life* magazine published an article titled "Wax and Wigglers: Life in Space?" The "wigglers" were Sisler's bacteria; the wax was the paraffin found by Nagy and his colleagues.

For about five years, Sisler said, he had been studying meteorites, concentrating on specimens provided by the Smithsonian Institution. "I have found either traces of organic matter or germs, or what appeared to be germs living or dead, in almost everything I have examined," he said.

His procedure with the two Murray specimens was as follows: The surface was sterilized by exposure to intense ultraviolet radiation, from all sides, for half a day. It was soaked in hydrogen peroxide to rid it of loose dirt. It was then held briefly over a flame and was plunged into a germicidal solution. Finally it was placed in a germ-free chamber at the National Institutes of Health in Bethesda, Maryland. Such chambers are tanklike containers within which everything, including the air, is sterile. By remote control Sisler ground up the samples with a mortar and pestle, then placed the material in a clear fluid rich in nutrients. The fluid sometimes became cloudy, although it often took several months to do so, which suggested to Sisler that the sparks of life in the meteorite might have been crippled in some way, requiring a long time to reconstitute themselves. He listed four possible explanations for his discovery of living bacteria: his analytical method may have been faulty; earth germs may have penetrated the specimen; there may have been life on the body whence this fragment came; or there may have been "autocatalysis" in which the meteorite material, immersed in a favorable growth medium, completed the final stage of chemical evolution and came to life —"something akin to spontaneous generation."

Sisler was not the first to report finding bacteria in meteorites. In 1932 Charles B. Lipman, in a publication of the American Museum of Natural History, described a series of experiments that resembled those of Sisler in that he took what he considered drastic measures to sterilize the exterior of each specimen and grind it up under what he believed were sterile circumstances, although he did not have at his disposal the elaborate facilities of the Germ-Free Laboratory in Bethesda.

Lipman was no fly-by-night crackpot. Having come to the United States from the USSR as a child, he had risen to be professor of plant physiology and dean of the graduate division at the University of California at Berkeley. From pulverized fragments of several stone meteorites (Pultusk, Johnstown, Modoc and Holbrook) he cultured a variety of bacteria, including cocci, chains of cocci, rods, chains of rods and chains of sausagelike bacteria. He conceded that they resembled earth bacteria, but said there was no valid reason that bacteria like our own could not have evolved elsewhere. He anticipated the argument that any bacteria from space would have been killed by the heat of the meteorite's flaming passage through the atmosphere. Actually, he said, the heat thus generated has time to penetrate only a fraction of an inch below the surface, an observation now well substantiated. For example, when a stone fell at Colby, Wisconsin, on a hot July afternoon in 1917, frost was forming on its surface by the time local residents reached it a few minutes later. The frigidity of space, within the heart of the stone, had already banished its surface heat.

Lipman said the evidence convinced him that meteorites "bring down with them from somewhere in space a few surviving bacteria, probably in spore form but not necessarily so."

Most scientists were skeptical, and in some cases indignant. Michael A. Farrell of the Department of Bacteriology at Yale University, writing in a subsequent issue of the museum's publication, said the newspaper publicity following Lipman's announcement "cannot but be disturbing to the minds of earnest searchers for truth, especially when the supposed findings fail of corroboration in other laboratories." Until more convincing evidence is brought forth, said Farrell, "Lipman's excursions into the field of life beyond this globe must be considered as a flight of imagination through space."

Sisler, more cautious than Lipman, readily conceded the possibility of contamination and, in fact, told the author that he believed this to be the most probable explanation. At that time (1963) he was still culturing his meteorite specimens in the belief that some might come to life only after several years. Also, he said, a large number of experiments were needed to show, statistically, whether or not the organisms were, indeed, contaminants. He stressed, however, that the possibility of their extraterrestrial origin had not been eliminated.

The publicity given Sisler's work sharpened the debate on the observations of Nagy and his colleagues in New York. Many skeptical voices were raised, but it was Edward Anders of the University of Chicago who emerged as spokesman of the doubters. His debate with the New York group began at a meeting of the American Geophysical Union in Washington soon after the 1961 session at the New York Academy of Sciences, and he again confronted Nagy's associate, Warren Meinschein, a few days later at the meeting where Sisler told of his live cultures.

Among the questions raised by Anders was whether the similarity of the Orgueil hydrocarbons to those in fuel oil might not be more than a coincidence. He noted that the meteorite is quite porous and hence "breathes" whenever there is a change in barometric pressure. The specimen had lain for almost a century in the American Museum of Natural History in New York and, Anders wrote in an analysis, "There are not many places on earth that burn and vaporize fossil fuels [coal, oil, gasoline, etc.] at a higher rate than New York City."

In other words, he said the meteorite had been breathing exhaust fumes since automobiles first began riding the streets of New York and some of the material must have stuck to the inner pores of the specimen. Only a very small amount would be needed, Anders said, to produce the characteristic spectrum of hydrocarbons. Meinschein and his colleagues countered by saying that they had found the meteorite a hundred times richer in hydrocarbons than could be explained by any such contamination.

What prejudiced many scientists against the arguments of Nagy, Meinschein and their colleagues was doubt that life could originate on a body in the asteroid belt. In England J. D. Bernal, who had pondered long on life's origin, wrote in *The Times Science Review* that the discovery of hydrocarbons in the Orgueil meteorite might be a sign of life on a former planet, were it not for the difficulty of finding a suitable home for this planet in the solar system. If it were inside the earth's orbit, the meteorite material would have been altered by the sun's heat. Such an alteration had clearly not taken place. If it were in the belt of asteroids beyond the orbit of Mars, it would be extremely cold. The asteroid belt, Bernal pointed out, receives only about one-tenth as much solar heat as does the earth. Hence he proposed that, unless the carbonaceous chondrites came from another solar system, the most obvious source of the hydrocarbons was the earth itself, meaning some form of contamination.

Bernal had earlier argued that such compounds, brought to Earth by meteorites, helped seed the primitive earth with material that figured in the evolution of life. He cited the proposal of Brian Mason that the carbonaceous chondrites are relics from the earliest stage of the solar system, their complex compounds having been synthesized by such stimuli as ultraviolet light and high-energy radiation from the youthful sun.

While the announcement of Nagy and his colleagues that they had found biological substances produced a stir, their report of a few months later created a sensation. In the issue of *Nature* for November 18, 1961, Nagy and George Claus of New York University Medical Center said that they had examined samples of four carbonaceous chondrites and in all of them had found microscopic particles resembling (but not identical to) fossil algae of the kind that live in water. Two other stony meteorites, examined for comparison, showed no such "organized elements," they said. They described five types of organized elements—that is, fossil life forms—that appeared in the carbonaceous samples. Some of them seemed to have perished in the midst of "cell-divisions," said Nagy and Claus, although they themselves put the expression in quotation marks. The "Type One" objects were small and circular with double walls. The material inside those walls stained diffusely in a manner typical of certain cells. The "Type Two" objects were like those of the first category except some were covered with spines or showed other appendages. "Type Three" objects were shaped like shields. "Type Four" objects were cylindrical, and a single "Type Five" object appeared to be six-sided with tubular protrusions on three sides, making it the most strikingly lifelike of the five types. The objects did not resemble any known mineral particles, the two men said. While they also found some obvious contaminants—well-known forms of terrestrial bacteria and algae—they believed their "fossils" must be indigenous to the meteorites. Except for the single specimen of Type Five, all the others resembled dinoflagellates or chrysomonads. The former, even though they are plants, propel themselves through the water. None of these forms live in the soil and hence they were viewed by Nagy and Claus as unlikely contaminants. Furthermore, whereas Orgueil fell in the south of France, Ivuna, another of those examined, fell in an arid, tropical region of central Africa. It seemed unlikely that similar contaminants would enter both. They said that they had interpreted the organized elements "as possible remnants of organisms."

Coming on top of all the other discoveries, this report impelled almost everyone with access to a carbonaceous chondrite to take a close look at it. Among them was Robert Ross at the British Museum in London. The museum possessed some fragments of Orgueil, and a number of the Type One forms described by Nagy and Claus were found in them, according to Ross. Likewise, he reported two with a mushroom shape, on a microscopic scale, and objects resembling collapsed spore membranes. He and his colleagues believed there was a strong indication that these were evidence of life somewhere in space, but they conceded that there was, as yet, no proof.

In Calgary, Alberta, Frank L. Staplin of Imperial Oil, Ltd., subjected a sample of Orgueil to the pollen analysis used by oil men to date layers of sedimentary rock drilled from deep in the earth. The fossil pollen grains are a clue both to the age of the rock and the climate in which it was laid down. Staplin found a number of objects that in size, texture and acid resistance "superficially" resembled some of the one-celled algae. He identified two entirely new genera, or general categories of plant, to which, according to custom, he appended his own name: *Caelestites staplin* and *Clausisphaera staplin*. The first name was clearly intended to denote its heavenly origin.

Another he identified as *Protoleiosphaeridium timofeyev*, a fossil pollen grain discovered in terrestrial rocks by Boris Vasilyevich Timofeyev, pollen specialist at a Soviet government institute in Leningrad (now St. Petersburg) that did research on oil prospecting.

Meanwhile, Timofeyev was himself looking at a carbonaceous chondrite—one that fell at Mighei, near Odessa, in 1889. He centrifuged a sample of its material and in the lighter fraction found a number of rounded objects that he thought resembled the oldest form of alga known, the Protosphaeridae, although they did not actually fall into any earthly classification.

Like Staplin, he hung his name on various of these presumed relics of life elsewhere, and when he heard of Staplin's work on Orgueil, he named one of the Mighei "fossils" *Prototrachysphaeridium staplini timofeyev*. In May 1962, at an All-Union Conference of Astrogeologists in Leningrad, Timofeyev told of his findings. A number of the Soviet scientists present were skeptical, but Timofeyev persisted in his view that the grains were evidence of life having existed elsewhere in the universe. When I visited Leningrad late in 1962, Timofeyev took me to his apartment to squint at microscopic preparations of Mighei material. The Soviet scientist bristled with enthusiasm, clearly convinced of the validity of his finds.

Confirmations of the Nagy-Claus report seemed to be coming in from right and left. P. Palik at Eötvös Lorand University in Budapest found in a bit of Orgueil six filament-like structures reminiscent of algae. She thought some of them "may possibly be indigenous to the meteorite." *Nature*, which traditionally opens its pages to controversy, devoted part of its March 24, 1962, issue to a symposium of articles on this subject, with an introduction by Harold Urey.

Urey made the startling suggestion that the life forms observed under the microscope of Nagy and Claus lived on the earth but had reached the New York laboratory via the moon. He cited his proposal that most of the stony meteorites are fragments of the moon kicked free by the impacts of larger bodies and pointed out that the meteorite "fossils" resembled water algae. Might it not be, he wrote, that the splash produced when a huge meteorite hit the earth carried some of these algae to the moon, where they soaked into the porous rocks and remained until another meteorite kicked them back again?

Urey pointed out that several other scientists had proposed that we may find microscopic life forms on the moon. In 1959 Carl Sagan, then at Yerkes Observatory, said the moon must once have had an atmosphere similar to the original atmosphere of the earth. If not replenished, it would have diffused off into space in about a thousand years, but presumably the lunar crust steadily gives off gases, and this may have been sufficient, he said, to stretch the lifetime of the lunar atmosphere to 100 million years. This would have been long enough for the synthesis of a considerable quantity of organic compounds in the manner postulated for the primitive earth. If life never emerged from this process and the organic compounds were protected, by meteoritic dust, from violent temperature changes and solar radiation, then these early compounds, extinct on earth, await the first scientists to reach the moon.

However, Sagan, who hoped life would be found in all kinds of places, also raised the possibility that there might be a habitable region of the moon, far enough below the surface to be uniform in temperature, warmed by underground radioactivity and moistened by water released from the rocks. If life evolved rapidly when the moon still had an atmosphere, it might still lie hidden beneath the surface, he said. Although this was "admittedly very speculative," he added, the idea that there may be something akin to life on the moon "must not be dismissed in as cavalier a manner as it has been in the past."

The burden of his paper, read to the National Academy of Sciences on November 16, 1959, was a warning against contamination of the moon by nonsterile spacecraft. Bacteria from the earth, he said, might find abundant compounds on which to feed, with a consequent "biological explosion" that would quickly destroy the precious, primitive substances. Or the microscopic invaders from earth might find living natives and devour them before they could be discovered and studied. Such possibilities, he said, are remote, but sufficiently real to make it imperative to sterilize space vehicles destined for the moon. To the dismay of some biologists, it was later decided to relax moon sterilization procedures in the interest of economy (and, presumably, to improve the chances of reaching the moon before the Soviets). It was argued that any bacteria surviving the trip would almost certainly be killed by ultraviolet radiation or other stresses on the unprotected lunar surface, but it was generally agreed that contamination of Mars by organisms protected within a spaceship would be intolerable.

An even more "far-out" hypothesis discussed by Urey in his introduction to the *Nature* symposium was that advanced by John J. Gilvarry, a British-born scientist who had taught at Princeton University and who when he sent his proposal to *Nature* in 1960 was with the Research Laboratories of the Allis Chalmers Manufacturing Company in Milwaukee, Wisconsin.

Gilvarry assumed that the moon, to begin with, had as much water as the earth or any other body of the solar system. If the moon had the highlands that we see there today, they would have been surrounded by seas 2 kilometers deep. The moon should have kept its oceans at least a billion years, he said. He argued that a number of the lunar craters, such as the huge bowl of Mare Imbrium, must have been formed by impact explosions under water. The smooth floors of the "seas" on the moon are oceanic sediments, he said, and are dark with the vestiges of "a primitive form of life" that existed in the vanished oceans.

Urey's reply was based on the generally accepted view that in its earliest infancy the earth had no atmosphere or oceans, its envelope of gases having been swept away. The present air and surface water thus must have come slowly from within the earth through volcanic action and the breakdown of surface materials (or, as it was later proposed, by multiple comets). The moon, with its weak gravity, was unable to retain such water and other gases; they diffused away as fast as they were liberated.

Among contributions to the 1962 symposium in *Nature* was an analysis of the reports of Nagy and his New York associates, written by Anders with two of his colleagues at the University of Chicago: Frank Fitch of the Department of Pathology and Henry P. Schwarcz of the Enrico Fermi Institute for Nuclear Studies. They said that in specimens of the Ivuna and Orgueil meteorites they, too, had found the spherical and oval forms described by the New York group. "Meteorites have long been notorious for containing structures resembling fossils," they said. They cited the discredited report published by Otto Hahn at Tübingen in 1880 describing a great variety of "organisms" that he had found in chondritic meteorites. While the work of Claus and Nagy had been done "with much greater care and competence," the Chicago group said, "the decision whether a certain form is of biological or inorganic origin is . . . quite subjective." In fact, the Chicagoans asserted that the rounded particles were strikingly similar to supercooled droplets of sulfur and hydrocarbon produced by experiments in their laboratory.

Nagy, Claus and Hennessy, far from capitulating to the objections of the Chicago group, said the latter's droplets of sulfur and hydrocarbon bore no relationship to the objects of their own study. They produced additional pictures of lifelike objects extracted from Orgueil and Ivuna and reported finding further samples in two other carbonaceous chondrites: Alais, the first of this kind to be discovered, and Tonk, which had fallen in India in 1911. A specimen of the most lifelike form of all—the hexagonal Type Five—was obtained from Tonk.

In a commentary forming part of the symposium, Bernal admitted to being impressed by the slides shown him by Claus. He believed the claims of the New York group merited serious consideration, despite the questions that they raised. What troubled him in particular was the suggestion that life which arose elsewhere would display the same chemistry as life on earth. He termed the alternatives equally unlikely: Either life, disregarding all other possible paths of development, always evolves the same chemical structure no matter where it arises, which he considered "inherently improbable," or life on earth and all life elsewhere has a common ancestry, which, he said, "strains the imagination." Perhaps, he added, Haldane was right and our spark of life came from elsewhere in the galaxy, "or indeed from other galaxies."

By now the debate had developed to the point where the New York Academy of Sciences, with Nagy as a prime mover, decided to call another meeting on the subject, to be held May 1, 1962. Most of the leading protagonists came and Harold Urey served as chairman. New supporters of the New York group came forward, including A. Papp of the Department of Paleontology at the University of Vienna and Pierre Bourrelly of the Department of Cryptogamic Botany at the Museum of Natural History in Paris. Bourrelly said the objects in the New York slides were definitely organisms and were not contaminants, although he found them more like earth forms than one would expect in life that had evolved elsewhere. Papp believed their resemblance to aquatic life on earth did not mean they had been aquatic in their distant home.

Anders and Fitch of the University of Chicago presented an array of counterarguments. Some of the meteorite extractions that had a more lifelike appearance, they said, were probably pollen grains that had infiltrated the specimen on earth.

By now the debate was not entirely academic. Claus, Nagy and Dominic L. Europa of the Department of Pathology at Bellevue Hospital in New York were clearly piqued at the implication that during preparation their specimens had become needlessly contaminated. In examining slides of meteorite material prepared in the Chicago laboratories, they also found fragments of earth life, showing, they reported acidly, "that experience is essential in preparing uncontaminated samples." Later, in the published version of these papers, the sharpest of the comments were deleted.

In summary they claimed they had by now studied 400 microscopic preparations of meteorite and related materials, disclosing thirty distinct types of organized elements. More, they said, had been found by other researchers. None were identical to earth species, though some were similar. "Full proof" of their extraterrestrial origin was lacking, they stated, "but the indications seem to be strong."

A year after the New York Academy of Sciences meeting, a devastating blow was dealt to Claus, Nagy and their coworkers. As noted earlier, of all the "organized elements" that they had found, by far the most convincing were those of the six-sided Type Five variety. They were comparatively rare. Two, as well as a few fragments, had been found in Orgueil and four in Tonk. When stained, their features came into clear relief: tufted protuberances from alternate faces of their hexagonal structure as well as other symmetrical characteristics.

On June 7, 1963, Fitch and Anders of the Chicago group fired their salvo in a report published in *Science*. They had stained ragweed pollen in the manner used by Claus's group and found that the grains thus treated were indistinguishable from the Type Five elements. The reason, as some had suspected, was that the staining distorted the grains.

This convinced most neutral observers that the Type Five elements were, in fact, pollen that had entered the samples while on earth. Such grains are a notorious contaminant of New York City air during the hay fever season.

Despite their Type Five setback, the New York group had a powerful ally in Harold Urey. Not that Urey was convinced that the meteorite objects were fossils, but he thought the proposals of Claus, Nagy and the others were not being given a fair trial. "It should be realized," he said after the New York Academy meeting, "that enthusiastic people may misclassify artifacts of one kind or another, or may mistake a contaminant for an indigenous form; in fact, it would be surprising if this did not occur in such a study. On the other hand, enthusiastic critics also make mistakes." If it can be shown, said Urey, that these "organized elements" are the residue "of living organisms indigenous to the carbonaceous chondrites, this would be the most interesting and astounding fact of all scientific study in recent years."

The debate continued with new tests, arguments and counter-arguments. The Orgueil meteorite has probably been more elaborately studied than any other chunk of material on earth, but the arguments for contamination won the day and by the 1990s little was being heard about "organized elements" in meteorites.

The debate had been a classic example of scientific discussion become personal, emotional and enmeshed with professional pride. The talents and ingenuity of participants were directed toward proving their case, rather than seeking out the truth.

Scientists became doubly aware of the danger of contamination in meteorites, since the specimens had lain on the ground, had been stored in museums, and had been frequently handled. When such life-related substances as amino acids were found in carbonaceous chondrites, skeptics noted that they were suspiciously similar to those in terrestrial organisms. Attempts were made to collect specimens immediately after their fall, as when a fireball hit Texas on the night of September 9, 1961. A large part of the population in the Dallas–Fort Worth area was watching the Miss America contest on television, since Miss Texas was a leading contender. Suddenly, as one viewer put it, "the television picture went all to pieces." Thousands of viewers got to their feet and tinkered with the TV controls. It was a Saturday night and the drive-in movies were full. The picture on the outdoor screen vanished in the blinding light of the fireball, which had left a trail of ionization that interrupted the TV signals. The weight of the original object was estimated at many tons, but Hurricane Carla drenched the landscape soon after the fall and only a half dozen fragments, totaling about 11 ounces, were found. Because the material of carbonaceous chondrites looks so much like dirt, they may escape the attention of even the most experienced meteorite-hunter.

When, on September 28, 1969, a carbonaceous chondrite fell near Murchison, 80 miles north of Melbourne, Australia, NASA and associated laboratories were primed for elaborate analysis of newly collected moon rocks. In particular they were armed with a technique that showed whether or not the structure of an amino acid was of the left-handed variety typical of those in proteins on earth.

Murchison was a Type Two carbonaceous chondrite—one of those suspected to be carriers of amino acids—and a sample was sent to NASA's Ames Research Center in California for analysis by an eight-person team led by Keith Kvenvolden. The results were unambiguous and spectacular. Among the different kinds of amino acid detected, fifty-five had no terrestrial counterparts. Also found were eight of the twenty-three occurring in proteins on earth, but most, if not all, of the meteorite amino acids were clearly not contaminants, for their molecular structure was not all left-handed in the manner characteristic of living material. Those from Murchison were almost evenly divided between the left-handed and right-handed forms. The analysis also showed what appeared to be purines and pyrimidines (components of the genetic molecules) as well as a variety of hydrocarbons. Similar substances were also found in a specimen from Antarctica. The analysts concluded that all carbonaceous chondrites of this type may have had a similar cargo when they landed, but part of it decomposed or was otherwise altered before the specimen could be analyzed. It is now widely believed that these and other prebiological substances in meteorites, comets and celestial dust grains may have been synthesized in the solar nebula before the earth was formed. While the seeds of our ancestors may not have fallen from the skies, the most basic components of living things seem to have done so.

11

Life's Origin—
Freak or Inevitable?

Speculation on the origin of life began in antiquity and flourished during the centuries when spontaneous generation was in vogue. Then, in 1871, Charles Darwin speculated that there must have been some form of chemical evolution leading to life, but at that time Pasteur convinced the world that outside the protective envelope of living cells, the basic components of life's molecules such as the amino acids forming proteins could not long survive in today's oxygen-rich atmosphere. It seemed as though life could have been created only by a miracle. Then, in the 1920s, two scientists proposed independently that the original atmosphere may have been fundamentally different from that of today. Both of them had studied the writings of Friedrich Engels, cofounder of communism, who a half century earlier had approached the subject from a materialistic point of view, and both doubted any sort of miracle. One, Alexander Ivanovich Oparin, was in Russia. The other, J.B.S. Haldane, was that most unconventional Englishman whose "astroplankton" theory was discussed in chapter 9.

Oparin and Haldane accepted the argument that in an oxygen-rich atmosphere like that of today any of the biological building blocks would have been destroyed by reaction with oxygen—they would have oxidized. Haldane also believed that much of the carbon now tied up in coal deposits and other remains of former life had, before the origin of life, been part of the atmosphere as carbon dioxide. He therefore proposed that the original atmosphere was chiefly carbon dioxide, ammonia and water vapor. When ultraviolet was shined on such an atmosphere, he noted, "a vast variety of organic substances are made, including sugars and apparently some of the materials from which proteins are built

up." That the original atmosphere lacked oxygen was further indicated by the presence today of certain primitive forms of bacteria that are killed by oxygen, such as those causing tetanus and
gas gangrene. Before the appearance of an oxygen atmosphere,
he said, the building blocks of life must have accumulated "till
the primitive ocean reached the consistence of hot dilute soup."
His lifeless ocean was murky with organic chemicals of enormous
variety, heaving under a lifeless sky and beating upon lifeless
shores until an oil-coated droplet formed with all of the necessary
ingredients, including the ability to duplicate itself. The organism
would have spread like wildfire until checked by exhaustion of its
food supply or the evolution of competitors.

The scenario proposed by Oparin was similar except that his
original atmosphere also included methane (chief constituent of
the gas used in our homes) and hydrogen. With little or no oxygen
this would have been what chemists call a "reducing" atmosphere.
Oparin, who until his death in 1980 was the world leader in speculation on life's origin, recognized that a long period of chemical
evolution preceded it. "It must be understood," he wrote, "that
no matter how minute an organism may be or how elementary it
may appear at first glance it is nevertheless infinitely more complex than any simple solution of organic substances." It is, he said,
"founded upon a harmonious combination of strictly coordinated
chemical reactions. It would be senseless to expect that such an
organization would originate accidentally in a more or less brief
span of time from simple solutions or infusions." The tiny bacterial rods, spheres and corkscrews that we see in a microscope may
look simple, but they are not.

In 1932 Harold C. Urey and Stanley L. Miller, his twenty-three-
year-old graduate student at the University of Chicago, decided to
test the Oparin-Haldane hypothesis. Two years later Urey was to
win a Nobel prize for his identification of "heavy hydrogen,"
whose atoms have two nuclear particles (a proton and neutron)
instead of only a single proton. In his test Miller used what he and
Urey assumed had been components of the original atmosphere
in which hydrogen was mated with the basic atoms of life: carbon
(in methane), nitrogen (in ammonia), oxygen (in water) and hydrogen (in hydrogen gas). The idea that Earth might once have
had such an atmosphere, in which any free oxygen quickly combined with hydrogen to form water, had been strengthened by
discovery that the atmospheres of Jupiter and Saturn are rich in

methane and ammonia. Methane had also been detected in the atmospheres of the other big planets: Uranus and Neptune.

Miller realized that some sort of energy input would be needed to break up the original molecules so that they could recombine into more complex forms. While on the young earth energetic wavelengths of ultraviolet sunlight could have done the job, it was easier to use electric sparks simulating lightning. Miller erected a system of flasks and tubes including a reservoir in which water was boiled, producing vapor that mixed with the methane, ammonia and hydrogen. The resulting mixture was subjected to a 60,000-volt high-frequency spark. The vapor that condensed was returned to the water-filled flask, boiled again and recirculated. This cycle was continued for a full week.

Within a day the water in the flask had become pink. By the end of the week it was deep red. Miller's analytical method (paper chromatography) revealed glycine, simplest of the amino acids, and alanine, another important amino acid, as well as lactic acid, acetic acid, urea and a large amount of formic acid. In one of his experiments, 15 percent of the available carbon (in the methane) went into production of organic compounds, a result which seemed to give strong support for the Oparin-Haldane hypothesis.

More recent observations of comets, meteorites and interplanetary dust have shown that such synthesis of prelife molecules has occurred throughout the solar system, very likely before the earth was formed. Experiments have shown that gas mixtures like that used by Miller have produced a wide range of biological molecules including components of the nucleic acids that determine the continuity of life. It was found that most protein amino acids could be formed from such simple molecules as formaldehyde without the aid of any enzymes—the class of proteins that today plays a universal role in stimulating life's chemistry.

Nevertheless, uncertainty persists regarding the original atmosphere. A 1990 report published by the National Academy of Sciences said it may have consisted primarily of carbon dioxide, nitrogen and water vapor. The report, "The Search for Life's Origins—Progress and Future Directions in Planetary Biology and Chemical Evolution," was drafted by a committee of the Space Science Board. Several hypothetical atmospheres had been tested, including one by Melvin Calvin at the University of California at Berkeley in which his laboratory examined the effect of high-energy radiation on an atmosphere of carbon dioxide and water

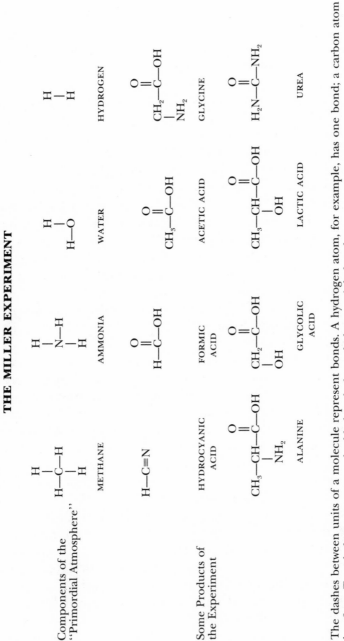

THE MILLER EXPERIMENT

The dashes between units of a molecule represent bonds. A hydrogen atom, for example, has one bond; a carbon atom has four. Two dashes represent a double bond; three dashes, a triple bond.

vapor. It was assumed that such radiation was coming from space or from radioactive rocks. Simple organic compounds such as formaldehyde and formic acid resulted, but the production was not as plentiful as in Miller's experiments, which had used abundant hydrogen. Perhaps the original atmosphere was not rich in hydrogen, as had been assumed in the Urey-Miller experiments, because the hydrogen had escaped into space.

Nevertheless, a continuous source of hydrogen, once oceans had formed, has now been proposed and tested in the laboratories of A. Graham Cairns-Smith at the University of Glasgow and David C. Mauzerall of Rockefeller University in New York. Ultraviolet sunlight shining on a sea enriched in ferrous iron (iron that lacks two of its electrons) would have separated the water molecules into hydrogen and oxygen. Those early seas may have been made iron-rich by remnants of the Great Bombardment during the earth's earliest history, but, said Cairns-Smith, their waters would not have been very hospitable to life. That the seas were full of iron is indicated by banded iron formations in layers of ocean sediment laid down between 1.5 and 3.8 billion years ago and found in the rocks of all continents. The subsequent creation of nearby coal deposits, as in Scotland, helped set the stage for the industrial revolution.

In view of uncertainty as to the original atmosphere, said the 1990 report, the manner in which the building blocks were first synthesized "is far from settled." Equally uncertain is how the basic organic molecules became organized into the far more complicated and interdependent ones associated with even the most primitive bacteria. It is generally assumed that this occurred on earth, rather than being introduced from elsewhere, as in the concept of panspermia or in Hoyle's life-in-comets hypothesis. Jacques Monod, a French Nobelist, in his 1970 book *Le Hasard et la Nécessité* (published a year later as *Chance and Necessity*) said it was highly unlikely that what he termed the conspiracy of improbable steps and coincidences needed to produce the miracle of life on earth could have occurred anywhere else. Hoyle has said that formation on earth of a living organism was no more likely than the assembly of a Boeing 747 airliner by a tornado tearing through a junkyard. Harold P. Klein of Santa Clara University in California was quoted in *Scientific American* as saying: "The simplest bacterium is so damn complicated from the point of view of the chemist that it is almost impossible to imagine how it happened." Yet Klein's task, as chairman of the National Academy of Sciences committee that prepared the 1990 report, was to attempt just that. It was a formidable job, but not hopeless.

Biologists point to evidence in the rocks at Isua in southwest Greenland that primitive life was already well established 3.5 billion years ago. Even more ancient "yeast-like" fossils, 3.8 billion years old, were reported found at Isua in 1978 by a scientist who put his own name on the "new species," but his report now appears spurious. Apart from the Isua fossils the oldest are from western Australia and South Africa. One form consists of many-layered mounds built by organisms which themselves have vanished but whose structures resemble some called stromatolites built 3.5 billion years later by their descendants. The others are chains of fossil cells resembling the cyanobacteria, or blue-green algae, of today.

It is clear that primitive life originated very soon after the Great Bombardment during the earth's earliest history. Some, in fact, argue that it originated several times and was wiped out by catastrophic impacts before it became permanently established. Clues to the bombardment timetable have been provided by studies of the moon, particularly by the dating of rock samples collected there by astronauts. Mare Orientale, a vast impact crater surrounded by a bull's-eye series of concentric rings, was formed an estimated 3.8 billion years ago and is believed to be the youngest crater more than 200 miles wide. It therefore seems approximately to date the end of the Great Bombardment. It was preceded by the giant impact that produced Mare Imbrium, a "sea" 750 miles wide, the determination of whose age (3.85 billion years) is regarded as one of the most reliable on the moon.

As noted earlier, the moon itself was probably formed earlier when an object as big as Mars struck the earth, producing an explosion that boiled off the oceans and vaporized much of the earth's upper mantle. The iron core of the impacting body melted and sank into the earth's own core and the moon condensed from the material thrown up by the blast. This is taken to explain why the density of the moon, having little or no core, is less than that of the earth. Such a cataclysm of course would have destroyed any organisms on earth, but it is remarkable that life appeared very soon after the last great impacts on the moon—and hence on the earth. It seems to have been far easier for primitive life to get started than for it to evolve into the many-celled organisms that

THE STRUCTURE OF ATP

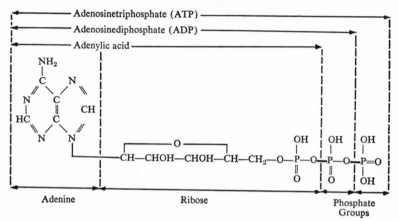

Heavy dashes indicate the energy-storage bonds between phosphate groups.

we see around us, for these organisms do not begin to appear in the fossil record until less than 1 billion years ago.

Meanwhile, evolutionary paths for the building blocks of life seemed clear. Components of the most important molecules of all—those of DNA (deoxyribonucleic acid), the architect of living things, and RNA (ribonucleic acid), the builder—had been synthesized in simple ways. The sugars bound into the skeletons of those molecules (ribose and deoxyribose) had been produced by merely shining ultraviolet light onto formaldehyde, a molecule assumed to have been generated from the primordial atmosphere. At least three of the five pyramidines and purines that spell out the messages in DNA and RNA (adenine, guanine and uracil) had been synthesized, although some researchers believed the method used with uracil was too artificial. NASA's Ames Research Center in Mountain View, California, had produced ATP (adenosine triphosphate), a key molecule in all forms of life from single bacteria to human beings. It performs the essential function of transporting energy between parts of the cell. Among those working on the problem was Cyril A. Ponnamperuma, a native of Sri Lanka (formerly Ceylon). He had worked with Melvin Calvin on the synthesis of adenine, a key ATP component, under electron bombardment. Another participant was Carl Sagan, then at nearby Stanford University and destined to become perhaps the world's most eloquent advocate of the search for distant life.

The Ames group, to test its results for the presence of ATP, exploited the fact that it causes the lanterns of fireflies to glow. The group was unable to find any such flies in California, but obtained dehydrated firefly tails from a supplier of exotic laboratory materials in Mount Vernon, New York. When tested, the tails glowed happily.

Another line of inquiry looked at the possibility that life elsewhere might be based on freight-train chains of silicon atoms, instead of carbon chains as on earth. Haldane, never averse to novel ideas, discussed whether, inside the earth, there might be life based on semimolten compounds of silicon. This he deemed "unlikely," but he also argued that a whole system of organic chemistry could take place based on liquid ammonia instead of water. This, too, seemed improbable, since ammonia is liquid only within a narrow range of frigid temperatures, making it a poor substitute for water as a medium for life. George Wald, professor of biology at Harvard University and a Nobel prize winner, noted that water is remarkable in that its solid state is less dense than its liquid form. In other words, ice floats. Wald and his students found nothing in the literature describing tests of ammonia's density as a liquid or solid, so they did the experiment themselves. In an ammonia bath blocks of frozen ammonia sank to the bottom. They reasoned that in an ammonia ocean, ice formed on its surface would sink, allowing a new layer to freeze and sink until the entire ocean was frozen. An ammonia world did not seem a likely place for life. In their 1930s analysis Urey and Miller wrote: "We know enough about the chemistry of other systems, such as those of silicon, ammonia and hydrogen fluoride, to realize that no highly complex system of chemical reactions similar to that which we call 'living' would be possible in such media."

By the 1960s experimenters were particularly excited at the evidence that out of all combinations of atoms that might have been produced by their tests, the ones most readily produced were those in life's building blocks, such as amino acids, sugars, purines, pyrimidines, and fatty acids. Optimism ran high. As Melvin Calvin wrote: "We can assert with some degree of scientific confidence that cellular life as we know it on the surface of the earth does exist in some millions of other sites in the universe."

Finding out, however, how these components organized themselves into the complex molecules of life has proved extremely difficult. As Miller himself has put it, "The problem of the origin of life has turned out to be much more difficult than I, and most other people, envisioned." A variety of ways have been proposed in which this could have first occurred, in particular the ability of an organism to duplicate itself. Jacques Monod had won his Nobel prize for his work on the mechanisms controlling bacterial reproduction and evolution, showing how complex reproduction was for even the simplest kinds of life now in existence. It depends on extremely long molecules of DNA, serving as blueprints, and various forms of RNA that help implement the genetic message carried by the DNA in supervising formation of proteins. An early proposal on how these long molecules could have formed was set forth in 1951 by the biophysicist J. D. Bernal, who said the building blocks of proteins (amino acids) and of nucleic acids could have originally stuck to layers of clay long enough for them to join into chains. This would require special clays whose presence on the early earth has been questioned by critics. It has, nevertheless, been kept alive by the Scottish chemist Graham Cairns-Smith, who believes such clay layers could have not only produced chains of amino acids and nucleotides but reproduced them. For a picture of first life, he has written, do not think about cells, "think instead about a kind of mud."

Nevertheless, more widespread today is the view that the original reproductive machinery was far less complicated and only remotely similar to the DNA-RNA system. All traces of that early chemical evolution seem to have been wiped out, but may be preserved in the sediments of Mars.

The 1990 Academy report singled out such sediment as most likely to explain how the machinery of life first evolved on our own planet. On Mars it may have gone through the early stages of evolution during the period when flowing water was scouring that planet's surface, as evident in spacecraft pictures. Or at least the stages of chemical evolution that preceded life are likely to have occurred. Such records have been lost on earth because its surface has repeatedly been "plowed under" by plate tectonics (continental drift), volcanism and erosion. On Mars there is one giant rift valley and a huge volcano, but there does not seem to have been global overturn of the surface through plate tectonics. Nor has Mars's surface been bombarded from space as violently as that of the earth, since its gravity is only one-third as strong, making for less high-speed impacts.

The Academy report noted that early stages in the evolution of biochemistry also seem to have occurred on Titan, the largest moon of Saturn. Titan is enveloped in a pinkish smog. Its atmosphere, like that of the earth, is predominantly nitrogen, but there is from 1 to 6 percent methane, producing on the surface an air pressure somewhat greater than that on Earth. Most remarkable, its smog is known to contain at least six hydrocarbons (ethane, propane, acetylene, ethylene, diacetylene and methylacetylene), three nitrogen compounds (hydrogen cyanide, cyanoacetylene and cyanogen), and two oxygen compounds (carbon dioxide and carbon monoxide). The smog almost certainly contains other, more complex substances. The ethane may have fallen to the surface, forming an ocean. Because of the great distance from the sun, the temperature is extremely low, so if there are continents they are formed of ice. Titan's extensive organic compounds may have evolved through stages that preceded the appearance of life on earth. Is there evidence, asked the report, of any unexpected pathway for such evolution? "Are unlikely reactions catalyzed in some unforseen way? Are compounds of biological interest, such as amino acids or adenine, produced? Are any of these results relevant to the events that preceded the origin of life on Earth?" To explore this possibility, the *Cassini* spacecraft under consideration by NASA and the European Space Agency would drop one or more probes through Titan's atmosphere and into its hypothetical ocean. NASA was a cosponsor of the Academy study and some of its scientists have proposed that because the surface of Europa, the second closest to Jupiter of its satellites, shows few impact craters, it may be sufficiently heated from within to have a water ocean beneath a thin crust of ice. "There may be life under the ice of Europa," said Robert Shapiro of New York University in his 1986 book, *Origins—a Skeptic's Guide to the Creation of Life on Earth.*

Life forms have traditionally been divided into two basic categories: those whose cells have no nucleus (prokaryotes) and those whose genetic material (usually DNA) is concentrated in a nucleus (eukaryotes), the latter including all higher plants and animals. Some biologists, however, divide what seem the most primitive surviving prokaryotes into two categories: the archaebacteria and eubacteria. They also believe the eukaryotes stem from a line much older than that of multicellular life, placing them in the same ancient category as the archaebacteria and eubacteria. Possibly the most ancient surviving eukaryote is the single-celled parasite *Giardia lamblia*, a cause of intestinal disorders, but knowledge and classification of the tiniest creatures is growing rapidly, particularly the "picoplankton" or microscopic sea organisms whose study during the past decade has shown they include most, if not all, the basic life forms, although, being so small, they are hard to culture and study.

Most remarkable about the surviving classes of ancient bacteria is that despite their superficial differences, their basic chemistry has much in common. In that sense there is only one form of life on earth, indicating that all must have had a common ancestor. This long-lost species from which all life stemmed is called the "progenote," and learning about it could explain how life came to be. Biologists point out that the chemical machinery evolved by the progenote may represent only one alternative route. Life on other worlds may be fundamentally different.

Reconstructing how our own earliest classes of organisms evolved and diverged depends to a large degree on deciphering the clues hidden in their chains of nucleic acids and associated proteins. While NASA and the National Science Foundation have been the chief explorers of life's origin, the National Institutes of Health are also playing a role through development of rapid ways to sequence those chains. Indeed, in terms of "dead genes" within them, the history of Homo sapiens may be hidden inside us, awaiting the fruits of such "molecular archaeology." DNA within our body cells contains many sections of unknown function, and some may be useless residues of our most ancient ancestry.

Of particular interest are clues to the history of ribosomes that are components of all cells, including the most primitive varieties. They are the "assembly plants" that carry out the protein-making instructions carried to them by RNA from the master DNA archive. A common bacterium may contain 15,000 ribosomes, each of which is a structure of more than 50 proteins and 3 specialized RNA's. How they do their protein-making job is still poorly understood.

Other clues may come from two cell components that long ago seem to have been free-living bacteria, but then found a permanent (and essential) home in parent cells. One group are the mitochondria, which do energy-processing for their host cell but bear a resemblance to some of the more primitive free-living bacteria. The other group consists of chloroplasts, whose chlorophyll enables them to perform the carbohydrate-making that is the essential function of plants. Such mating of organisms, or "symbiosis," occurs widely, as in the *Escherichia coli* that inhabit our intestines, helping our digestion, or those bacteria that enable ruminants to digest plants.

Biologists have long puzzled over a "chicken versus egg" paradox: proteins can seemingly be made only on instructions from nucleic acids, yet nucleic acids cannot perform without the help of catalytic proteins. Which came first? The answer may be that the manner in which they interact may have evolved from a simpler and quite different process. One proposal is that proteins were originally able to replicate themselves, then "invented" nucleic acids, and finally were enslaved by them. Another is that nucleic acids were first able to multiply without protein help.

In the 1960s Leslie Orgel and Francis Crick at the Salk Institute in La Jolla and Carl Woese of the University of Illinois began wondering whether RNA might have served as both chicken and egg, having produced both DNA and proteins. This idea gained momentum in the early 1980s when two molecular biologists, Thomas R. Cech of the University of Colorado and Sidney Altman of Yale University, found that certain RNA's could act as enzymes, snipping RNA molecules into parts that could then recombine, a finding for which they shared a Nobel prize. Manfred Eigen at the Max-Planck-Institut für Biophysikalische Chemie in Göttingen, another Nobelist, proposed that during that early period, RNA could not only reproduce itself but could evolve through replication errors, setting the stage for evolution of the more efficient DNA-RNA machinery. Thus was born the concept of an "RNA world" in which, for a long time, that nucleic acid was dominant, without need for DNA. The idea is not universally popular, since it does not explain whence came the original RNA. The latter has never been synthesized under plausible early-earth conditions.

Ever since the Urey-Miller experiment showed how amino acids could have been produced in quantity, some experimenters have tried to show that they could independently form protein chains that could reproduce themselves. Oparin had pointed out that when a variety of proteins and other large molecules are mixed, they sometimes form into droplets with peculiar properties. Some of the molecules migrate to the surface of the droplet, forming a skin, and there is also segregation of material inside the droplets. Oparin reported that such "coacervates" can absorb certain substances from surrounding fluid while rejecting others—an important property of living cells. The droplets swell with ingested material to a certain size, then divide, much like a drop of water.

The most persistent investigator of independent protein synthesis has been Sidney W. Fox, most recently head of the Institute for Molecular and Cellular Genetics at the University of Miami. In the 1950s he began heating mixtures of the amino acids constituting proteins to very high temperatures. When the resulting "proteinoid" material was dissolved in hot water, then allowed to cool, it formed into a multitude of microscopic spheres looking much like the cocci that produce pneumonia. According to an enthusiastic assessment by Robert Shapiro of New York University, Fox's spheres show many lifelike properties, such as catalysis of chemical reactions, surfaces that resemble membranes and, above all, an ability to proliferate, evolve and produce nucleic acids. Fox's work has been followed up by others, such as Cyril Ponnamperuma, that veteran explorer of how life's chemistry may have evolved, but the idea that Fox's protein spheres are analogues of life has been harshly criticized by such leaders in the field as Urey and Miller.

The possibility of protein-based replication may, however, be demonstrated by the agent causing an extraordinary family of ailments including scrapie (a disease of sheep), "mad cow" disease, kuru (perpetuated, until recently, by cannibalism), and possibly Alzheimer's disease. Scrapie's name derives from the compulsion of infected sheep to scrape off their wool. In France it is called *la tremblante* because of trembling and loss of coordination before the sheep collapses and dies. In 1935 French workers showed that scrapie could be transmitted by inoculation. The agent, however, proved perplexing. It passed readily through filters fine enough to stop all bacteria. It was apparently not a virus, since it seemed to contain none of the DNA that enables a virus to proliferate, surviving such DNA-killing measures as severe heating, lethal ultraviolet or formalin.

Kuru was first described medically by Vincent Zigas and D. Carleton Gajdusek of the National Institutes of Health, who in 1957 had found it among the Fore tribe of New Guinea. By tactful inquiry they learned that the tribesmen traditionally ate the brains of those who died, hoping thereby to preserve the spirit of the deceased. It is now clear that all diseases of this class are concentrated in the brain and seem caused by one or more very slow-acting agents. They are suspected to be proteins that can multiply inside brain tissue and have been named "prions" by one of their investigators, Stanley B. Prusiner of the University of California at San Francisco. The same type of disease seems to have infected cattle. In 1986 British cows were fed meal containing tissue from sheep that were apparently infected with scrapie, and an epidemic of "mad cow" disease (bovine spongiform encephalopathy) ensued. To stem the epidemic at least ten thousand cattle were killed and burned. Relatively rare human ailments, such as Creutzfeldt-Jakob disease and Gerstmann-Strassler syndrome, are attributed to the same or similar agents. A relationship to Alzheimer's disease has been proposed but is doubtful.

Gajdusek, who in 1976 won a Nobel prize for his studies of such diseases, and other scientists have been able to transmit the infectious agent to laboratory animals, making it possible, for example, to continue studying kuru after the outlawing of cannibalism. During a visit to Gajdusek's laboratory near Washington, D.C., he showed me a kuru-infected chimpanzee which pathetically hung on my neck.

As pointed out by Robert Shapiro in his 1986 book *Origins—a Skeptic's Guide to the Creation of Life on Earth,* since the scrapie agent and other prions seem able to proliferate in the brain, perhaps proteins can replicate on their own. If so, it would "strongly support the idea that during the course of evolution a system of this type preceded the one based on nucleic acids."

A very different theory for life's origin depends, as did the synthesis of Fox's "proteinoids," on extremely high temperatures. It began to receive attention after the deep-diving submersible *Alvin* discovered an entirely new way of life, or ecosystem, surrounding hot, metal-rich geysers where the Pacific floor is being torn apart along the East Pacific Rise. The geysering water contains hydrogen sulfide, methane and ammonia. As I know from personal experience, an *Alvin* dive at any site is memorable, but those on the volcanic ridges in midocean have revealed a whole new world. Living there, under 8,000 feet of water, are groves of rigid tube worms, 10 feet tall, whose digestive systems are packed with reddish bacteria. These perform a classic symbiotic role, deriving energy for the worm by digesting hydrogen sulfide from the vents. There are giant clams that, when I opened them, were red with a similar form of bacteria. Some of these organisms seem to thrive in water as hot as 360 degrees Celsius (680 degrees Fahrenheit). They are forever cut off from sunlight, depending instead for energy on digesting sulfur compounds from the hot vents. John B.

Corliss of NASA's Goddard Space Flight Center, who took part in the vent discoveries, had proposed that, not being dependent on photosynthesis (and chloroplasts), they could have supported the earliest life and, in their deep-sea environment, remained relatively protected from cataclysmic impacts. Carl Woese of the University of Illinois, a supporter of this hypothesis, argued that archaeobacteria, which seem to have evolved little since the time of their origin (assumed to have been billions of years ago), prefer hot environments, some of them surviving temperatures as high as 120 degrees Celsius (248 degrees Fahrenheit). A related proposal has been made by Christian de Duve of Rockefeller University who, in 1974, won a Nobel prize for his work on cell function. He proposed that life was originally based on sulfur compounds called thioesters that, he believes, were synthesized in the hot, highly acidic vents and performed many of the functions of living systems, particularly the transport of energy within each cell. This is now done throughout the plant and animal worlds by ATP (adenosine triphosphate). De Duve believes it was first done by thioesters.

A persistent mystery in biology is the origin of one-sidedness in the structure of such basic substances as the amino acids within proteins (all left-handed) and the sugars within nucleic acids (all right-handed). Some scientists have tried to relate this to an asymmetry in physics in which there is a slight but unmistakable tendency for a certain form of radioactivity to eject electrons more often in one direction than the other, but this explanation has not been generally accepted.

Perhaps the most essential element in the development of life may be the tendency of biological systems to organize themselves, as opposed to the progress of nonliving systems toward disorder. Since, thanks to outside sources of nourishment and energy, living organisms can survive indefinitely (apart from the effects of aging), they are immune to the thermodynamic law that requires progression toward disorder, the phenomenon known as entropy. A leader in the so-called "Brussels school" of thought on self-organization is Ilya Prigogine, a Russian-born theorist who heads the Center for Statistical Mechanics and Thermodynamics at the University of Texas. In 1977 he won a Nobel prize for his theoretical work. The deviation from thermodynamics concerns the tendency of large, seemingly chaotic systems, including large biological molecules, to organize themselves. Such a phenomenon has also been proposed by Manfred Eigen, the Nobel laureate from Göttingen, Germany. A striking feature of a human being, or even the tiniest organism, is the complexity of its constituents. In humans it is highly improbable that genetic information in the fertilized egg specifies every detail in the intricate structure of the end product—such as the countless tiny capillaries, the multitudinous cells of the nervous system, and those of the various organs. Self-organization has been at work.

Life, said Prigogine in a 1971 issue of UNESCO's *Impact of Science on Society*, "no longer seems to be a precarious miracle, a struggle against a universe which rejects it. With the generalization of thermodynamics, we come to realize that entropy is not forcibly synonymous with disorder and death; in certain particular conditions, it is organization and, finally, life which are the rule."

This, he said, had been ignored by Jacques Monod in his book arguing the improbability of life. "We must leave behind the rigid dualism of chance and necessity and go beyond this contradiction in principle. The living being is not a strange product of chance, nor the improbable winner in a huge lottery."

It is curious that so many winners of the Nobel prize have written hopefully about the possibility of life elsewhere, such as Melvin Calvin, Harold Urey and George Wald. Perhaps as members of that scientific elite they feel immune to ridicule. One of them, Christian de Duve, ended his recent book on an admittedly optimistic note: "Life belongs to the very fabric of the universe. Were it not an obligatory manifestation of the combinatorial properties of matter, it could not possibly have arisen naturally."

He could not, he said, conceive of the universe "as not bound by its very nature to give birth somewhere, sometime, perhaps in many places and at many times, to beings capable of enjoying beauty, experiencing love, seeking truth, and apprehending mystery."

Assuming that life originates on a planet, what are the chances that it will evolve into higher organisms and, finally, human beings? It is remarkable that on earth it took so long for multicellular life to progress from the most primitive species. The subsequent history of evolution was marked by both spurts and long unproductive periods. Although life originated some 3.5 billion years ago, it did not begin to evolve rapidly until between 800 and 1,000 million years ago. Then, in the early Cambrian Period some 570 million years ago, after nucleated cells had evolved, a "revolution"

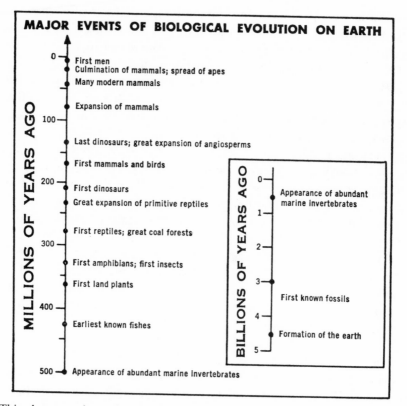

This chart was drawn by Huang to illustrate the brief portion of evolutionary history during which man has existed.

or "evolutionary frenzy" occurred. According to Ernst Mayr of Harvard the animal kingdom developed into at least forty phyla, or basic subdivisions. The diversity of life during that period is dramatically represented by imprints of marine organisms in the Burgess shales of British Columbia, the Cheng-jiang fauna of China, the Sirius Passet fauna of Greenland and the Emu Bay shales of South Australia. Many exotic species evolved, including one whose fossil tracks, 20 centimeters wide, resemble tire tracks on a wet mud flat. Had some large animal experimented with crawling from the sea? Despite "valiant attempts at reconstruction, nobody has any idea what this animal looked like," reported Henry Gee in a 1992 issue of *Nature*.

Then, in the Middle Cambrian, some sort of disaster struck. As many as 80 percent of marine animals became extinct. A large percentage of the phyla disappeared. As pointed out by Mayr, "there is no good evidence for the origin of any new phylum after the end of the Cambrian," 500 million years ago, which seems difficult to explain.

The great catastrophes, at intervals of a few hundred million years, cleared the decks for bursts of evolution, including the one that killed off the dinosaurs and opened the way for development of the mammals and, finally, of the hominids. Mayr argued that such a history depended on so many chance events that it is unlikely to have occurred elsewhere. Certainly there has nowhere been an exact repetition, but we are surrounded by demonstrations that, once the spark of life is ignited, it will evolve in whatever direction is open to it, and do so over and over.

12

Mars

No other object in the heavens has been the subject of such bitter controversy as the planet Mars. It has figured prominently in debates about extraterrestrial life, for if life had evolved there, this would encourage the belief in many inhabited planets in other solar systems.

Not very long ago it was still widely suspected that even intelligent life might exist on Mars, or might once have done so. In the 1960s Iosif Samuilovich Shklovsky, a prominent (and unorthodox) Soviet astrophysicist and proponent of the search for intelligent life in other worlds, proposed that Phobos, one of the two moons of Mars, might be an artificial satellite launched by an extinct civilization. He based this on an observation at the United States Naval Observatory (later found erroneous) that Phobos was slowly dropping out of orbit.

Another Soviet talked of bringing back Martian cattle of extraordinary usefulness. In 1962 Frank B. Salisbury, professor of plant physiology at Colorado State University, proposed in the authoritative journal *Science* that since seasonal color changes in regions of Mars seemed to indicate plant life, there might be animals and even intelligent organisms. He cited not only the peculiar behavior of Phobos but reports of what seemed eruptions or explosions on its surface. "Was this volcanic activity," he asked, "or are the Martians now engaged in debates about long-term effects of nuclear fallout?" He even suggested that when a robot lands on Mars and extends a mechanical arm to grasp a sample of the surface, the results radioed to earth may be perplexing: "At least I can imagine how I might react if such an apparatus landed in my back yard and started grabbing for my apple tree, the cat, and maybe me!"

Discovery of the essential features of that planet began in the seventeenth century, when Huygens made drawings that showed a white cap on one of its poles. Others observed the periodic changes in its dark markings, and in the eighteenth century Sir William Herschel noted that these included alterations of color. The white polar caps were of snow and ice, like those on earth, he said, and he argued that the "inhabitants" of Mars enjoy an environment much like our own. Herschel observed the planet on its close approaches, or "oppositions," of 1777, 1779, 1781 and 1783. Of all the planets, Mars is the most easily observed. Its orbit around the sun is the next outward from the earth, but because both orbits are elliptical, the distance between the earth and Mars, at opposition, is highly variable. The "close oppositions," when Mars comes within 35 million miles, occur every fifteen to seventeen years. Venus is nearer, but clouds completely hide its surface and, because its orbit lies inside that of the earth, its near side is dark at its closest approach.

It was two observations in the year 1877 that drew particular attention to Mars. One was by Giovanni Virginio Schiaparelli of Milan Observatory in Italy, who saw what he called *canali*. The other was the discovery by the United States Naval Observatory of the two Martian moons, one of them with a rather peculiar orbit.

Schiaparelli's *canali* would probably have best been translated as "channels," but instead they became known as canals, implying construction by Martian engineers organized within some superior, global society. Schiaparelli himself was noncommittal: "Their singular aspect, and their being drawn with absolute geometrical precision, as if they were the work of rule or compass," he wrote, "has led some to see in them the work of intelligent beings, inhabitants of the planet. I am very careful not to combat this supposition, which includes nothing impossible."

Soon thereafter word reached a Bostonian named Percival Lowell that Schiaparelli's eyes were failing, and Lowell decided to pick up the Italian's torch. He was the brother of Amy Lowell, the poet, and of Abbott Lawrence Lowell, for twenty-four years president of Harvard. Percival was primarily an Orientalist, having lived in Japan and Korea, written extensively on the Far East, and served as foreign secretary to the Korean Special Mission to the United States. Now, in his thirties, he became caught in the excitement of Schiaparelli's observations and decided to become an astronomer.

In 1894 Mars was to pass in close opposition, and Lowell went to Arizona to observe it through the clear, dry air of that region. "A steady atmosphere is essential to the study of planetary detail," he wrote, "size of instrument being a very secondary matter. A large instrument in poor air will not begin to show what a smaller one in good air will. When this is recognized, as it eventually will be, it will become the fashion to put up observatories where they may see rather than be seen."

His prediction has been dramatically fulfilled in Arizona with the establishment of some of the largest observatories in the world. The fruit of his own efforts, the Lowell Observatory at Flagstaff, was the first well-equipped station devoted primarily to study of the planets.

From Lowell's observation of the oppositions in 1894 and 1905 he calculated that Mars's atmosphere was only half as dense as that on the summit of Mount Everest, and that Mars "is very badly off for water," depending for its meager supply on the melting of its polar ice in spring. The canals would then have been built to carry water into the desert. Lowell realized that, for them to be visible from earth, they would have to be many miles wide. Hence it was proposed that what is seen are bands of irrigated land, rather than the canals themselves.

Already in Lowell's day there were vehement doubters. Astronomers of repute looked at Mars and said they could see no canals at all. Alfred Russel Wallace, who independently devised a Darwinian theory of evolution, said flatly that Mars is lifeless and that "water-vapour cannot exist" there. The latter statement was, however, proved incorrect in 1963 when California's 100-inch telescope on Mount Wilson, overlooking Pasadena, recorded the telltale emissions of water vapor in the Martian air.

Another skeptic was Svante Arrhenius, author of the concept of panspermia, who explained the changing colors of Mars in terms of hygroscopic, or water-hungry, salts deposited on the bottoms of shallow lakes. These, he said, dry up in winter, the moisture being deposited on the winter pole as ice crystals. When they melt in the spring, the airborne moisture is absorbed by the salts, which thus are darkened.

With even the most powerful telescopes, gazing at Mars through the constantly changing distortions of the earth's atmosphere is like looking through flowing water. The mind compensates for this, to some degree, by storing the ever-changing images and averaging them into an integrated picture. This, however, is somewhat subjective. Once the canals had been reported, there was a strong temptation for observers to see them. That they did so may also be related to a tendency of the brain, in poor visibility, to "see" as straight lines what are really strings of dark spots.

Excitement at the possibility of a Martian civilization soared after the writings of H. G. Wells and Edgar Rice Burroughs. "Those who have never seen a living Martian," wrote Wells in his 1898 *War of the Worlds*, "can scarcely imagine the strange horror of its appearance." The Burroughs books, written on the eve of World War I, had such titles as *A Fighting Man of Mars, The Warlord of Mars* and *A Princess of Mars.* They rivaled in popularity those he wrote about Tarzan of the Apes. In 1920 Guglielmo Marconi reported that some of his radio stations on either side of the Atlantic had for many years been hearing strange signals, and one of his associates said Marconi suspected Mars might be the source.

With Mars due to make a close opposition in 1924 there were a variety of suggestions on how to communicate. The Sperry Gyroscope Company, which had become a preeminent searchlight maker for air defense during the war, said that its new, high-powered model, if clustered, could be seen by the Martians. David Todd, who had been head of the Astronomy Department at Amherst College in Massachusetts, with the backing of a wealthy collaborator, proposed to convert an abandoned mine shaft in Chile into a mammoth telescope. The shaft, almost 60 feet in diameter, was so located that Mars would pass directly over its mouth. As a mirror at the bottom of the shaft, Todd proposed to spin a basin of mercury 50 feet in diameter fast enough that the mercury would assume the proper concavity (a principle that by the 1990s was being used to cast very large telescope mirrors from molten glass in a rotating oven). The telescope, he claimed, would bring the surface of Mars, which was to pass at a distance of 34,630,000 miles, to within less than 2 miles of his instrument. Astronomers at Harvard and Yale denounced the plan as "preposterous" and "a foolish, wild scheme." They pointed out that such magnification would only produce a meaningless blur.

Todd's next proposal was to persuade all radio stations on Earth to shut down during the close passage and listen for signals from the other planet. The possibility of a civilization on Mars superior to our own was still sufficiently in men's minds for Todd to have some success. On August 21, 1924, the Chief of Naval Operations of the United States Navy sent a dispatch to the twenty most powerful stations under his command, from Cavite in the Philippines to Alaska, the Canal Zone and Puerto Rico, telling them to avoid unnecessary transmissions and to listen for unusual signals. A similar order was sent to the Army stations, and the executive officer of the Army Signal Corps announced that William F. Friedman, chief of the code section in the office of the chief signal officer, was standing by to decipher any messages received. He expressed confidence that Friedman could do so. (Friedman later led the cryptologists who, by 1940, had broken enough of "Purple," the top-secret Japanese code, to decipher many of the messages to Japanese embassies that anticipated the attack on Pearl Harbor.) Elsewhere a man who had been experimenting with radio transmission of photographs tuned up his equipment in case Mars wanted to send us pictures. There was, of course, considerable skepticism in the government and in other circles. Only one or two commercial broadcasting stations seem to have closed down for the proposed periods of five minutes each hour during the passage. The more conventional astronomers were aghast at the whole thing. *The New York Times* predicted editorially that if a great many listeners tried to pick up Martian signals, some were sure to hear them on the basis of the old adage that the wish is father to the thought. Indeed there were such reports. Some from Vancouver, British Columbia, proved later to have come from new United States radio beacons. "Harsh dots" were reported by one of the most powerful receivers in Britain, but were quickly dismissed as being of natural origin. The bubble of excitement collapsed, and it was not until the start of space exploration that the search for life on Mars became realistic. The first Soviet and American space flights were projected for the International Geophysical Year of 1957–58, but when American pride was humiliated by Soviet launching of the first *Sputnik*, visible after dusk to most of the world's inhabitants as a coasting point of light in the sky, President Kennedy pledged in 1961 that "before the decade is out," Americans would be landed on the moon and brought safely home. Seven times, between 1969 and 1972, Apollo spacecraft made manned landings, and less dramatic were a series of unmanned flights to Mars.

In 1965 *Mariner 4*, as it flew past that planet, observed enough of its surface to show that the canals were a myth. *Mariner's* instruments also found atmospheric pressure on the surface of Mars to be even lower than had been supposed—only two-hundredths that on the earth. This meant that liquid water would evaporate immediately. There could be no lakes, ponds or rivers, much less oceans. Two more fly-bys, *Mariners* 6 and 7, confirmed these results. *Mariner* 7 flew over one polar cap, and on August 7, 1969, at a press conference in the auditorium of Caltech's Jet Propulsion Laboratory, which was conducting all the unmanned flights on behalf of NASA, Gerry Neugebauer of Caltech reported that infrared emissions from the polar cap showed its temperature to be −123 degrees Celsius (−180 degrees Fahrenheit). This indicated that it was covered primarily with frozen carbon dioxide, rather than water ice. George Pimentel, a distinguished biologist at the University of California at Berkeley, however, reported that according to spectroscopic measurements from *Mariner* 7 the temperature at the edge of the ice cap was compatible with water ice and that there seemed to be gaseous methane and ammonia in the atmosphere above that region, hinting at the presence of life. Those of us in the auditorium could hardly suppress our excitement, which was soon conveyed to the world at large. Public relations pressure for quick results had overcome caution. Within weeks it was shown by laboratory tests that the spectral features on which Pimentel had based his report could just as well be produced by carbon dioxide, which by then was known to be by far the major constituent of the Mars atmosphere, and Pimentel had to withdraw his report.

The first comprehensive pictures of the Martian surface began coming in late 1971 when *Mariner 9* went into orbit around the planet, sending extensive images of its features. At first a giant dust storm hid the surface, but as it cleared, two particularly striking features began to appear. One was a giant volcano, now known as Olympus Mons, more than three times as high as Everest and with a base as broad as the distance from Virginia to New York. Also coming into view was a canyon 2 to 3 miles deep and 3,000 miles long, many times the length of the Grand Canyon. To produce such features Mars at some time must internally have been violently active. Nothing comparable to these features has been recorded elsewhere in the solar system.

Harold Masursky of the U.S. Geological Survey, leader of the *Mariner 9* imaging team, wanted to lower the spacecraft's orbit to get a closer look, but, according to Masursky, Joshua Lederberg "turned ashen" at the suggestion. Lederberg in 1958 had won a

Nobel prize for his work in genetics and had been chairman of a national committee considering ways to search for life in space. Working with him now as his protégé was Carl Sagan, a kindred soul in that respect. They had also played a leading role in establishing international commitments to avoid biological contamination of other celestial bodies. If inadvertently a single microbe was carried to Mars, the moon or some other such object, it could either overcome any life native to that body or be mistaken for a native by future explorers. Of even greater concern was the possibility that a spacecraft returning from there might carry an organism perilous to life on earth, a fear that led to elaborate and costly measures to quarantine returning lunar astronauts and their specimens. Lederberg feared that lowering the *Mariner 9* orbit would create a greater danger of its ultimately crashing and contaminating Mars.

The possibility that Mars was once very different from now, with abundant liquid water, was evident in *Mariner 9* images that showed networks of channels like those formed on earth by the drainage of rainy areas. There were also regions like the "scablands" of Washington State, including Grand Coulee. In Washington they had been produced by catastrophic floods when an ice dam broke at the end of the last ice age, draining a giant lake in a matter of hours. On the floors of the Martian channels were many impact craters, indicating that they had been subjected to bombardment for millions or billions of years and were therefore very old. But if Mars once had plentiful liquid water, Lederberg reasoned, it could have supported life—and still might.

During the period of the *Mariner 9* mission a series of Soviet craft were sent to the planet. One of them, *Mars 1*, a lander, reached the surface but sent no data. *Mars 3* landed and sent twenty seconds of television pictures, but they were featureless. This was the time of the global dust storm that greeted *Mariner 9*'s arrival, but activities by the Soviet craft had been programmed in advance and could not be altered to wait out the dust storm. All told, the Soviets said they sent eight missions that included two more landers, *Mars* 6 and 7, but neither were successful.

Meanwhile Voyager, an ambitious, $4-billion American program to land a robot on Mars, was being planned, and several of those destined to supervise the ultimate landing worked on life-detection schemes. It was argued that biological tests on the first landing were critical, for a bacterial stowaway on that spacecraft,

no matter how stringent the efforts at sterilization, could proliferate across Mars in time to confuse experiments by the next
lander to arrive. Norman H. Horowitz, professor of biology at Caltech, and Gilbert V. Levin, who had developed a novel way to
detect microorganisms in drinking water, designed what they
called "Gulliver." It contained three baby cannon that would fire
25-foot sticky strings across the Martian landscape. Soil would ad-
here to the strings, which would then be reeled in and soaked in
a nutrient broth. If bacteria fed on it, since carbon in the broth
was radioactive, measurable amounts of radioactive carbon dioxide would be released.

Another device, "Multivator," was designed by Lederberg and
his associates at Stanford University. Martian soil would be blown
into fifteen compartments and tested for a variety of biological
substances. A "Wolf Trap," designed to suck in dust and culture
it in a broth, was proposed by Wolf Vishniac, professor of biology
at the University of Rochester. The costly Voyager project was
killed by President Johnson in 1967, but from its ashes rose the
historic Viking Project, in which Horowitz, Lederberg and Vish-
niak played prominent roles.

Two Viking spacecraft were to be launched, each in two parts:
one to remain in orbit around Mars, the other to be detached and
land on the planet. Their launching was to be timed so that after
their eleven-month flight, they arrived two months apart. In that
way, if by some amazing good fortune both were successful, those
at JPL (the Jet Propulsion Laboratory) monitoring their streams
of radioed data could concentrate on one at a time.

Despite the discouraging reports from the Mariner spacecraft,
there was still hope that the two landers might find traces of life,
past or present. The critical soil-sampling experiments were to be
fed by extending a 10-foot boom with a little shovel at the end.
Two cameras atop the lander were to scan the scene, looking, a
few die-hards hoped, for large animals or other signs of life, but
more realistically to enable those at JPL to guide the remotely
controlled boom to a propitious area for soil samples, including
those to be tested for microorganisms. After the boom had been
retracted it was to dump the specimens of Martian soil into each
of three hoppers, where they would be ground, sieved and fed to
the experiments. Three of them, fed from the same hopper, were
to look for signs of life. Samples from the second hopper were
destined for a gas chromatograph mass spectrometer to be ana-

lyzed for organic compounds, including both those associated with life and ones delivered by meteorites and comets. The third hopper fed an X-ray fluorescence spectrometer to analyze the soil for inorganic material. There was also a seismometer to detect tremors, whether from Marsquakes or the footfalls of giant beasts.

Heroic efforts were made to check out this elaborate and complex assembly of devices—the most sophisticated biological payload ever sent into space. Prototypes of the landers were tested in the desert. The detectors were tried in Antarctica's dry valleys, whose extreme cold and aridity come as close to Martian conditions as could be found on earth. So much was at stake, in hopes, complexity and cost, that conservative participants wanted to do further testing and, perhaps, first send simpler prototypes to Mars. But there were many pressures, political as well as scientific (including fears of Soviet rivalry and of new dust storms on Mars), for an early launch.

Viking 1 and its attached lander were sent into space in August 1975, after the lander had been sterilized by eighty hours of baking. As an added precaution it was enclosed in a "bioshield" to isolate it until it was free of the earth's bacteria-laden atmosphere. Eleven months later *Viking 1* began orbiting Mars, and the debate began on where to land. Some of us were allowed to stand around the wall of the JPL room where the debate was taking place, and the division into two camps became obvious. From the *Mariner 9* pictures the biologists and geologists had identified sites of prime scientific interest, such as the remnant river valleys and gouges, whereas the engineers wanted the smoothest regions. Another fear was dust. The world's largest radio antenna, at Arecibo, Puerto Rico, had been probing the Martian surface with radar, and its director, Frank Drake, pioneer searcher for extraterrestrial life, said the reflections from parts of Mars indicated it was dusty. As reported by Henry S. F. Cooper, Jr., in his 1980 book *The Search for Life on Mars: Evolution of an Idea*, Drake warned that the Viking lander could possibly "sink up to its eyeballs."

The radar was most effective when its beam bounced off that region of Mars directly facing the earth. Chryse, a low-lying plain in the northern hemisphere, was the favored landing area, but the Arecibo radar could not get a direct shot at it until June 1976, a week before the scheduled landing on July 4. By then Harold Masursky of the U.S. Geological Survey, who was now head of the site selection team, had gotten his first look at pictures of the projected landing area taken by the Viking orbiter. They were much clearer than those from *Mariner 9* and showed that the planned site in the Chryse plain was cut by deep channels. The lander was tolerant of considerable tilt, but it only cleared the ground by 9 inches. Boulders could be fatal. The landing was postponed and the site shifted 60 miles onto a flatter part of the plain.

Finally, on July 29, 1976, a month after *Viking* had reached Mars and gone into orbit, it was decided to fire the explosive bolts holding the lander to the orbiter. Compressed springs would push the two craft apart and the lander would begin its descent. As soon as this occurred it was obvious something had gone wrong. Years of planning, ingenuity, designing, testing and hoping hung in the balance. In my science-writing career I can think of no period as tense as the 192 minutes until the lander was supposed to reach the surface.

We were crowded into JPL's auditorium on the foothill slopes of California's San Gabriel Mountains. Television screens were mounted so that the media and many of those from the laboratory could see the first close-up pictures of Mars—a glimpse of reality after centuries of speculation.

It had been expected that signals from the descending lander would be relayed via the orbiter's much more powerful communications, but none were being received. The only indication that the lander had not been destroyed were its own faint signals, and these said nothing about its situation. Detachment of the lander had apparently changed the orientation of the orbiter so that its large radio antenna was no longer aimed at the earth. Did this mean that the lander was aimlessly careening off into space?

Traveling at 10,000 miles an hour and temporarily enclosed in an "aeroshell" for protection during entry into the Martian atmosphere, the lander was programmed to enter at a shallow angle, minimizing the shock and heating. When it had slowed to 560 miles an hour, 20,000 feet above the planet's surface, it was to shed the aeroshell and deploy a parachute. Despite the very thin air the chute was expected to have a braking effect. Then, at 4,000 feet and with only forty seconds to go before touchdown, it was to jettison the chute and fifty-four little jets were to reduce its speed to six miles an hour before it landed. The jets were designed to avoid plastering the surface with chemicals that might confuse the sampling experiments.

Seconds after the scheduled landing time there was a joyous

shout over the public address system. Signals from the lander indicated that it had touched down and was functioning. Then, on the left margin of each video screen in the auditorium, a white line appeared. Beside it a second line formed, then a third. Gradually the screen was being covered by lines that were painting a picture of a gravelly, rocky terrain.

Thomas A. ("Tim") Mutch of Brown University, a tall, courtly gentleman in charge of the imaging team, was in front of a microphone, ready to broadcast a running commentary. He became almost hysterical. "It works!" he cried. It seemed a miracle that the system was doing so after a multitude of pitfalls, test malfunctions and budget constraints. What came into view, after a glimpse of the lander's feet, was a rust-colored, rocky, partly sandy landscape with no sign of life. The reddish color of Mars is evident from earth even with the naked eye, and, as suspected, its surface proved to be rich in iron oxide—that is, rust.

Because of the vast distance between the earth and Mars it took eighteen minutes for a signal from the lander to reach JPL and a comparable time for any JPL commands to reach it. Hence the lander carried to Mars enough instructions to fulfill at least the early phases of the mission without further help, although, unlike the Soviet landers, JPL could override the instructions if necessary. High priority was assigned to the photographs. Then, after eight days, the sampler arm was programmed to reach out and scoop up the first specimens. This allowed ample time for the JPL team to study the photographs and pick a suitable sampling spot. The shovel reached out, dug up a sample, and dumped a load of Martian soil into the biological hopper.

In light of the earlier evidence that the surface of Mars could not retain liquid water, the prospects for life were considered poor by many of the Viking scientists, but three days later the head of the biological team, Harold P. Klein of NASA's Ames Research Center, told the assembled press that "important, unique and exciting things" had been found. He referred to the experiment of Vance I. Oyama, his colleague at Ames, in which Martian soil was exposed to water vapor to see whether that stimulated biological activity, thereby releasing telltale gases. Samples of the resulting gas were drawn off periodically and analyzed. Then, if there were no indications of activity, a mixture of nutrients was to be added, containing such substances useful to terrestrial life as vitamins, amino acids, purines and pyrimidines.

The surprise was that two and a half hours after the start of the experiment, long before nutrients were to be added, analysis of the gas showed twenty times as much oxygen as could be explained by Martian air in the chamber. The thought that this might be biological diminished the next day when the increase in oxygen gradually tapered off.

Also incubating was soil in the experiment of Gilbert Levin, the expert at detecting microorganisms in water supplies. As nutrients, the soil was supplied with seven synthetic amino acids and carbohydrates in each of which ordinary carbon had been replaced by radioactive carbon 14. It was assumed that, if something in the soil "ate" and digested one of the nutrients, that would release carbon dioxide or, possibly, some other gas containing the carbon 14. The first radioed report came less than a day after the nutrient had been added, and almost immediately radioactivity was detected above the 500 counts per minute assumed to be coming from other devices in the lander. In nine hours it reached 4,500 counts, and it eventually climbed to 10,000. This, said Levin, was "very much like a biological signal," and Oyama reported that his own result showed "very active" surface material. Nevertheless, Klein told the press these results must be viewed "very, very carefully." The memory of Pimentel's erroneous report from the *Mariner 7* fly-by was in many minds.

Klein pointed out that some form of plant life could not be expected to have released the burst of oxygen in Oyama's experiment, since the sample was in total darkness. The suspicion grew that when the extremely desiccated soil was moistened by the nutrient, some substance with an extra oxygen atom, such as peroxide or superoxide, was releasing that atom as soon as it came into contact with water.

It was proposed, but without Levin's concurrence, that such oxidants could also explain the release of carbon dioxide in his experiment. While Levin's apparatus could not detect the oxygen that might have been released by the moistening of oxidants, the oxygen could have reacted with carbon in Levin's nutrients, accounting for the observed release of radioactive carbon dioxide.

The experiment prepared by Norman Horowitz was the only one that tried to mimic the hostile environment of Mars. A sample of Martian soil was exposed to an imitation Martian atmosphere, but one whose carbon dioxide and carbon monoxide were radioactive. The soil was incubated for 120 hours under imitation sunlight, whereupon the synthetic atmosphere was removed and the soil processed to see whether plantlike organisms in it had consumed any of the radioactive gases. The first run was startling. While the number of radioactive emissions was well below that obtained in similar tests of soils on earth, they were well above the number from sterile test samples.

Ever since *Mariner 4*, in 1965, had ruled out the possibility of liquid water on the Martian surface, Horowitz, who had previously championed the possibility of life on Mars, had become a leading skeptic. Now, of all the Viking experiments, his seemed to show the strongest evidence for life. In presenting his results to the press he said: "I want to emphasize that we have not discovered life on Mars—*not.*" His data were "conceivably of biological origin," he said, but could also be explained in other ways. All told, his test was repeated nine times, six at Chryse and three by *Viking 2* when, after an equally miraculous descent, it landed on the Utopia Plain, almost halfway around the planet. In no case were there as many radiation counts as in that first sample. The explanation, he said later, may have been a reaction between the radioactive carbon dioxide and the iron-containing minerals that are abundant on the Martian surface. This could have formed substances that contained radioactive carbon and remained after the carbon dioxide had been withdrawn. In his 1986 book *To Utopia and Back: The Search for Life in the Solar System* Horowitz wrote that the reactions are still not completely understood, but added: "the chance that their source was biological seems negligible."

It had been assumed that even if organic molecules from existing or former organisms were not found, *Viking* might detect such material from the stages of chemical evolution that preceded life. It was also suspected that the Martian surface would be coated with organic material from meteorites, comets and other space debris. The search for such molecules, conducted by a gas chromatograph mass spectrometer, found no organic molecules whatsoever. They had been wiped out by the pervasive oxidants and probably cannot be found anywhere on Mars except by excavation.

In 1978 the *Viking 2* orbiter was the first to go off the air, but its lander on the Martian surface continued to function until 1980. For more than six years the *Viking 1* lander sent pictures documenting the cycle of seasons and local dust storms that produced drifts visible to its cameras. It finally expired in November 1982. Some of the prime figures in Viking have perished, including Tim

Mutch, Wolf Vishniac and Harold Masursky. Carl Sagan, who for much of his career argued for life on Mars, as he has for life throughout the universe, now concedes that the signs of present or past life, as well as organic molecules from impacting material, could not have survived the "witches' brew" of lethal oxides on the planet's surface. That Mars was not always so is obvious from the signs of ancient water courses, and Sagan, like many others, believes that fossils may still be found.

The failure to find life on Mars is disappointing but, in retrospect, should not have been surprising. With only half the earth's diameter and only 38 percent of its surface gravity, Mars was unable to retain a substantial atmosphere and its fate has little bearing on that of more earthlike planets throughout the universe.

13

The Uniquely Rational Way

On September 19, 1959, the British journal *Nature* published a proposal that startled the scientific world. The authors, two physicists of impeccable reputation, suggested that at this very moment intelligent beings in some distant solar system might be trying to communicate with us. Furthermore, they presented an argument pointing to one precise radio frequency as the uniquely logical channel for such communication. They proposed that radio telescopes be aimed at certain nearby stars in search of the hypothetical signals.

Actually the discussion of possible methods for communication with other worlds long predates the discovery of radio waves. During the many decades when even the most conservative scientists believed intelligent life might exist on Mars—or even the moon —there was talk of ways to reveal the presence of intelligence on earth. In the last century the mathematician Karl Friedrich Gauss is said to have proposed that broad lanes of forest be planted in Siberia, forming a gigantic right-angled triangle. The inside of the triangle would be planted in wheat to give it a uniform color. A modification of this plan was to erect squares on each side of the triangle, forming the classic illustration of the Pythagorean theorem.

The astronomer Joseph Johann von Littrow, who became director of the Vienna Observatory in 1819, suggested that canals be dug in the Sahara, forming geometric figures 20 miles on a side. At night kerosene was to be spread over the water and set on fire. In France during the 1870s Charles Cros urged that the government construct a vast mirror to reflect sunlight toward Mars. A related scheme was for a system of mirrors that could be rearranged, periodically, in a form of slow-speed semaphore.

The discovery of radio waves offered a far more realistic means of interplanetary communication, and this possibility excited two of the men who pioneered in this field, Marconi and Tesla, to the point where both suspected that they had heard signals from another world. Although far less well known than Edison, Nikola Tesla was an inventive genius whose contributions to the application of electrical energy were of major importance. He seems, for example, to have been the first to propose an effective way to use an alternating current—a scheme strongly opposed by Edison, but one that made possible the harnessing of the power of Niagara Falls. Tesla invented a multitude of dynamos, motors, transformers and the like. He performed some of the earliest experiments in radio communications, apparently being the first to use transmitting and receiving antennas tuned to the same frequency.

He was also given to grandiose schemes and strange ideas. In 1917, for example, he said: "We will deprive the ocean of its terrors by illuminating the sky, thus avoiding collisions at sea." One of his grandest ideas was to set the electric field of the entire planet aquiver. He believed the world had an electric charge and that he might be able to generate oscillations within it. To this end, and with financial aid from John Pierpont Morgan, he set up a laboratory in Colorado Springs, Colorado, in 1899, equipped with a 200-foot transmission tower and high-voltage equipment. One night, when he was alone in the laboratory, Tesla observed "electrical actions" that, he later reported, appeared to be signals:

> The changes I noted were taking place periodically, and with such a clear suggestion of number and order that they were not traceable to any cause then known to me. I was familiar, of course, with such electrical disturbances as are produced by the sun, Aurora Borealis and earth currents, and I was as sure as I could be of any fact that these variations were due to none of these causes. . . . It was some time afterward when the thought flashed upon my mind that the disturbances I had observed might be due to intelligent control. . . . The feeling is constantly growing on me that I had been the first to hear the greeting of one planet to another.

When Tesla was asked by the American Red Cross, at the turn of the century, to predict likely developments of the next hundred years, he replied on January 7, 1900, with a reference to these apparent signals:

> Faint and uncertain though they were, they have given me a deep conviction and foreknowledge, that ere long all human beings on this globe, as one, will turn the eyes to the firmament above, with feelings of love and reverence, thrilled by the glad news: "Brethren! We have a message from another world, unknown and remote. It reads: one . . . two . . . three . . ."

The century that followed is now almost at an end. It is noteworthy that the 1959 proposal for a listening program envisaged that intelligent beings in another world would use just such a device for attracting attention—a series of pulses representing such numbers as one, two, three, etc. Had it not been for Tesla's personality, his report might have made more of an impression. He was an eccentric figure who allegedly believed in mental telepathy and displayed a strange affinity for pigeons, in particular a certain white bird that, in his later years, he spoke of with peculiar passion. He implied that he had had a closer rapport with this pigeon than with almost any human being and that when it died he could no longer work creatively. He fed the pigeons in Bryant Park, in the heart of Manhattan, with devoted regularity and, toward the end, when too weak to go himself, paid a Western Union messenger to do so.

Tesla was seized upon by the spiritualists, although he does not seem to have been one himself, and was the hero of a book published in 1959 under the title *Return of the Dove* (his pigeon thereby being transformed into a somewhat different species). The book, by Margaret Storm, described him as having been born aboard a space ship en route from Venus to the earth in 1856. He was deposited, it said, in a remote mountain province of what became Yugoslavia. Tesla was, in fact, a Croatian by birth.

While Tesla was experimenting in Colorado, the Italian physicist Guglielmo Marconi demonstrated the capabilities of radio by sending messages between England and the Continent. By 1901 his radio waves had spanned the Atlantic and a revolution in communications was under way. Then, at the approach of World War II, two developments made communication with other worlds seem more realistic: the birth of radio astronomy and the explosive evolution of radar.

In the early 1930s it was proposed that emissions at the extreme short-wave, or "microwave," end of the radio spectrum, where it approaches the even shorter wavelengths of light, might be focused into an invisible, penetrating beam that could search out enemy targets. In 1935 the British Air Ministry authorized the secret construction of five stations along the east coast of England, designed to sweep microwave beams across the sky to warn of any surprise attack by Germany's new Luftwaffe. Two years later another fifteen of these stations were installed, and during the war that followed, this equipment, which came to be known as radar, proved so valuable that there was a massive research effort in electronics. Transmitters, receivers and antennas of ever greater efficiency were developed until, by 1946, it was possible to bounce radar signals off the moon. By 1959 ten groups in six countries had studied the moon with radar echoes, observing, for example, its surface properties in a manner impossible through optical telescopes.

The use of radar for such astronomical work in the United States was a by-product of the effort, largely financed by the Air Force, to develop equipment that could give the country as much warning as possible of a missile attack. One of the most powerful instruments in this program was the parabolic, or dish-shaped antenna, 84 feet in diameter, mounted on a tower atop Millstone Hill, near Westford, Massachusetts. It was operated by the Lincoln Laboratory of the Massachusetts Institute of Technology, which in February 1958 used it for the first time in an effort to obtain radar echoes from another planet. It was aimed at Venus, our nearest neighbor beyond the moon, and prolonged analysis of the tape-recorded results seemed to show an echo. President Eisenhower sent a congratulatory message, and, although the analysis is now thought to have been erroneous, the experiment awakened scientists to the growing power of the world's transmitters. When Venus came close to the earth again in 1959, new attempts were made, both by Millstone Hill and by what was then the world's largest antenna at Jodrell Bank in England. Again the results were unconvincing, but when Venus made its close approach in March and April of 1961, a number of antennas in England, the Soviet Union and the United States, including an 85-foot dish at Goldstone, California, obtained echoes.

Two such antennas at Goldstone, 7 miles apart in the Mojave Desert, constituted the central unit of the Deep Space Instrumentation Facility, operated for NASA by the Jet Propulsion Laboratory. The station was one of three located around the world to keep continuous track of American space vehicles. After several weeks of preparation and testing, the equipment was ready on March 10, and the two antennas were aimed at Venus, one to transmit and the other to receive. Some six and a half minutes after the transmitter was turned on, the stylus recording incoming noise on a moving roll of paper moved over slightly—indicating the arrival of a weak echo.

Both antennas were controlled by gears that kept them aimed squarely at Venus, despite the planet's motion across the sky. After a half hour, to find out whether or not the incoming echoes were genuine, the motion of the transmitting antenna was halted and Venus slowly moved out of its beam. Six and a half minutes later the stylus moved back to the intensity level it had shown before the experiment. The interval was that required for radar waves to travel to Venus and bounce back.

The transmissions were at 2,388 megahertz per second with a power of 12,600 watts. About 10 watts of this total hit the surface of Venus. It was calculated that 9 watts were absorbed and 1 reflected back into space. Of this, one-hundredth of a billionth of a billionth of a watt returned to strike the receiving antenna, yet this was still ten times the background noise, making it observable.

Within weeks other observatories, including Millstone Hill, were obtaining echoes, and the latter was able to reexamine earlier records and discover, hidden there, true echoes from the Venusian surface.

Thus did earth's inhabitants send their first signals to another planet and receive "answers." Among the factors which made this possible was the use of radio frequencies in which the sky is comparatively quiet. Another was the availability of two revolutionary new devices for amplification of returning echoes. If you turn up the volume of an ordinary radio, the hum generated by its amplifiers becomes so loud that it drowns out any weak signals that might otherwise be received. The new devices, known as the maser and the parametric amplifier, eliminate almost all of this receiver noise. Both were used with the Goldstone antenna.

Meanwhile, radar echoes had been obtained from the sun and other bodies of the solar system, but still more dramatic evidence of the capabilities of radio for long-range communications came from radio astronomy. It was shown that radio waves traverse the vast reaches of the universe even more readily than light waves, and eventually this new science pointed to what seemed the uniquely rational wavelength on which to search for signals.

Radio astronomy came into being by accident through the efforts of Karl G. Jansky, of the Bell Telephone Laboratories, to track down the high-frequency static that was interfering with his company's transoceanic communications. At a field station in Holmdel, New Jersey, he built an antenna array on a wooden frame 100 feet long which rode on four wheels salvaged from a Model T Ford. A motor rotated this array, which he called his Merry-Go-Round, once every twenty minutes. He began his observations in August 1931, on a wavelength of 14.6 meters (20,600 kilohertz) and soon tracked down the sources of two types of static. One came from the lightning in nearby thunderstorms. The second he attributed to distant storms whose radio emissions were apparently bent back to earth by the ionized regions of the upper air. A third form was quite different, producing a loudspeaker hiss whose intensity slowly changed during the day. He reported in 1932, in the *Proceedings of the Institute of Radio Engineers*, that the direction from which this hiss came traveled "almost completely around the compass in twenty-four hours." In the previous December and early January, he said, its direction coincided generally with that of the sun, his array not being able to pinpoint the source. Then he noticed that its direction was shifting and, as of his writing, on March 1, he said, "it precedes in time the direction of the sun by as much as an hour."

Nevertheless he thought the source was probably some effect of sunlight on the earth's atmosphere. The shift in direction might be related, he said, to the seasonal rise of the sun to higher elevations in the sky. However, an astronomically minded colleague suggested that Jansky rework his data, using celestial coordinates (the "latitude and longitude" of the stars on the heavenly sphere) to designate the apparent direction of the source each day. He made the exciting discovery that it remained fixed among the stars. At first there seemed to be a jump in his record until he realized that he had forgotten to allow for the shift from Standard to Daylight Saving Time.

At a meeting of the American Section of the International Scientific Radio Union on April 27, 1933, he announced his conclusion. The radio emissions, he said, appeared to be coming from beyond the solar system. They might be from a single source, he added, or "from a great many sources scattered throughout the heavens," the direction at which he observed maximum intensity being simply the center of this activity. He noted that this direction, toward the constellation Sagittarius, the Archer, was that of the center of our galaxy. Both of his suggestions proved correct. The direction that he identified is that from which intense radio emissions indicate incredibly fierce activity in the core of the galaxy now known as Sagittarius-A. But it was also soon shown that radio emissions are coming from the entire galaxy, as well as from the most distant objects in the universe of which we have knowledge, now called quasars.

Jansky's report of signals from beyond the solar system was front-page news, even though he discounted the possibility that they were of artificial origin. The astronomers, apart from two young men at Harvard (Jesse L. Greenstein and Fred L. Whipple), gave it the cold shoulder. It was for an amateur astronomer and radio "ham," Grote Reber, to make the next big step. Reber was a radio engineer and, at his own expense, built a parabolic antenna in his backyard at Wheaton, a Chicago suburb. Being 31 feet in diameter, the dish crowded his yard. It was the first such antenna turned on the heavens, and Reber's results, published in 1940 and thereafter, showed virtually the entire Milky Way to be a source of radio "noise," with several areas of intense emission.

World War II, which impeded further progress in radio astronomy, also gave it a new stimulus. In February 1942, British Army radar operators complained of a new form of German jamming. The Germans had been trying to nullify the British warning system by flooding the radar receivers with signals on their echoing frequencies. The new interference was referred to J. S. Hey of the Army Operational Research Group, who noticed that the direction from which the jamming came did not point to one or two transmitters on the coast across the Channel. Instead, all the directions pointed to a single source: the sun. It happened that there was at this time a very large sunspot, and Hey guessed that it was the villain, not the Germans. In that same year solar emissions were also identified in the United States, but anything to do with such high-frequency research was considered a military secret, so radio observations of the sun did not come into their own until after the war.

Meanwhile, in 1944, word of Reber's observations had reached the German-occupied Netherlands and had excited the interest of astronomers at the Leiden Observatory. Unlike American scientists, who were preoccupied with the war effort, the Dutch were free to dabble in pure science, and Jan H. Oort, director of the observatory, held a seminar on the implications of Reber's observations. Since radio waves are an extension of light waves into longer wavelengths, the discovery that parts of the sky "shine" in the radio portion of the spectrum meant that an entirely new window had been opened on the heavens. Until then all of our knowledge of the universe, apart from a few clues culled from meteorites or cosmic-ray particles, had been acquired through observation of that narrow band of the electromagnetic spectrum which we call light.

However, Oort pointed out, one drawback to the radio spectrum seemed to be that, unlike light, it lacked the sharp emission (or absorption) lines that had proved so useful in astronomy. These lines, the by-products of various atomic processes, could be used to detect relative motion toward and away from the observer and many other phenomena.

As though picking up this challenge, one of Oort's students, Hendrick Christoffel van de Hulst, came up with the proposal that clouds of individual hydrogen atoms, as opposed to the paired atoms of hydrogen gas, should emit radio waves at a wavelength of 21 centimeters. Since it was suspected that such clouds exist widely in space, there should be a sharp augmentation of cosmic radio noise at that wavelength—in other words, an emission line.

The hydrogen atom normally consists of an electron and a proton, both in effect spinning and thus acting like tiny bar magnets. Because like repels like, in neighboring magnets, the most natural alignment of these particles is with their magnetic poles pointed in opposite directions. It therefore takes a bit of energy to flip the electron over so that its positive pole is aligned with the positive pole of the proton. When such a flip has occurred, the atom has a slight reserve of energy. Ultimately the electron flips back, emitting this energy as a radio wave. The wave oscillates at a characteristic frequency of 1,420,405,752 times a second (1,420 megahertz), which corresponds to a wavelength of 21 centimeters. In the lonely reaches of space, where hydrogen atoms are free from strong energy inputs and magnetic influences, they still receive nudges of energy from collisions and radiation. Van de Hulst proposed that hydrogen clouds throughout the universe must be shedding this energy as radio waves at 21 centimeters. By good fortune, this wavelength passes freely through space and the earth's upper atmosphere. Longer waves are apt to be absorbed in the upper air.

The next step was to look for such emissions. For a number of years Edward M. Purcell at Harvard had been studying the radio frequency resonances of atomic nuclei. In 1945 he devised a technique that made it possible to measure the utterly tiny magnetic fields generated by the spin of these nuclei, using a high-frequency coil that could be tuned to match the resonance frequency of the nucleus under study. For this he and Felix Bloch, who had done similar work at Stanford University, shared the Nobel prize in physics for 1952.

It was in the previous year that Purcell, with his Harvard colleague Harold I. Ewen, sought to detect the 21-centimeter hydrogen emissions predicted by van de Hulst. At a cost of $400 a university carpenter built a horn antenna on the laboratory roof. The electronic equipment, said Purcell, "was all scrounged," and on March 25, 1951, when they turned on the receiver, the predicted emissions were there. Less than two months later the Netherlands astronomers obtained similar results and both groups found that the emissions, instead of showing up as a single, narrow line at 21 centimeters, were spread out in wavelength, with several peaks of intensity.

What they were seeing were hydrogen clouds moving at different velocities toward and away from the receiving antennas, with consequent modification, or "Doppler-shifting," of the wavelengths. These relative motions were produced by the earth's spin, by our planet's orbital flight around the sun, by the sun's flight around the core of the galaxy and by inherent movements of the hydrogen clouds themselves. The relative contributions of all these factors could be determined and distances to the various clouds could be indirectly estimated. Soon the Dutch scientists were mapping spiral arms of our galaxy, hidden from us by dust clouds but revealed by their 21-centimeter emissions. The dust that closed much of the universe to our telescopes had suddenly become transparent.

The discovery of the 21-centimeter emissions gave a tremendous boost to radio astronomy. Large antennas already under construction or planned had opened up many lines of investigation. They signaled eruptions on the sun, even in cloudy weather; they indicated the surface temperatures of the moon and nearby planets; they disclosed the existence of atomic particles trapped and gyrating furiously within distant magnetic fields, as they do in the radiation belts surrounding the earth and Jupiter or in the turbulent gas clouds of the Crab Nebula.

In England the mammoth radio telescope at Jodrell Bank was under construction when word was received that the 21-centimeter emissions had been observed. The diameter of the British dish is 250 feet (a football field is 300 feet long), and it is mounted on a frame that can be tilted and rotated to aim the antenna at any point in the sky. To achieve rotation, the array, weighing 2,000 tons, rides a circular, double railroad track. To tilt the 800-ton dish assembly the British installed a drive taken from one of the main battery gun turrets of the battleship *Royal Sovereign.* Protruding 62.5 feet from the apex of the dish is a mast that holds the receiving equipment at the focal point. The parabolic shape of the dish, like that of the mirror in an optical telescope, reflects all radio emissions from the direction of its aim to this focal point. The dish was designed for wavelengths so long that a coarse wire mesh was sufficient to reflect them to the focus. The plans were hastily revised to provide a surface of solid metal sheets that would reflect waves as short as 21 centimeters.

In the United States the only large dish capable of observing at that wavelength was the one, 50 feet in diameter, on the roof of the Naval Research Laboratory, across the Potomac River from Washington's National Airport. Hence it was decided that the first big antenna at the projected National Radio Astronomy Observatory in West Virginia should be designed for wavelengths of 21 centimeters and even less.

This is how matters stood when the two physicists began formulating their proposal that there be a search for intelligent signals from another world. They were Giuseppe Cocconi and Philip Morrison, both of them professors on the faculty of Cornell University in Ithaca, New York. In the spring of 1959 Cocconi was working on a paper that he was to present at an international conference on cosmic rays to be held that summer in Moscow. It was his thesis that some celestial objects, such as the Crab Nebula, should be strong emitters of gamma rays as an extreme form of what is known as synchrotron radiation. If one should be so unwise as to peek into one of the very-high-energy circular electron accelerators used in physics research, one would see a strange and brilliant light. This is synchrotron radiation and represents energy given off by the electrons because they are forced, by magnets, to fly in a circle. The radiation is given off throughout much of the electromagnetic spectrum, including visible light, radio waves and gamma rays, and differs from other radiation in that it has no spectral lines.

Both the light and the radio emissions from the Crab Nebula lack such lines, showing that they are products of synchrotron radiation, one of the less controversial predictions of Iosif Shklovsky. This nebula is one of the most awesome sights of the sky, for in photographs it looks like a snapshot of an explosion. That, in effect, is just what it is, for the Crab Nebula is the remnant of a supernova first seen in 1054, when it was visible by day as well as by night. The gases and other debris are still flying outward at great velocity, and within these clouds are magnetic fields that have trapped gyrating electrons, producing the radiation.

One of Cocconi's points was that the gamma rays from this nebula should stand out boldly against the sky because such emissions must be a rarity in the heavens. He proposed a gamma-ray detector that could, in a rough way, determine the direction from which such rays were coming.

One evening, as he discussed this with his wife, Vanna, herself a physicist at Cornell, it occurred to them that the scarcity of gamma-ray emitters in the sky pointed to these rays as a promising vehicle for interstellar signaling. If an unexpected source of gamma rays were observed, would it not be worth examining very closely?

The Cocconis had often discussed with Morrison the possibility of intelligent life in other worlds. Morrison had been a group leader at Los Alamos during the effort there to produce the first atomic bomb. He jumped at the suggestion that it might be a good idea to look for signals, rather than try sending any. However, he argued against gamma rays as difficult to generate and receive. A synchrotron adequate to generate a beam of gamma rays would probably cost millions of dollars, whereas radio transmitters and receivers are so cheap that taxis all over the world carry them.

The radio frequencies that travel readily through space and through the atmospheres of planets like our own lie in the range between 1 and 10,000 megahertz. The most powerful instrument for receiving signals in this general region was that at Jodrell Bank, now in operation more than a year. From CERN, the international atomic research center in Switzerland, Cocconi wrote to Sir Bernard Lovell, founder and director of the great radio astronomy observatory, with a proposal that Sir Bernard at first dismissed as "frivolous." The letter, dated June 29, 1959, began as follows:

Dear Dr. Lovell,

My name is probably unknown to you, so let me start by saying that I am now at CERN for one year, on leave from Cornell University, where I am professor of Physics.

Some weeks ago, while discussing with colleagues at Cornell the emission of synchrotron radiation by astronomical objects, I realized that the Jodrell Bank radio telescope could be used for a program that could be serious enough to deserve your consideration, though at first sight it looks like science fiction.

It will be better if I itemize the arguments.

(1) Life on planets seems not to be a very rare phenomenon. Out of ten solar planets one is full of life and Mars could have some. The solar system is not peculiar; other stars with similar characteristics are expected to have an equivalent number of planets. There is a good chance that, among the, say, 100 stars closest to the sun, some have planets bearing life well advanced in evolution.

(2) The chances are then good that in some of these planets animals exist evolved much farther than men. A civilization only a few hundred years more advanced than ours would have technical possibilities by far greater than those available now to us.

(3) Assume that an advanced civilization exists in some of these planets, i.e., within some 10 light years from us. The problem is: how to establish a communication?

As far as we know the only possibility seems to be the use of electromagnetic waves, which can cross the magnetized plasmas filling the interstellar spaces without being distorted.

So I will assume that "beings" on these planets are already sending toward the stars closest to them beams of electromagnetic waves modulated in a rational way, e.g. in trains corresponding to the prime-numbers, hoping in a sign of life.

Cocconi then went on to point out that planets in orbit around even the nearer stars would still appear so close to them, as seen from the earth, that signals originating from such planets would be lost in emissions from the star itself unless the wavelength was in a part of the spectrum where the star was comparatively quiet.

This, he argued, limits the suitable wavelengths to those in the radio portion of the spectrum and to the short-waved end of the gamma rays. Of these the radio waves seemed the logical choice because they are so much easier to send and receive.

The transmitted signals, he pointed out, must be strong enough to stand out clearly against the background emissions of the sky as a whole and must also be powerful enough to survive their very long journey to the solar system. Their loss of strength, in this respect, is determined by the so-called inverse square law, which applies to any form of electromagnetic radiation, be it light, radio waves or gamma rays. The law says that the intensity falls off according to the square of the distance from the source. Thus if you increase fourfold your distance from a light, it becomes sixteen times dimmer.

In view of these considerations, said Cocconi, if a radio dish the size of that at Jodrell Bank were used to transmit from a distant planet and the British instrument were on the receiving end, the power required to transmit detectable signals would lie beyond the reach of our present technology. "But I want to have faith," he wrote, "and will assume that they have larger mirrors and more powerful emitters and can do it." What he therefore hoped Lovell would do, he said, was "a systematic survey of the stars closest to us and spectroscopically similar to the sun, looking for man-made signals."

"As I said before," he wrote in conclusion, "all this is most probably fiction, but it would be most interesting if it were not.

"I leave to you the judgment on the feasibility of such a search."

Lovell's reply, on July 14, was brief and, in Cocconi's words, "rather disappointing." Lovell said such a search would be "difficult," and added that among current tasks of the radio telescope was "a survey of certain flare and magnetic stars." When that was done, he said, "perhaps we shall have time to look at a few others." As the years passed, Lovell warmed to the idea and even published Cocconi's letter as an appendix to one of his books. In *The New York Times Magazine* of December 24, 1961, he wrote that he "would still find it difficult to justify the diversion of any of the world's present radio telescopes to such speculative work." Nevertheless, he continued, "during the past two years or so the discussion of the general problem of the existence of extraterrestrial life appears to have become both respectable and important."

The cool reception that Lovell initially gave to the proposal did not discourage Morrison and Cocconi. Both men went to the cosmic-ray meeting in Moscow, and by this time Morrison had an idea that opened the way for a far more precise suggestion. He had discussed the problem with colleagues at Harvard, where excitement at discovery of the 21-centimeter line in the radio spectrum was still running high. The weakness of Cocconi's original proposal was the vastness of the spectrum that had to be searched. If you spin your radio dial, a powerful broad-band signal is easy to spot, but the transmitted energy is spread over a wide range of frequencies. The narrower the band, the farther away is a signal audible above the background noise, assuming that the receiver is correctly tuned. But the narrower the band the more difficult it becomes to find the signal, particularly if you do not know what frequency to look for.

Searching the entire range of possible frequencies, as Morrison said later, was like trying to meet a friend in New York without having a prearranged rendezvous. It would be absurd to wander the streets in the hope of a chance encounter. The sensible approach was to put yourself in your friend's place and guess what logical meeting place he or she might select, such as the information booth in Grand Central Terminal. What had to be found, Morrison said, was the "uniquely rational way" for interstellar attention-getting.

Fortunately, he pointed out to Cocconi, there is such a rendezvous in the radio spectrum, and this became the kernel of the joint proposal that the two men discussed in Moscow and then sent off to *Nature* from Cocconi's new home in Switzerland. "Just in the most favored radio region," they wrote, "there lies a unique, objective standard of frequency, which must be known to every observer in the universe: the outstanding radio emission line at 1420 Mc/sec [21 centimeters]." It has come to be known as the "magic frequency," for it seems by far the best for interstellar communication. It is that part of the radio spectrum that most readily penetrates the earth's atmosphere. It has the least competition from sources elsewhere in the galaxy, apart from the whisperings generated by drifting clouds of hydrogen. It is subject to relatively little interference from a frequency-dependent effect known as quantum noise. And finally, since it reveals positions and motions of the vast hydrogen clouds that form a major component of the galaxy (including its otherwise invisible far side), it would be scanned by radio telescopes everywhere and be an obvious meeting place.

Copies of the Morrison-Cocconi proposal, distributed in early August of 1959, went to many of those who, they thought, would be receptive. In particular they sent it to P.M.S. Blackett, the noted British physicist, in the hope that, as a friend of Morrison, he could see that it was published in *Nature* despite its sensational content. *Nature* seemed the logical choice, since it was so often the vehicle for bold new ideas. The article was published in the issue of September 19 and precipitously brought the debate into the open.

After repeating some of the initial arguments in Cocconi's first letter, the two physicists pointed out that while the lifetimes typical of civilizations elsewhere in the universe are unknown, it is not unreasonable to suppose that some endure "for times very long compared to the time of human history, perhaps for times comparable with geological time." Such beings must have achieved levels beyond the reach of our most imaginative speculators and should be awaiting the appearance of intelligent life in our solar system. "We shall assume," they wrote, "that long ago they established a channel of communication that would one day become known to us, and that they look forward patiently to the answering signals from the sun which would make known to them that a new society has entered the community of intelligence."

What would be the nature of signals from such an inquiring civilization? Cocconi and Morrison suggested that they might come as pulses, transmitted at a rate of about one per second, with prime numbers or "simple arithmetical sums" as a device for attracting attention. Because we have to look out fifteen light years to find seven stars with luminosities and expected lifetimes comparable to that of our sun, we must think in terms of travel times of many years for messages between civilizations. One light year is the distance traveled by light waves or radio waves in a year, at roughly 186,000 miles a second, and therefore it would take ten years for a radio message to reach a star ten light years away and twenty years for an exchange of messages.

In this situation a message from a distant civilization would logically be extremely long, taking perhaps years for its transmission. As elaborated by Morrison in subsequent discussions, such a lengthy message would be interspersed with attention-getting signals and "language lessons." The bulk of the message, however, would be a detailed description of the transmitting civilization—enough to keep scholars at the receiving end busy during the years required for a reply and further message exchanges. As a possible timetable he suggested one-minute periods for the calling signal, followed by ten minutes for a language lesson and then fifty minutes of information comparable to that found in an encyclopedia, whereupon there would be another minute of calling, ten minutes of language instruction and fifty minutes of the encyclopedia. The "language" would, of course, not be spoken but would be a system of symbols designed to convey information and ideas. The cycle of language lessons might be completed every ten hours, whereupon it would repeat, but he said it could take fifty years to run through the encyclopedia. Interspersed within the transmissions might be occasional instructions on how to build better receivers, transmitters and so forth. The transmitting civilization would presumably tape-record its messages and bring them up to date from time to time.

The signals would have to be directed at many stars for enormous periods of time before being likely to bear fruit. Cocconi and Morrison said they did not believe a highly advanced society would find it too burdensome to beam signals at, say, 100 nearby suns that seemed promising abodes of evolving life. An antenna like the huge one projected by the Navy for Sugar Grove, West Virginia, they said, would be able to send messages to planets ten light years away, assuming a receiving antenna of the same size at the other end, and using transmitters no more powerful than those already available on earth.

The dish then under construction at Sugar Grove, some 30 miles northeast of Green Bank, was to be fully steerable, like that at Jodrell Bank, but it was to be 600 feet in diameter compared to the 250-foot width of the British antenna. One of its tasks, as later reported in the press, was to eavesdrop on Soviet domestic radio traffic reflected back to earth by the moon. The project later ran into seemingly insuperable engineering difficulties. To be effective, such an antenna must retain its parabolic shape to within about one-sixteenth of the wavelength to be observed. This permits almost no sag at all, as the dish is swung and tilted to scan various parts of the sky, and demands a frame of great strength. It was finally found that a movable array weighing some 36,000 tons would be necessary to meet these requirements, yet the supporting structure already being built was not adequate for such a weight. Hence, the project was finally abandoned, much to the dismay of those who looked forward to its use as an extremely powerful receiver for scientific research—including, perhaps, the search for signals from other worlds.

Morrison and Cocconi pointed out in their article that, if another civilization had already made contact with sister worlds elsewhere in the galaxy, it would be far more persistent and patient in its search for emerging technologies.

The two men sought to narrow the search further by selecting suitable targets for a listening program. They calculated that there are some 100 stars of appropriate type within fifty light years. Of the seven that are within fifteen light years, three (Alpha Centauri, 70 Ophiuchi and 61 Cygni) are viewed from the earth against the backdrop of the Milky Way so that the 21-centimeter emissions coming from beyond them are forty times more intense than in other portions of the sky. Hence signals from near those stars, on that wavelength, could be observed only if they were extremely strong. The other four (Tau Ceti, O_2 Eridani, Epsilon Eridani and Epsilon Indi) are suspended against less noisy backdrops in the southern half of the sky. Some astronomers believed that at least two in the first group were poor candidates on other grounds. Alpha Centauri, third brightest of all stars and our nearest neighbor, is in a three-part system, one of whose components is much like our sun, but the system is probably too young for life to have evolved. The other, 61 Cygni, is also in a mutiple-star system, where planetary orbits might not be stable enough for life to evolve.

By a remarkable coincidence, at the National Radio Astronomy Observatory in West Virginia a search at 21 centimeters of two of the best candidate stars was already in preparation. Unaware of the Morrison-Cocconi proposal, Frank D. Drake had tuned his antenna to that wavelength as the most promising one in radio astronomy, but he had also become intrigued by the possibility that beings in other worlds might be trying to communicate. His effort is described in Chapter 15, "False Alarms."

Morrison and Cocconi pointed out that one phenomenon distinguishing signals generated on a planet from the general din of 21-centimeter emissions might be the ubiquitous Doppler effect, produced by the planet's orbital motion. They calculated that this, comparable to the change in pitch of the horn on a passing car, could alter the frequency of signals from a planet in orbit around its mother star by as much as 300 kilohertz, up or down. The earth, for example, travels around the sun at some 67,000 miles an hour. If a distant civilization monitored our radio signals from directly above or below the solar system, there would be no Doppler effect, any more than there is relative motion between one's eyes and the surface of a spinning globe when one looks down at it from above. However, if one views the globe sideways, from over the Equator, one side is coming toward you and the other is receding, making the relative motion of the edges maximal. The chances are small that the orbit of a distant planet would be so oriented as to eliminate all Doppler effect.

In conclusion Morrison and Cocconi urged that the readers of *Nature*, who include most of the world's scientific elite, not dismiss this analysis as a form of science fiction.

We submit, rather, that the foregoing line of argument demonstrates that the presence of interstellar signals is entirely consistent with all we now know, and that if signals are present the means of detecting them is now at hand. Few will deny the profound importance, practical and philosophical, which the detection of interstellar communications would have. We therefore feel that a discriminating search for signals deserves a considerable effort. The probability of success is difficult to estimate; but if we never search, the chance of success is zero.

14

Other Channels

Although the 21-centimeter radio band, and neighboring wavelengths, have continued to be the most popular for those seeking emissions from other civilizations, there have been very different proposals. Ronald N. Bracewell, a leading radio astronomer, argued in a 1960 issue of *Nature* that radio signals were uneconomical. He noted that the search on 21 centimeters was aimed at those nearest stars not disqualified by factors inimical to life. But, he said, "do we really expect a superior community to be on the nearest of those stars which we cannot at the moment positively rule out? Unless superior communities are extremely abundant, is it not more likely that the nearest is situated at least ten times farther off, say, beyond 100 light years?"

If that is so, it means that after ruling out unlikely candidates, we must still scan one thousand stars to find the highly advanced civilization that we are looking for and that civilization, in turn, must direct its calling signal at one thousand stars in the hope, ultimately, of finding someone else. "Remember," Bracewell said, "that throughout most of the thousands of millions of years of the Earth's existence such attention [to our planet] would have been fruitless."

It was more logical to assume, he said, that superior civilizations would send automated messengers to orbit each candidate star and await the possible awakening of a civilization on one of that star's planets. Such a messenger would resemble some of our space probes in being powered by light from the sun to which it was assigned. However, it would be far more sophisticated in design, equipped, as Bracewell explained later, with a miniaturized computer perhaps no larger than a man's head, yet with some characteristics of an intelligent being.

"If we contemplate the resources of biological engineering, which we have not begun to tap yet," he said in a 1962 lecture at the University of Sydney, "it is conceivable that some remote community could breed a subrace of space messengers, brains without bodies or limbs, storing the traditions of their society, mostly to be expended fruitlessly but some destined to be the instruments of the spread of intragalactic culture."

Such a messenger, said Bracewell in his *Nature* article, "may be here now, in our solar system, trying to make its presence known to us." This dramatic suggestion could not be dismissed as crackpottery, for Bracewell was a respected scientist, coauthor of a standard text on radio astronomy and now an associate professor on the faculty of Stanford University. Because a messenger might have to wait millions of years before performing its task, it would have to be heavily armored against radiation damage and meteor impacts, Bracewell said. It would be placed in an orbit lying within the zone surrounding the target star suitable for life. The messenger would keep watch for narrow-band emissions indicating that a nearby civilization had begun to use radio for communications. The observed frequency was clearly one that had passed readily through the atmosphere surrounding the sending planet and therefore could be used in the probe's response. It would simply repeat the detected emission, producing an echo that would excite the interest of the original sender. If the latter then repeated the transmission, the probe would know it was recognized and begin communicating. "Should we be surprised," said Bracewell, "if the beginning of its message were a television image of a constellation?"

He pointed out that strange radio echoes have been observed in the past. In 1927, 1928 and 1934, Carl Störmer of Norway and others reported hearing radio signals from Holland repeated from three to fifteen seconds after their original transmission. Some thought they were being reflected from clouds of ionized gas scattered around the solar system.

Then, in 1972, after Bracewell had published his proposal and drawn attention to the delayed echoes, a young Scotsman said the seemingly random variations in echo times were actually a message from an orbiting probe, describing whence it had come. The Scotsman, Duncan A. Lunan, a graduate student at Glasgow University, said it showed its origin to have been the double star Epsilon Boötis. Lunan's account was published in *Spaceflight*, the journal of the British Interplanetary Society. He also sent his material to Bracewell, who said the young man had "hit on a clever idea." It is, he said, "very suggestive, but now requires confirmation." There has been none. Nevertheless, Bracewell has urged that peculiar radio transmissions of any sort not be ignored as were, for a time, the remarkably powerful emissions from Jupiter. The latter, he noted, represent an output of 1 billion watts per megacycle, yet they were not recognized until 1955, when Bernard Burke and Kenneth Franklin of the Carnegie Institution first reported the phenomenon. The source has proved to be electrons gyrating in Jupiter's huge radiation belts.

It was also possible, Bracewell said, that the probes, "sprayed" toward nearby stars by a superior civilization, might merely report back when they heard signs of intelligent life, using a star-to-star relay system for more efficient communications. If such a super-world is 100 light years away, the announcement that we have reached the communicative stage has completed only part of its journey. Since radio waves travel at the speed of light, the 100-light-year trip will take 100 years. The message may include samplings of what was heard, giving the recipients a strange picture of our civilization, including such early radio shows as *Amos 'n Andy*, Caruso singing, or newscasts telling of global war, which might discourage further monitoring on the assumption that our civilization was doomed.

Bracewell said we should not expect that many worlds are seeking to make contact with us, since supercivilizations are probably already linked in a "Galactic Club." This federation of intelligent beings in many solar systems would allocate sectors of the galaxy for search by individual civilizations: "Our impending contact cannot be expected to be the first of its kind; rather it will be our induction into the chain of superior communities, who have had long experience in effecting contacts with emerging communities like ours."

Less than a year after the appearance of Bracewell's proposal, *Nature* published one by two authors arguing for light signals. It has been estimated by Frank Drake that, to cover the immense distances between solar systems, the energy of radio signals would have to be concentrated within a band of frequencies not much wider than 100 hertz, or cycles, per second. One of the great advantages of radio waves for communications is that, by means of various kinds of oscillator, they can be generated in very narrow bandwidths. However, until the invention of the maser and its optical counterpart, the laser, there was no known way to do this in that portion of the spectrum occupied by light. Even the most brilliant lamp emitted an inconsequential amount of energy within any narrow part of the spectrum, and therefore the use of light for long-distance signaling was out of the question.

All our conventional light sources, be they electric bulbs, arc lamps or candle flames, produce light by the heating of some material. What happens on the atomic level is that heat stirs up the electrons which, at random times, shed their excess energy in the form of light waves. This haphazard process in billions of atoms produces a wide range of wavelengths. Furthermore, the waves are not "coherent"; that is, they are chaotically jumbled; there is no symmetrical train of waves that could be modulated to carry information. It is because the waves generated by a commercial radio station are coherent that its output can be modulated to carry a Beethoven symphony or a human voice.

Discovery of the maser and laser grew out of speculation by a number of scientists, including Charles H. Townes at Columbia University in New York, N. G. Basov and A. M. Prokhorov in the Soviet Union, and Joseph Weber at the University of Maryland, concerning possible ways to "organize" the excitation of atoms and their subsequent radiation of energy. An atom or molecule can become excited either in response to a wave at its resonant frequency or by the action of heat-induced collisions. Once it is excited, there are two ways in which it can radiate a wave at its characteristic frequency. One is by spontaneous emission, at an unpredictable time of the atom's own choosing, so to speak. This is the manner characteristic of electric lights. The other is by "stimulated emission," which occurs when an excited atom or molecule encounters a wave at its resonant frequency. Since the atom is already excited, the new wave jars loose from the atom another wave of that frequency and the two waves go sailing off, hand in hand. More scientifically, they are "in phase," like soldiers marching in step. They are therefore the stuff of which a smooth wave-train is made—a coherent wave suitable for message-carrying.

The first maser was developed in 1954 when Townes, James P. Gordon and Herbert J. Zeiger at Columbia University found a way to assemble excited ammonia molecules in a resonant, metal-walled chamber. This could be done because excited ammonia molecules are repelled by an electrostatic field, whereas such molecules in the lower energy state are attracted by it. The experimenters sent a stream of low-pressure ammonia gas through an electrostatic field, segregating the excited molecules, which were then directed into the metal chamber. There they waited, aquiver with excitement, for a wave of the right frequency. (For ammonia, it is 24,000 megahertz, which lies in the microwave, or radar, region of the spectrum.) When such a wave entered the chamber and encountered one of the molecules, it released an identical wave, thus producing two waves (or photons) from one. These, in turn, interacted with other molecules, releasing more waves, which bounced between the walls of the chamber, constantly producing more and more waves on that same frequency. It was "microwave amplification by stimulated emission of radiation," and the device came to be known, from the initial letters of that formidable title, as the "maser."

While such early masers were able to amplify microwave signals to an extent never previously possible, they were useful only in a very narrow frequency range centered on 24,000 megahertz. The next discovery, thanks largely to the work of Nicolaas Bloembergen at Harvard, was that similar amplification on a number of frequencies could be achieved by increasing the energy states of free electrons in a crystal structure. For example, some masers were made with a silicon crystal containing occasional atoms of phosphorus. The latter fit into the crystal lattice so that one electron from each atom is free and available for excitation by microwaves on frequencies other than that to which these electrons are resonant. Once excited, the electrons will amplify waves on their own frequency.

In 1958 Townes and Arthur L. Schawlow showed how it would be possible to apply the maser principle to light waves and thus produce a unique kind of light. Writing in *The Physical Review*, they proposed that a resonant chamber be built with mirrors facing each other at each end, the distance between them being thousands of times greater than the wavelength to be generated, but nevertheless a multiple of half that wavelength, thus permitting light to resonate between the mirrors. Light waves generated along the axis of the maser would travel back and forth between the mirrors, constantly growing in intensity as more electrons in the central crystal gave off their stored energy. A partial transparency of one mirror would allow some of these waves to escape in a beam that was not only very directional and intense, but was concentrated in a narrow frequency band and was coherent. For helping create a whole new field of optics, both Schawlow and Bloembergen, like Townes, became Nobelists.

The first maser to operate within the frequency range of visible light was dependent on a pencil-like rod of synthetic ruby whose ends were machined with great precision and silvered to provide mirrors facing each other along the axis of the rod. Around the rod was a coil of xenon flashtube that generated an intense light, serving to excite electrons within the crystal. Ruby consists of aluminum oxide with a few of the aluminum atoms replaced by chromium atoms. The more of the latter there are, the deeper the red of the ruby. Within the lattice of interlocking atoms that form a ruby crystal, these chromium atoms have extra electrons that are free of the lattice and hence can be excited to higher energy states by the action of ordinary light. Light at the characteristic wavelength of a particular ruby crystal, usually near 7,000 angstroms, produces a cascading of the excited electrons and a shower of additional waves of precisely the same color. (An angstrom is 1/100-millionth of a centimeter.) In the first lasers, only those waves directed along the axis of the rod could survive and gain in intensity. Those aimed obliquely flew out of the crystal and were lost.

The successful production of such a device was announced in July 1960 by Theodore H. Maiman of the Hughes Aircraft Company, and in rapid succession it was found that other crystals would produce light at various wavelengths. It was characteristic of these early devices that they had to operate in pulses, but in February 1961, Bell Telephone Laboratories announced the development of a continuously operating type that used a mixture of helium and neon gas instead of ruby. Such devices, whether dependent on a crystal or a gas, soon came to be known as lasers (for "light amplification by stimulated emission of radiation").

There followed a scramble to use them for a wide variety of purposes, including pinpoint destruction of tissue in delicate surgery and various forms of communication. The available frequencies spread across the infrared and visual parts of the spectrum. On May 9, 1962, a laser beam generated by a ruby crystal the size of a 6-inch pencil was used by a group at the Massachusetts Institute of Technology to illuminate a small patch of the darkened moon. The stark lunar landscape glowed sufficiently for the re-flected light to be detectable by instruments in Massachusetts.

When astronauts landed on the moon, they left behind special mirrors so that laser beams later reflected back to the stations on earth that originated them could be used to show the extremely slight changes in the relative positions of continents and islands as they move with respect to one another.

The increase in laser capabilities stimulated the thought that such a device might be chosen by those in other worlds seeking to attract our attention. It is not surprising that this was first pro-posed in a 1961 issue of *Nature* by Charles Townes, "father" of the maser and laser, with a colleague, Robert N. Schwartz. Both were by then at the Institute for Defense Analyses in Washington, where Townes was director of research. He was soon to become provost of MIT.

It seems to have been a "historical accident," they said, that lasers were not discovered before radio as a means of long-range communication. Another civilization "might have inverted our own history and become very sophisticated in the use of optical or infra-red masers rather than in the techniques of short radio-waves." In our case, however, lasers were at a rudimentary stage of development. "No such operating device was known a year ago," they wrote, yet the technology in this field promises to ad-vance at great speed, with major gains in the next decade. Already, they added, we are not far from the development of lasers able to communicate on wavelengths of visible light, or in adjacent por-tions of the spectrum, "between planets of two stars separated by a number of light years." Furthermore, they said, it may be pos-sible that our present-day telescopes and spectrographs might be able to pick up signals from another civilization not much more capable than we are.

They discussed two methods of interstellar communication by laser which they designated System A and System B. The former would pass the beam of a single laser through an optical system with a 200-inch reflector, comparable to that of the telescope on Mount Palomar in California. This would produce a beam that, to start with, was 200 inches wide, instead of the very narrow beam typical of a laser; but the spread of the beam would be sufficiently reduced so that millions of miles away it would be much narrower than if such an optical system were not used.

The other scheme, System B, would use a battery of twenty-five lasers without any optics. They would be clustered to produce a beam initially only 4 inches wide; but the spread of the beam, slight as it was, would be five times that of System A. However, unless the laser and the huge telescope of System A were lifted above the atmosphere, the advantage of its well-disciplined beam would be lost to the light-scattering effects of the air. Both systems would use 10-kilowatt lasers generating a wavelength of 5,000 angstroms.

Schwartz and Townes pointed out that a few years ago the idea of operating a laser from a space platform or from a high-altitude balloon would have seemed out of the question. Now, orbiting observatories were in preparation and balloon-borne telescopes were coming into their own. In 1963, Stratoscope II, a 3-ton instrument with a 36-inch mirror, demonstrated that such remote-controlled telescopes could be operated with great precision when suspended from balloons. However, it was a big jump from a 36-inch telescope to one of 200 inches—of which there was only one in existence. Even though System A could deliver to the targets light 100 times as intense as System B, assuming the equipment was carried aloft, the two scientists conceded that at present System B was the only practical one for earth's inhabitants.

Under favorable seeing conditions, they calculated, a person could view the signals from System A at a distance of 0.1 light years with the unaided eye. With binoculars, the range would be some 0.4 light years. For System B the distances would, in each case, be one-tenth as large. If, however, a telescope like that on Mount Palomar was in use at the receiving end, System A could be observed visually at 10 light years and System B, at the same range, would show up on photographic plates exposed for 1.5 hours. Thus they demonstrated that even our primitive laser technology was nearing the point where it could send observable beams to the nearer candidate stars.

What, then, should we look for? At a distance of 10 light years, the earth and the sun would appear only one-half second of arc apart. Most telescopes could not separate objects so close together, even if they were above the turbulent atmosphere.

Hence, the transmitting civilization must see to it that its laser signals are not hidden by the brilliance of its nearby sun. One way to avoid this, Schwartz and Townes pointed out, would be to select a wavelength of light that is absorbed by stellar atmospheres.

When one looks at the sun, even through a pocket spectroscope, its brilliant spectrum of colors is subdivided by a series of black lines. It is as though some physics teacher had drawn the lines there, with precise neatness, to aid students. But the lines are "drawn" by nature. As light from the sun passes through the solar atmosphere on its way to the earth, gases in that atmosphere absorb the light at their characteristic wavelengths, leaving black lines in those parts of the spectrum. Two of the most prominent lines, in this respect, are those absorbed by calcium in the violet (at wavelengths of 3,933 and 3,968 angstroms). Each line is several angstroms wide and, at those wavelengths, the light of the sun is dimmed some tenfold. If a laser transmitted light in an extremely narrow line at the center of such a broad "black" line in the star's spectrum, astronomers on a distant planet, scanning the light of various stars, might recognize it as artificial. They could then examine this light more carefully to see whether it was coded, in some way, to carry a message.

The narrower the band of the emitted light, the brighter and more obvious it would be, so long as astronomers at the other end were capable of detecting such narrow bands of emission. Our present capabilities in optics limit us, in general, to the detection of spectral lines representing a spread of at least 0.01 angstroms in wavelength. One problem pointed out by Schwartz and Townes was that the wavelength spread of a laser beam, though narrow to begin with, could be broadened, and the signal therefore dimmed, by shifts in observed wavelength caused by changing relative motion between transmitter and receiver. The most likely cause of such Doppler effects would be the spin of the home planets of both the receiving and sending civilization. The effect could become significant when film was exposed for several hours to pick up faint lines in the spectrum. It was therefore proposed that the transmitting civilization might adjust its wavelength to allow for the spin and orbital motion of its planet and that those on the receiving end might make similar allowances for their own motion.

The two scientists in effect apologized for questioning the uniqueness of 21-centimeter radio waves as the logical channel. "It may be both an encouraging enlargement of possibilities, and at the same time an unwelcomed complication," they said. "The rapid progress of science implies that another civilization, more advanced than ourselves by only a few thousand years, might possess capabilities we now rule out." But in any case, they added, it might be appropriate "to examine high-resolution stellar spectra for which lines are unusually narrow, at peculiar frequencies, or varying in intensity."

One of the tasks of *Copernicus*, the orbiting astronomical observatory launched in 1972, was to scan three of the nearest candidate stars—Tau Ceti, Epsilon Eridani, and Epsilon Indi—at an ultraviolet wavelength at which those stars hardly shine at all. This, it was hoped, might allow a civilization orbiting one of those stars to announce itself by an appropriately tuned laser. No such radiation was detected.

Those who leaned toward radio waves as the best vehicle for interstellar signaling found drawbacks to the laser proposal. Among the radio proponents was Bernard ("Barney") Oliver, a big, ebullient man who, as vice president for research and development of the Hewlett-Packard Company, had become one of the most avid supporters of the search for extraterrestrial signals. The company, in Palo Alto, California, is one of the leading makers of high-performance electronic devices. In a 1962 issue of *Proceedings of the Institute of Radio Engineers* Oliver discussed the great variety of new capabilities that lasers have provided; but he concluded that before such devices can be useful for signaling to other solar systems, "a tremendous increase" in laser output must occur.

Frank Drake, who conducted the first search on 21 centimeters, came to a similar conclusion. He argued, as well, that mirrors used to collect light signals must be far more precisely made than those in radio telescopes, since the permitted error is controlled by wavelength. The production of a mirror 200 inches in diameter for the Mount Palomar telescope was a feat of optical technology, yet the dish of the Jodrell Bank radio telescope is 3,000 inches wide, and the one constructed at Arecibo, Puerto Rico, in a bowl-shaped valley, is 24,000 inches across. Drake sought to analyze the problem in terms of physical laws that are universal and would therefore influence engineers on any planet of the universe. Thus he argued that the choice of wavelength is partly controlled by the flexibility of metals and other solids that might be used in building mirrors for radio or optical waves. Such flexibility, determined by the inherent binding properties of solids, points to the longer, or radio, wavelengths as those for which the largest dishes could be built. The limitations we encounter in antenna structures, he said, "are not peculiar to ourselves, but are common to all civilizations."

Also universal are two of the mysterious "constants" that affect all forms of electromagnetic radiation, be it in light or in the radio parts of the spectrum. They determine the extent of the "quantum" noise that results from the very nature (or quantum behavior) of light and radio waves. The extent of this noise, at any frequency, is equal to the frequency multiplied by the Planck constant and divided by the Boltzmann constant. These constants will be described by other names in other worlds, but their combined effect must be known to every physicist in the universe, whether a hundred-legged blue sphere or a two-legged cylinder, like Frank Drake and the rest of us. The effect of the law is to point toward lower frequencies and longer wavelengths as the channels most free of quantum noise. Thus, as pointed out by Oliver, while a radio system has far more beam spread than one using light, at optical frequencies the quantum noise, per cycle, is thousands of times greater than at the proposed radio frequencies.

The optimum frequency for communication with any one star is also determined by the competing background noise from that part of the galaxy lying behind the star, as seen from the earth. Thus, while the consideration of quantum noise (the inherent noise in all electromagnetic radiation) favors low frequencies, the noise from various parts of the galaxy sets a lower limit on such frequencies. Drake calculated that the most favorable frequency for conversing with any star will lie between 3,700 and 9,300 megahertz (3.2 and 8.1 centimeters). We can already send radio messages out 1,000 light years, he said, bringing some 2 million stars within our range, whereas, at our present level of technology, laser signals could hardly be seen beyond 10 light years.

Such arguments have not impressed Townes and the other laser adherents. They point out that our present engineering and laser capabilities are a poor guide to those of others more advanced than we. Furthermore, such disadvantages as greater quantum noise, for light waves, may ultimately be outweighed by the narrower and more intense beam of a laser. The truth, according to Townes, is that we are not yet in a position to decide with assurance which is best.

One proposal by Bernard Oliver that at first seemed attractive was to look for signals of extremely short duration and broad bandwidth. He cited a peculiarity of radio signaling, namely, that the shorter the length of a pulsed signal, the broader the bandwidth necessary for efficient transmission and reception. This

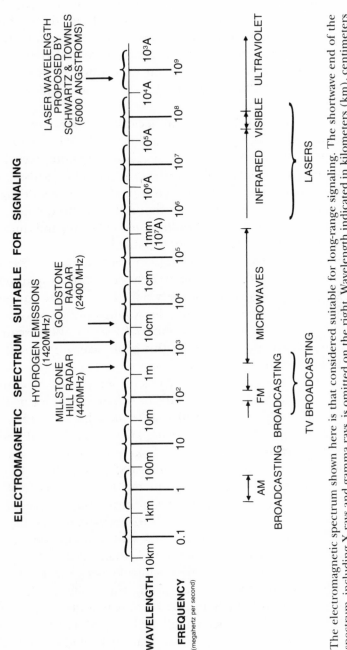

The electromagnetic spectrum shown here is that considered suitable for long-range signaling. The shortwave end of the spectrum, including X rays and gamma rays, is omitted on the right. Wavelength indicated in kilometers (km), centimeters (cm), millimeters (mm), and Angstroms (A).

seemed to offer a way to get around the problem of frequency selection. He calculated that if one could cram 100 million billion watts into a pulse a ten-billionth of a second long, the optimum bandwidth would span the entire radio spectrum to which the atmosphere is transparent. Assuming that we could detect such very short pulses (although at present we cannot), it would only be necessary to tune the receiver to this very broad bandwidth and aim the antenna at suitable stars.

Oliver later shelved this proposal because of the effect on such broad-band signals of electrons along their path. While space between the stars is almost a vacuum, there are still enough particles, including electrons, per cubic inch to affect the transmission speeds of radio waves. Through such a medium, waves of higher frequency travel slightly faster than those of lower frequency. The effect on a broad-band signal would be to smear out the duration of each pulse until it became undetectable.

Oliver then proposed that a huge, fan-shaped antenna 20 miles long be built across the equator so that the earth's spin would daily sweep its aim across almost every star in the sky. He conceded that the observing time on each star would only be about one second per day and that a large number of stars would be scanned simultaneously. His point was that this gross scanning might disclose a target worth closer scrutiny.

Oliver's most grandiose proposal was presented at a meeting of the American Association for the Advancement of Science in 1966. Build an "orchard" of 10,000 dish antennas, he said, each 100 meters (328 feet) in diameter, covering an area 6 to 10 miles wide. Together they would scan a patch of sky the size of the sun for artificial signals. Such a system might cost $5 billion, he said, "but let me remind you that this is about one-tenth the cost of the Apollo Project," the manned lunar landings. Only 10 percent of its time would be spent on the search for extraterrestrial life, the rest being devoted to research that, he said, would produce "far more fundamental scientific discoveries" than Apollo.

Five years later his scheme, the Cyclops Project, was the subject of an "educational program" for young university faculty members, sponsored jointly by Stanford University and NASA's Ames Research Center. It led off with a talk by Philip Morrison, followed by Oliver, Bracewell (now director of Stanford's Radio Astronomy Institute), Sebastian von Hoerner of the National Radio Astronomy Observatory, David S. Heeschen, that observatory's director, and others. Although NASA was a cosponsor, it was not a NASA proposal. Cyclops was, as Oliver wrote me, "a modest first look at the engineering problems and probable costs of engaging in the most exciting of all space programs—that of ending the cultural isolation of the human race."

In 1977 three Stanford men analyzed whether the ultimate search system should be on earth, in orbit above the atmosphere, or on the moon. If a Cyclops array was set up on the far side of the moon, it would be shielded from man-made radio noise and its dish antennas could fit snugly into lunar craters. If a single giant dish was built in earth orbit, a screen could be orbited between the dish and the earth as shielding. Spacecraft could then fly radio receivers into the focus of the dish.

An even more far-out proposal was published in 1977 by thirteen scientists at the Space Research Institute of the Soviet Academy of Sciences. Among them were such leaders of the Soviet search as Iosif Shklovsky and Nikolai Kardashev. Also participating was Roald Sagdeev, who, as head of the institute, became a bridge between the American and Soviet space programs and confounded his fellow Russians by later marrying Julie Eisenhower, granddaughter of the former American president. The proposal, titled "Infinitely Built-up Space Radio Telescope," spoke of large reflectors, 1 to 10 kilometers in diameter, orbiting as far apart as 10 astronomical units (one such unit equals the earth-sun distance of 93 million miles)! "Very Long Baseline Interferometry," or VLBI, was becoming one of the most powerful tools in radio astronomy. If VLBI antennas are hundreds of miles apart, they can distinguish details of distant targets as small as the fingers of a human hand. In 1969, to get even greater baseline, simultaneous observations were made at Green Bank, West Virginia, and the Crimean Astrophysical Observatory in the Soviet Union. At each site, for VLBI observations, the incoming waves must be timed with great precision, using atomic clocks. The records can then be brought together and matched.

By 1977, as pointed out by the Soviets, the use of VLBI on earth had reached its limit—the diameter of the planet. While using VLBI antennas in orbit has been discussed in the United States, the USSR proposal was extraordinarily ambitious, even though the Soviets claimed that building antennas 1 kilometer (0.6 miles) wide would be cheaper in space than on earth. Bigger ones, assembled in space from modules 200 meters (650 feet) in diameter, would be adjustable to maintain precision of their overall shape. The program would depend on space shuttles and space tugs to carry components into orbit and assemble them. If necessary, giant shields would protect the antennas from earth-made radio noise, but they could also be far from the earth, separated by the width of the solar system. Such an assembly, the Soviet scientists said, could be used for planet detection and the search for intelligent signals far more effectively than anything now contemplated. They estimated the cost at $1 billion to $10 billion. Development of the first module for such an enterprise, they said, could begin "now."

Out of the Cyclops study, in the search for a rendezvous in the radio spectrum, came the "water hole" concept. By then many other telltale emissions from atoms in space had been detected, in addition to those of hydrogen at 21 centimeters. While they were not as powerful as those of hydrogen, in some parts of the sky the hydrogen emissions almost drowned out any that might have been artificial. To define what came to be called "the water hole," attention turned to emissions from hydroxyl (at 1,612, 1,665 and 1,667 megahertz), which, like water, is a mating of oxygen and hydrogen, but with only one hydrogen atom, as opposed to the two in a water molecule. (Chemists refer to it as OH, as opposed to H_2O.) That part of the radio spectrum between the hydrogen and hydroxyl emissions (1,420 and 1,667 megahertz) is a region in which both the earth's atmosphere and space beyond are most transparent. Since hydrogen plus hydroxyl produces water, the region between those two frequencies came to be known as "the water hole" and has figured in many subsequent sky searches. "There, right in the middle," wrote Oliver, "stand two sign posts that taken together symbolize the medium in which all life we know began." In the galaxy different species might meet there, said Oliver, "just as different terrestrial species have always met at more mundane water holes."

By the 1990s the discovery that gas clouds in parts of the galaxy (such as clouds of methyl alcohol or hydroxyl) sometimes act as natural masers led James M. Cordes of Cornell University to revive a 1976 proposal by that perennial source of extraordinary ideas, Thomas Gold. This was that astronomical lasers may be used by other civilizations to make their signals powerful enough to cross the entire galaxy. By 1988 almost 190 masers producing water-vapor radio waves had been detected in the dust shells around some aging stars. The stars were shedding material as their lives neared an end. One report told of observing 60 celestial masers, 49 of which had not previously been detected. Most of them radiated at the 1,612 megahertz of hydroxyl. Light from the stars was apparently energizing hydroxyl in the dust shells, which then dumped the energy at the telltale wavelength. While the process is not as efficient as in a laboratory maser, it is on such a vast scale that the emission can be far more powerful than anything likely to be produced by a civilization. Even stronger emissions are generated in the nuclei of some galaxies, producing as much as 350 times the sun's luminosity at the single water-vapor wavelength. While any one maser emission is extremely directional and may be short-lived, suitable alignments recur often enough, according to Cordes, for long-range communication.

Another proposal would use the gravity of the sun to focus otherwise weak artificial signals from afar. A number of such "gravitational lenses" have been observed, in which a distant galaxy focuses light from an even more distant source. Even more provocative is the proposal of Philip Morrison that our distant friends might be using "Q" waves "that we are going to discover ten years from now!" Or, he said, a way may be found to handle neutrinos as message carriers. While these elusive and paradoxical atomic particles travel at the speed of light and can go right through the heart of a planet, no one knows an easy way to generate them, detect them, or modulate them to carry information.

There have been other suggestions. M.E.J. Golay of the Technische Hogeschool in Eindhoven, the Netherlands, cited the disadvantage of listening directly on 21 centimeters, since this involved competing with the natural hydrogen emissions. He suggested that better locations for a search were at half or double the hydrogen frequency. Drake then picked up this idea of examining multiples, or "harmonics," of the hydrogen frequency, which had also been suggested by Sebastian von Hoerner. Drake said our distant friends may have chosen the harmonic closest to the minimum-noise frequency applicable to their star. Here, in effect, was a refinement of the Cocconi-Morrison proposal. But was it too refined to be "uniquely logical"?

There are those who say: "But what if no one is calling and all are listening?" This does not necessarily mean that all is lost. It might still be possible to detect the more powerful radio emissions produced routinely by an advanced civilization—communication

192 We Are Not Alone

with its interplanetary vehicles, tracking them by radar, and so forth. For example, BMEWS, the Ballistic Missile Early Warning System in Greenland, transmitted radar pulses of 5 million watts, powerful enough, probably, to kill a man at short range.

The possibility of interstellar eavesdropping was analyzed at a 1961 symposium of the Institute of Radio Engineers by J. A. Webb of the Lockheed Georgia Company. The stray signals likely to reach us from a sophisticated society ten light years away, he found, would probably be undetectable by our present antennas. Matters would be different, however, if the antennas were above the atmosphere, either in orbit or on the moon. Ultimately, he said, antennas 10,000 feet in diameter might be erected in space and these could probably detect a civilization "tens of light-years" away.

Frank Drake, after the negative results of his Project Ozma (described in the next chapter), also leaned toward eavesdropping. He reasoned that we are unlikely to call other worlds if we have no real evidence that they exist. Why, then, should we expect them to call us? He proposed that a method analogous to the "coded pulse" technique used to detect weak radar signals be applied to this problem.

It was the coded-pulse method that enabled Millstone Hill to first obtain radar ranges on Venus. A series of pulses is transmitted toward the target and the returning signal is crosscorrelated with a record of the original pulses. Although the individual echoes may be lost in the background noise, when they have been aligned with the original pulses they can be superimposed, electronically, producing a composite signal that is evident.

Drake assumed that another civilization like ours would be using a multitude of frequencies for its television broadcasts, communications and other activities. None of these signals would, by itself, be detectable. However, he proposed that a sensitive receiver, aimed at the candidate star, record in segments of 100 cycles at a time, sweeping the 10 billion cycles of the radio spectrum available for such eavesdropping. This would produce records of signal strength in 100 million segments, or bands, of the spectrum. Then a second sweep would be made, producing a second set of records for the 100 million bands. These would be correlated with the first set. In those bands where artificial signals were concealed in the noise, such signals would be atop one another. The noise, being random, would not pile up in this way.

The superimposition of all 100 million bands would then produce a different effect than if the distant solar system were not using radio.

In 1978 Carl Sagan wrote to *Science* questioning the idea of eavesdropping. He cited the trend on earth toward cable television. In not very many years, he said, we will no longer be "wastefully leaking radio power to space," and the same can be assumed for others. By then Woodruff T. (Woody) Sullivan III of the University of Washington in Seattle had become a major eavesdropping proponent and replied that not only is the future of television technology uncertain, but major other radio sources have emerged. The Naval Space Surveillance System at Archer City, Texas, he said, radiates 14 billion watts in a bandwidth of only about 0.1 hertz. Any eavesdropper within 28 percent of the sky, out to a distance of 60 light years, would be able to detect it with an antenna like the one in Arecibo. With a Cyclops array, the range would be 600 light years, he said. Most detectable are space surveillance radars and the carrier signals of television stations, which transmit parallel to the earth's surface, some of them with great power. For a review paper in *Science* Sullivan mapped 2,191 TV stations around the world, most of them in Western Europe and the United States, but with clusters in Japan and southeast Australia. As the earth rotates, these emissions sweep the skies like clustered beacons, emitting signals detectable out to from 25 to 250 light years with Cyclops antennas. By listening to us in this manner one of our neighbors could learn a great deal about our planet and its inhabitants, Sullivan said (although detecting the carrier signal of a station would not produce a program).

A quite different group of proposals has been based on the assumption that a distant civilization would try to attract our attention by means of some "marker" or beacon. One idea, discussed by Drake and others, was to enclose one's sun in a cloud of material that absorbs some unusual—and hence clearly artificial—wavelength of light. As noted earlier, there are black lines in the spectra of the sun and other sunlike stars, denoting substances in their gas envelopes that absorb certain wavelengths. Suppose that in the light of a distant star we found an absorption line that we could not explain naturally? Drake cited the example of technetium. This element is not found on earth and only weakly in the sun, in part because it is short-lived, in its most stable form decaying radioactively within an average of 20,000 years. As the name of technetium implies, it is observed on earth only where produced artificially. Although seen in the spectra of certain unstable stars, it is not readily detectable in those resembling our sun. Therefore, if we saw evidence of technetium in the light of a sunlike star, it would be cause for further investigation. Drake estimated that such star-marking would require only a few hundred tons of a light-absorbing substance, spread around the star; but the scheme would still place heavy demands on a civilization's technology. Not only would the stuff have to be distributed uniformly, but it would have to be maintained in the face of the steady outflow of gas from the sun—the so-called "solar wind." The great attraction of such a scheme is that it would take advantage of the enormous output of radiation from a star, freeing the civilization from the need to generate such vast amounts of energy and letting the star do the work instead.

Philip Morrison discussed converting one's sun into a signaling light by placing a cloud of particles in orbit around it. The cloud would cut off enough light to make the sun appear to be flashing when seen from a distance, so long as the viewer was close to the plane of the cloud's orbit. Particles about a micron (one-millionth of a meter) in size, he thought, would be comparatively resistant to disruption. The mass of the cloud would be comparable to that of a comet—some 100,000 billion tons—covering an area of the sky 5 degrees wide, as seen from the sun. Every few months the cloud would be shifted to constitute a slow form of signaling, the changes perhaps designed to represent algebraic equations.

An inadvertent marker was inherent in the "far-out" proposal of Freeman J. Dyson, an Englishman at the Institute for Advanced Study in Princeton, New Jersey. In 1960 Dyson, as one of his highly imaginative proposals, discussed how distant supercivilizations may have used their technology. He argued that population growth in such worlds would continue to press the limits of available sustenance, as set forth in the eighteenth-century theory of Thomas Robert Malthus. The limiting factors would be available material and available energy. Both shortages could be met, Dyson argued, by dismembering planets of that solar system and using the material to build a shell completely enclosing the parent star. The entire energy radiated by the star would then become available—40,000 billion times that which falls on the earth.

He calculated that from the material of Jupiter such a shell could be built with energy equivalent to that radiated by the sun

The southern part of the diffuse nebula IC 1396 in the constellation Cepheus is full of dark and dense areas where stars are being born. This masked and enhanced monochrome photo was made at the European Southern Observatory Headquarters from a blue-sensitive plate obtained with the 48-inch Oshin telescope at the Mount Palomar Observatory.

The Keck Telescope atop
Hawaii's dormant volcano
Mauna Kea is the world's
largest optical-infrared telescope. The photo at right
was taken from a helicopter
by Roger Ressmeyer.
When the Keck, with its 33-
foot-wide segmented mirror,
is twinned in 1996, the two
telescopes, working in tan-
dem, should be powerful
enough to distinguish the
headlights of a car 16,000
miles away.
*(California Association for
Research in Astronomy)*

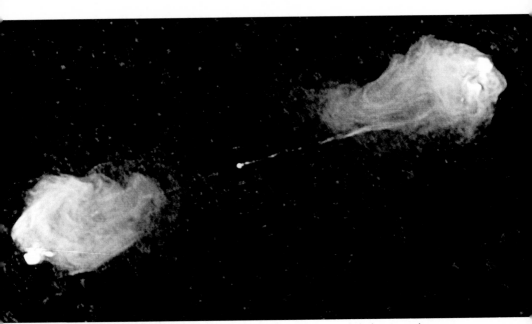

Jets emitted on either side of radio galaxy Cygnus A (central dot) are analagous on a huge scale to "T Tauri" stars, which may form solar systems from the remnants of similar jets, produced while they themselves are being born.
(National Radio Astronomy Observatory)

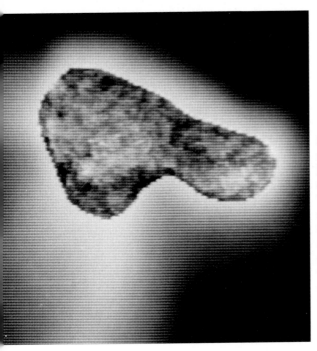

More than 30 "organic" molecules were identified by three spacecraft–Giotto, Vega 1, and Vega 2—that passed close to Halley's comet on its return to Earth's vicinity in 1986. This computer-processed photo was taken by Vega 2 on March 9.
(Institute for Cosmic Research of the Academy of Sciences of the USSR)

A small portion of the observation time of the Arecibo Observatory in Puerto Rico is devoted to searches for extraterrestrial intelligence. With a 300-meter dish, it is the world's largest radio telecope. Arecibo is part of the National Astronomy and Ionosphere Center, operated by Cornell University under a cooperative agreement with the National Science Foundation.

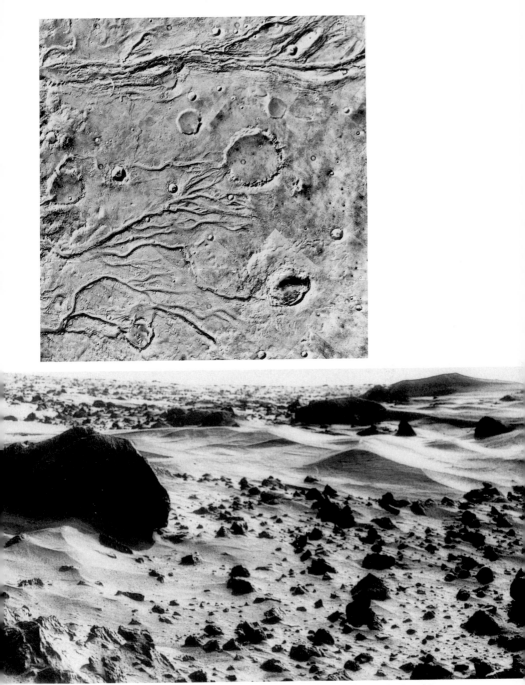

Meanders and stream beds on Mars (top), photographed by Viking Orbiter I in 1976, are suggestive of a massive flood across the cratered terrain. Shapes and textures of dunes in a photograph by Viking Lander I that same year indicated the general direction of recent wind storms was from upper left to lower right. *(Jet Propulsion Laboratory)*

Frank Drake, back row, second from right, and part of his Project Ozma team, in front of the Howard E. Tatel Telescope used in the first search, starting in 1960, for radio signals from extraterrestrial civilizations.

(National Radio Astronomy Observatory)

This screen illustrates how the NASA-SETI targeted search can distinguish pulsed signals from random noise. Each spot is either noise or a signal. Location on the screen, from left to right, represents frequency. In this test are shown, from top to bottom, the results of 90 rapid observations at millions of frequencies. Included are both noise and recurring pulses. Can you identify the latter? If not, turn the page. *(Nasa Ames Research Center)*

Carl Sagan at the inauguration of the SETI project on Columbus Day, 1992. Above is the lower edge of the 112-foot-wide dish of antenna used to conduct the search at NASA's Goldstone Station in California's Mohave Desert. *(Walter Sullivan)*

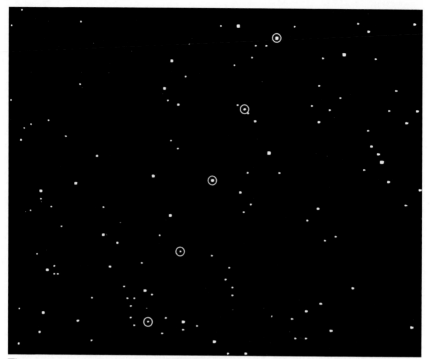

The computer took only 1.2 thousandths of a second to identify and mark the train of five regular pulses. The slope of the line shows that the frequency is gradually being altered by the Doppler effect. The computer is not only quicker but more reliable than humans at detecting patterns.
(NASA Ames Research Center)

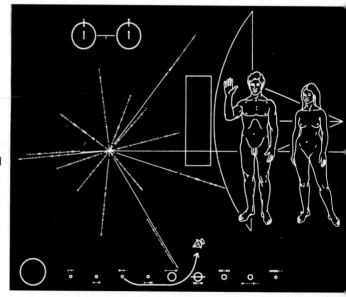

Metal plaque with message from Planet Earth to extraterrestrials. Attached to Pioneer 10, the first spacecraft to travel beyond the Solar System, its point of origin is indicated by lines showing the direction and distance of Earth to the center of the galaxy and 14 pulsars. *(NASA)*

in 800 years—a short period on the time scale under discussion. The shell would be 6 to 10 feet thick, surrounding the star at twice the distance of the earth from the sun. The outside of such a shell, because it was warm, would radiate just as copiously as the central star, he said, but in the infrared part of the spectrum, centered near a 10-micron wavelength, which easily penetrates the earth's atmosphere. Hence, he proposed a search for sources of radiation on that wavelength.

Readers of *Science* pointed out in letters that such a shell could not survive the gravitational stresses and the tendency of the material to move into the equatorial plane of the shell's rotation. Dyson replied that he envisioned a "loose collection or swarm of objects traveling on independent orbits around the star."

When I visited Shklovsky in Moscow, the Soviet astronomer developed Dyson's idea further. A civilization capable of such feats, he said, would think nothing of allocating 1 or 2 percent of its solar energy to a beacon radiating on a key wavelength, such as 21 centimeters, particularly if the beacon seemed to offer the only chance of drawing the attention of another civilization. Shklovsky was pessimistic as to the density of civilizations in space. On the average, they are probably at least 700 light years apart, he said, since we have not inadvertently stumbled upon the signals from one of them. If a civilization could divert a small percentage of its star's energy into a suitable wavelength, it would be a marker par excellence. To find such a beacon in our own galaxy, said Shklovsky, would probably be a long task, since, with the entire sphere of heaven to choose from, it would be hard to know where to start. Instead, he proposed to look at the great spiral nebula in Andromeda, the nearest galaxy like our own. Being 2 million light years away, the entire galaxy is encompassed within the field of view of a radio telescope. So brilliant a beacon should be evident even at this distance. If one was seen in Andromeda, it would mean there was a supercivilization there 2 million years ago and this, Shklovsky said, would then justify a patient search of our own galaxy.

One of the most startling proposals, in discussions of how other beings could communicate, was that of Leslie C. Edie of Bellmore, Long Island. In the issue of *Science* for April 13, 1962, he suggested that the long-chain molecules now being extracted from certain meteorites (the carbonaceous chondrites) might have been put there by some remote civilization and hurled toward us in great numbers. Might these long molecules contain coded information, he asked? Similarly two Japanese scientists in 1979 proposed that a message from afar may be enclosed in the genetic material of a harmless virus sent to earth. The virus could replicate itself, once it had infected an organism, such as the bacteria that normally live in the human gut. The virus would then spread through the populace like an epidemic until someone deciphered the message, encoded in that part of a virus's genetic material, or DNA, whose function was otherwise mysterious. There is, unfortunately, no evidence that such a message is among, or within, us.

15

False Alarms

In one of those remarkable coincidences that occur repeatedly in scientific history, at the time when Cocconi and Morrison were suggesting a search for intelligent radio signals on 21 centimeters, a plan to do so was quietly in preparation. Such is the logical flow of scientific thought that it leads to simultaneity in many such developments.

The search was being prepared by Frank D. Drake, then a young astronomer at the newly built National Radio Astronomy Observatory at Green Bank, West Virginia. From boyhood Drake had been captivated by the idea that there might be life in other worlds and when, as an undergraduate at Cornell in 1951, he heard Otto Struve lecture on astronomy and the slow-rotating stars as evidence for a multitude of solar systems, Drake found that such ideas were shared by a man of the highest distinction.

After being graduated in 1952 from Cornell, home base of Morrison and Cocconi, he fulfilled his three years' military service as electronics officer aboard the heavy cruiser USS *Albany*, acquiring practical experience that would stand him in good stead when it came to attempting longer-range communications than any contemplated by the Navy. After his discharge he took his doctorate in astronomy at Harvard. Concentrating more and more on the new and exciting field of radio astronomy, he took a job on the side, working for Harold I. Ewen, codiscoverer of the 21-centimeter emissions, who had gone into business on his own as a founder of the Ewen-Knight Corporation, one of the burgeoning electronics firms in the Boston area. Drake became intensely interested in the 21-centimeter signals as a wedge for prying into various astronomical problems, and he is credited with being the first to propose, on the basis of radio observations, that Jupiter has belts of trapped, gyrating particles like those girdling the earth, only 1,000 times more intense—sufficiently so, in fact, to produce strong synchrotron radiation.

These various lines of activity, at Harvard and then at Green Bank, repeatedly directed his mind toward the question of communication with distant worlds. When he arrived at Green Bank in April 1958 as one of the initial staff members of the observatory, he found it a scene of mud and snow with no permanent buildings. The observatory offices were in a farmhouse. Yet there was something remarkable about this national listening post on the universe, even in its formative stages. The site was in a valley partly sheltered from extraneous, man-made radio signals by flanking mountains. West Virginia had passed a Radio Astronomy Zoning Act to bar producers of undesirable radio interference from moving into the area. It was said to be the first law of its kind ever passed, protecting, as well, the nearby Navy site at Sugar Grove. It applied only to unlicensed radio activity. To curb licensed forms of interference a national "radio quiet zone" 100 by 120 miles in area was established under the Federal Communications Commission.

Another fear was aircraft interference. The danger was spelled out in the journal *Science* by Richard M. Emberson, director of construction at the site. The metal surface of an airplane, he said, can mirror man-made radio signals down into a radio telescope, even though the transmitter is hidden behind mountains. If the plane itself is transmitting, "the situation becomes quite serious."

The transmissions may not be directly on the frequency of the cosmic signals being observed, yet, he said, their side effects can mask the faint emissions from space. In particular, he noted, the 21-centimeter emissions from hydrogen "lie in a band assigned to aviation purposes. Thus, for example," he continued prophetically, "the spurious signals from aircraft flying in the neighborhood of Green Bank will interfere with this important astronomical observing band, even if the aircraft transmitter's primary frequency is carefully set away from the hydrogen frequency."

Eagerness to observe on the magic wavelength of 21 centimeters had been a major factor in persuading the government of the need for a national observatory, and the first big antenna at Green Bank was designed with such observations in mind. At about the time that this dish was completed, in March 1959, Drake went to a nearby grill facetiously called "Antoine's" for a quick lunch with Lloyd V. Berkner, acting director of the observatory and head of Associated Universities. The latter was a teaming up of nine northeastern universities whose first job had been to establish the Brookhaven National Laboratory on Long Island, home then of the world's most powerful atom-smasher. It now had the added task of setting up the National Radio Astronomy Observatory, the government's role in the two projects being justified on the ground that, while scientifically important, they were too costly to be undertaken by any one private institution. Berkner, who had gone to the Antarctic with Admiral Byrd and been a pioneer in radio science, had risen to world prominence as a science administrator. It was he, for example, who first proposed the International Geophysical Year of 1957–58. He was not one to shrink from unusual schemes, and Drake knew it.

Drake's reasoning, in his presentation to Berkner, was not unlike that set forth soon afterward by Cocconi and Morrison in their letter to *Nature*. For thousands of years man had speculated about the possibility of life in other worlds without any hope of verifying the hypothesis. The question had seemed utterly academic. But now, Drake pointed out, the situation was changing rapidly. The reception by Millstone Hill, near Westford, Massachusetts, of radar echoes from Venus (later shown to have been announced prematurely) demonstrated the enormous enlargement of man's capabilities. Such amplification devices as the maser and parametric amplifier had been increasing the sensitivity of radio telescopes at a rate of 50 percent a year. A maser had already been installed at the focus of the 50-foot dish atop the Naval Research Laboratory near Washington, resulting in a sensational increase in sensitivity. A similar device could clearly be used on the 85-foot dish towering above the valley where Drake and Berkner were munching hamburgers. The new dish stood on a three-legged structure that held it aloft like a gigantic robot warrior presenting its shield to heaven.

Drake proposed listening on 21 centimeters more on practical grounds than as a "uniquely rational" channel that, as proposed by Cocconi and Morrison, might be chosen by a highly advanced society. Observations on that wavelength were, as cited earlier, a prime goal of the 85-foot antenna, and one of Drake's colleagues hoped that the so-called Zeeman effect could be found in the 21-centimeter line of the radio spectrum. In 1896 a Dutch physicist named Pieter Zeeman had discovered that when light is emitted by atoms in a magnetic field, the lines of its spectrum are split, each part being polarized. The extent of the effect indicates the strength of the magnetism producing it; hence the Zeeman effect has been used to chart the magnetic fields of the sun and other celestial objects. Could the 21-centimeter line of the radio spectrum be exploited in the same way?

Because extremely sensitive receivers were needed for this conventional investigation, the added cost for equipment to listen for intelligent signals would be only about $2,000, although there would also be operating and manpower expenses, as well as diversion of the much-sought-after antenna from other work. Berkner knew an attempt to intercept signals from another solar system would be scoffed at by some conservative scientists. Nevertheless he encouraged Drake to go ahead, and soon thereafter, at a meeting of the observatory's trustees, he described the project to Otto Struve. A few weeks later Struve was named the observatory's first full-fledged director and the program, to a large degree, became his responsibility. It was a natural culmination to his lifelong interest in the possibility of extraterrestrial life. While growing up in Czarist Russia he had read, with his father, the speculations of Percival Lowell regarding life on Mars, and Struve's subsequent work on the slow-rotating stars had persuaded him that planets are plentiful in the universe. In 1960 he estimated that, in the Milky Way Galaxy alone, there are 50 billion solar systems. As to how many of them have produced intelligent life, he wrote in connection with Drake's project: "An intrinsically improbable single event may become highly probable if the number of events is very great." If the probability of finding intelligent life on a planet at one given time is substantially more than 1 in 10 billion, he said, "then it is probable that a good many of the billions of planets in the Milky Way support intelligent forms of life. To me this conclusion is of great philosophical interest. I believe that science has reached the point where it is necessary to take into account the action of intelligent beings, in addition to the classical laws of physics."

The idea that intelligence may play a role in shaping events in the universe, alongside the conventional laws of nature, was indeed a revolutionary one. In fact Struve believed that we are on the threshold of a new view of the universe, as remarkable in its departure from the past as was that of the Renaissance and the Copernican revolution.

Drake decided to call his effort Project Ozma, "for the princess of the imaginary land of Oz—a place very far away, difficult to reach, and populated by exotic beings." He and his colleagues were hesitant to say anything publicly about their sensational plan, and, although the project was initiated in April of 1959, they kept quiet about it until after publication of the Cocconi-Morrison proposal in *Nature* that fall.

When word did get out, there were hoots as well as cheers from the scientific community. It was somewhat embarrassing to Struve, even though, as former president of the International Astronomical Union and one of the world's most respected astronomers, his reputation was almost invulnerable. Project Ozma, he wrote, "has aroused more vitriolic criticisms and more laudatory comments than any other recent astronomical venture, and it has divided the astronomers into two camps: those who are all for it and those who regard it as the worst evil of our generation. There are those who pity us for the publicity we have received and those who accuse us of having invented the project for the sake of publicity."

He conceded that the chances of intercepting signals from beings near the first two or three sunlike stars that are scanned were "almost zero," but he said the instruments developed for the project would benefit all radio astronomy, and he added in conclusion: "There is every reason to believe that the Ozma experiment will ultimately yield positive results when the accessible sample of solar-type stars is sufficiently large."

Thus Struve did not envisage Ozma as a one-time effort, but as a continuing search for life in other worlds, carried out, perhaps, with pauses for technological improvement and reconsideration of methods, but with determination to persist until contact is made.

Drake stated his case for Ozma in *Sky and Telescope*, the magazine published at his alma mater, the Harvard College Observatory. By the most optimistic reckoning, he said, we may expect communication signals from a quarter of all stars. At the other extreme he cited the "extremely conservative" estimate of Harlow Shapley that intelligent life may exist near but one star in every million. The truth, he suggested, lies somewhere in between. He predicted that within fifty years we will have developed radio technology to the point where further improvements will no longer influence our ability to communicate with other worlds. The limitations, henceforth, will be background noise in space and other natural factors. Since our radio technology was fifty years old, he said, this means that civilizations characteristically jump, in the brief span of one century, from having no capacity for interstellar communication to having the maximum capability. On an astronomical time scale, he said, "a civilization passes abruptly from a state of no radio ability to one of perfect radio ability. If we could examine a large number of life-bearing planets," he continued, "we might expect to find in virtually every case either complete ignorance of radio techniques, or complete mastery."

Thus the chances that any of our neighbors are in our state of transition are negligible. As Drake told an assembly of physicists on the eve of his listening effort, "In view of the continuous formation of stars, there should be a continued emergence of technically proficient civilizations." Since it has taken life several billion years to evolve to our present level and the life expectancy of our planet and sun, in their present form, is another 5 billion years or more, the present transitional stage from a primitive state to one of unpredictably sophisticated technology is but a moment on that time scale. Therefore, Drake said, we must expect most societies that have crossed the threshold of civilization to be more advanced than our own.

Shortly after making these remarks, at a symposium on radio astronomy at Colgate University in Hamilton, New York, Drake was ready to begin his experiment. As targets he selected two of the stars most often discussed as candidates: Tau Ceti and Epsilon Eridani. Tau Ceti is in the constellation Cetus, the Whale. Epsilon Eridani is part of Eridanus, the River. Although both stars are about eleven light years away, Drake calculated that signals from planets near them should be observable with the 85-foot dish if they were generated by a million-watt transmitter operating through a 600-foot antenna like the one then being built by the Navy (and later abandoned) at nearby Sugar Grove. To be detected under these circumstances, the signals would have to be concentrated within a very narrow band of frequencies.

In an effort to eliminate background interference, such as man-made emissions and natural "noise" sneaking in from other parts of the sky, two receiving "horns" were rigged at the focus of the great 85-foot dish. One horn was aimed to receive signals from the target star (and its planets, if any) reflected to the focus by the parabolic dish. The other horn was aimed away from the star and did not catch the reflected radio waves, yet, like the target horn, it picked up background noise and stray emissions sneaking in from other parts of the sky. An electronic switch, operating at very high speed, alternately connected the two horns to the receiver. All emissions entering the receiver from the horn aimed away from the star were assumed to be interference and this amount of noise was subtracted, automatically, from the emissions arriving via the other horn. What remained, it was hoped, would be any signals coming from the distant solar system, plus "cosmic noise" reflected to the focus from the small region of the galaxy lying directly behind the star.

Another arrangement, designed to eliminate the cosmic noise, was based on Drake's assumption that artificial signals, for maximum efficiency, would be concentrated into a band of frequencies no more than 100 cycles in width. The receiver listened simultaneously on a broad bandwidth, which was assumed to be almost entirely cosmic noise, and on a narrow bandwidth. The broadband noise was then subtracted, electronically, from the narrowband reception, leaving exposed any signals from life elsewhere.

Of the two new devices for amplifying very weak radio signals, the maser was the more sensitive but also the more expensive, and had to be kept at temperatures of several hundred degrees below zero, whereas the parametric amplifier was cheaper, simpler, and could be used at room temperature. Drake therefore installed a parametric amplifier to boost the net sensitivity of the horns. This key amplifier was donated by Dana W. Atchley, Jr., president of Microwave Associates, Inc., maker of such equipment in Burlington, Massachusetts. Atchley was brimming with enthusiasm about the project and later was one of the few invited to the secret conference in Green Bank on the problem of establishing contact with other worlds.

Down the line from the parametric amplifier was an oscillator that had to remain constant, in frequency, to within one part in a billion. This was done by using a special quartz oscillator housed inside an oven set for 100 degrees Fahrenheit. This oven was itself within another oven, the purpose of the twin ovens being to protect the crystal from even slight variations in temperature.

The equipment was tuned continuously to observe each 100 cycles of bandwidth for one minute, covering, altogether, about 400,000 cycles of the radio spectrum, centered on the hydrogen frequency of 1420.4 megahertz (1,420,400,000 cycles per second).

When Drake and his associates took aim at their first star they

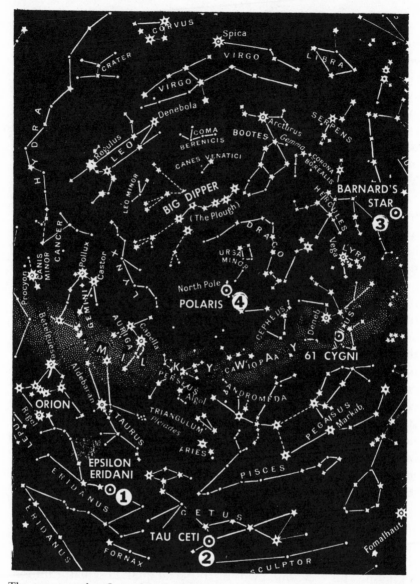

The two stars that figured in Project Ozma were Epsilon Eridani (1) and Tau
Ceti (2). The report of a planet circling Barnard's star (3) is now questioned.
Although it is one of the closest stars to the earth, it cannot be seen with the
naked eye. The North Star, Polaris (4), is a variable with a 4-day cycle.

knew the chances of success were almost nil; yet they could not suppress a certain excitement, if only because this was the start of a search that ultimately could lead to the most revolutionary discovery in the history of mankind.

At roughly 4 A.M. on April 8, 1960, the dials were set in the control building under the great shieldlike antenna, aiming it at Tau Ceti, then newly risen above the mountainous horizon to the southeast. The clockwork was adjusted to keep the antenna pointed at the star as it climbed across the dome of heaven. The receiver was switched on and, for the first time, the two-legged animal that we call man sought to intercept signals from beings of unknown physiognomy in another world.

Although daylight made the star optically invisible, its radio emissions should not have been affected. No unusual emissions were observed and, as the star began setting in the west, Drake decided to shift to the other target, Epsilon Eridani. The output of his equipment was observed in two ways. It was fed into a loudspeaker and also was recorded on moving paper. Shortly after the antenna had been aimed at Epsilon Eridani they heard a loud noise from the speaker and, on the moving paper, the needle went off the scale. Some very strong signal was coming in! The volume was turned down and the needle wrote a series of high-speed pulses at roughly eight per second, so uniformly spaced that they could only be the product of intelligent beings.

"We were so surprised and so unprepared for it," Drake told fellow astronomers later, "we didn't know what to do. Everybody just looked at each other." He ran across the room and began checking the connections and wires, seeking to find out whether the equipment was somehow to blame. Such possibilities as a car with faulty ignition or a nearby defective electric razor were quickly eliminated. The signal, said Drake, "was very close to what everyone had been predicting, a pulsed signal coming from cosmic space."

Drake, however, remembered an earlier experience while he was a twenty-six-year-old graduate student at Harvard. With the university's new radio telescope near the village of Harvard he was measuring the spectrum of emissions from the Pleiades star cluster when he began detecting signals at a wavelength close to the "magic" frequency of 21 centimeters. In fact it was offset from that frequency to an extent matching the motion of those stars relative to the earth. "I could barely breathe from excitement,"

Drake reported later. But when he aimed the antenna slightly to one side, the signals were just as strong. They were obviously made by man and not extraterrestrials.

Just as Drake and his Project Ozma colleagues started to make the same test on Epsilon Eridani, the signal abruptly stopped of its own accord. It had been received for a total of five minutes. The plan had been to avoid publicity unless they were sure of their results, but one of the crew phoned a friend in Ohio; news of the detection became public, and they were deluged with press inquiries. Meanwhile, each day they looked at Epsilon Eridani at the bandwidth of their original observation, hoping the signal would come back.

"After a few days we realized there was a good chance it was really some kind of interference, and so we actually added a completely separate receiver that had just a little horn stuck out the window. Sure enough, about ten days later, the signal came back and we saw it not only with the 85-foot telescope but also with equal intensity coming in the horn." It was clearly local. "We have never known what it was, but it had all the earmarks of being an electronics countermeasure system, probably airborne." It could thus have been a device to jam hostile radar. The warning of Richard Emberson, the observatory's construction chief, that 21-centimeter reception would be particularly vulnerable to aircraft interference had been borne out. During 200 hours of listening from May to July, 1960, the two stars were scanned without further causes for excitement.

As noted earlier, Iosif Shklovsky, the USSR's leading proponent of the search for life beyond the earth, argued that a supercivilization in another galaxy might use part of its star's energy for a beacon radiating on such a wavelength as 21 centimeters, and early in 1964 one of his associates, Nikolai S. Kardashev, proposed in the scientifically conservative *Astronomical Journal* of the Academy of Sciences of the USSR that two heavenly sources of radio emission might, in fact, be such beacons. They had been catalogued as CTA-21 and CTA-102 by the California Institute of Technology and had been singled out the previous year in a survey of 160 such "discreet sources" (or radio stars) published in the *Monthly Notices* of Britain's Royal Astronomical Society. What had struck the authors, R. G. Conway of Britain's Jodrell Bank Observatory, K. I. Kellerman of Caltech, and R. J. Long of the Mullard Observatory at Cambridge University, was the peculiar shape of the radio-frequency spectra of these two sources, compared to the others. In both cases their emissions were strongly concentrated toward 900 megacycles. The emissions, Kardashev pointed out, are thus centered in that part of the spectrum best suited to communication across the cosmos. Furthermore, he cited the calculation of his colleague, V. I. Slish, that, unlike the other radio sources, these two were virtually pinpoints in the sky.

To support his argument Kardashev proposed that three levels of technology are possible: One is comparable to that of our planet, which, as a whole, can generate 4,000 billion watts of power. He classed it a "Type One" civilization. Then, he said, there are those postulated by Dyson with energy at their disposal equivalent to that radiated by a star—400 million billion billion watts. With the earth's annual 3 to 4 percent increase in power production, ours could become such a "Type Two" civilization in about 3,200 years. Finally he spoke of "Type Three" civilizations that dispose of energy comparable to that generated by an entire galaxy. His proposal was that peculiar radio sources, such as CTA-21 and CTA-102, might be beacons of Type Two or Type Three technologies.

What created a worldwide sensation was the report by another of Shklovsky's colleagues at the Sternberg Astronomical Institute that he had observed rhythmic fluctuations in the signals from CTA-102. This first appeared quietly on February 27, 1965, in *The Information Bulletin on Variable Stars*, published under the auspices of the International Astronomical Union. The author, Gennady B. Sholomitsky, said that for several months he had repeatedly compared the strength of emissions from CTA-102 with those from a presumably unchanging source (3C-48). He did the same with CTA-21. The emissions from the latter proved to be steady, but those from CTA-102 fluctuated in what appeared a smooth "sine curve," with peaks at roughly 100-day intervals.

In his report Sholomitsky did not speculate as to the reason for the fluctuations, but on April 12, 1965, the Soviet news agency TASS announced that he had found CTA-102 to be the beacon of a "supercivilization." It was evidence, the agency said, "that we are not alone in the universe."

This created such a furor that Shklovsky, Kardashev and Sholomitsky held a hurriedly summoned press conference the next morning. Shklovsky criticized the TASS report and said it was "a little premature'' to say the signals were artificial. However, he maintained that this was still at least possible.

Soon thereafter the idea that CTA-102 might be a beacon was dealt a severe blow. Mount Palomar reported that a peculiar, blue-shining object had been found in the precise location of CTA-102. Furthermore, Maarten Schmidt of that observatory had discovered that the spectral lines of its light were so radically shifted toward the red end of the spectrum that the object must be receding from the solar system at some 114,000 miles a second. In our expanding universe this speed can be used as a rough index of distance. It showed CTA-102 to be one of the most remote objects ever observed—apparently one of the newly discovered bodies known as "quasars," whose brilliance makes them visible, both in light and radio waves, far beyond anything else. It was hard to believe such emissions could be artificial, and so the flurry subsided.

Then, in 1967, a false alarm that developed into one of the most important discoveries in modern astronomy occurred at Cambridge University in England. A graduate student, Jocelyn Bell, like many students, seemed hopelessly behind in her work— a half kilometer behind, for that was the length of chart paper with three-track recordings of radio signals waiting to be analyzed. And the recorders of the receiving system, which she herself had helped build, were pouring out more chart paper at 100 feet a day.

The purpose of the project was to identify quasars by the extent to which their radio emissions scintillated, or "twinkled," since twinkling indicates that the source is very pointlike. Stars twinkle, but the nearer, brighter planets do not. And so, if radio sources scintillated, it was likely that they were very far away and probably quasars. Bell's thesis adviser, Antony Hewish, had concluded that the scintillation of radio sources—which is much slower than star twinkling—was not, as in the case of stars, caused by turbulence in the atmosphere but by clouds of ionized gas moving rapidly out from the sun—the "solar wind."

The observing program was ambitious. A 4.5-acre field, large enough to accommodate fifty-seven tennis courts, was covered with 2,048 dipole antennas (each consisting of two aligned rods cut to fit the wavelength being observed). Jocelyn Bell, aged twenty-four, had been responsible for wiring them into a coordinated receiving system. By July 1967, all had been ready, and recording began. With the earth's rotation sweeping the field of view across the heavens it still took several days to scan the entire sky visible to the system, whereupon the scanning was repeated. In this way fluctuating sources that were man-made could be eliminated. Only if, week after week, the signals were observed coming from the same spot among the stars would they be considered quasar candidates.

Bell did the data analysis, as her interpretation of the recordings would be her doctoral thesis. In examining those made on August 6 she noticed what she later called "a bit of scruff"—a wavering signal a half inch long on the 400 feet of chart representing one full coverage of the sky. At the time when the "scruff" had been recorded it was midnight, with the antenna aimed directly away from the sun—the direction shadowed from the solar wind by the earth itself. Quasar scintillation therefore seemed unlikely. The observation was dismissed—and almost forgotten—as of local origin.

In September Bell saw it again. "I began to remember that I had seen this particular bit of scruff before, and from the same part of the sky," she said later. "It seemed to be keeping pace with twenty-three hours, fifty-six minutes—that is, with the rotation of the stars."

It was a similar argument that gave birth to radio astronomy. When Karl Jansky first recorded radio noise from the sky in the early 1930s he thought it was of solar origin, and it was only after the sun, in subsequent months, changed its position relative to the stars that he realized the source had not moved with the sun but remained fixed among the constellations (whose passage overhead occurs every twenty-three hours and fifty-six minutes, rather than every twenty-four hours).

By the end of September the "bit of scruff" had shown up on six (though not all) occasions as the constellation Vulpecula, the Little Fox, passed through the antenna beam. Hewish thought it might be one of the stars that periodically erupts or "flares," so he and Bell decided to operate a fast recorder as it passed, observing its fluctuations in greater detail. For weeks it virtually disappeared. Then late in November, according to Hewish's account, "Miss Bell undramatically announced: 'It's back.' " A high-speed recording, made on November 28, showed clocklike pulses at a repetition rate of slightly more than one per second.

When Jocelyn phoned Hewish with this astonishing result, he replied: "Oh that settles it, it must be man-made." No rhythmic phenomena known to astronomers occurred at so fast a tempo—

no spins, no orbits, no hypothetical vibrations. The shortest variable-star periods were about eight hours.

"We considered and eliminated radar reflected off the moon into our telescope, satellites in peculiar orbits, and anomalous effects caused by a large, corrugated metal building just to the south of the four-and-a-half-acre telescope," Bell said. Even though the source seemed fixed among the stars, Hewish believed it must be in the vicinity of the earth. "After all," he said later, "seasoned radio astronomers do not make the mistake of supposing that every queer signal on their records is truly celestial; in ninety-nine cases out of a hundred peculiar 'variable radio sources' turn out to be some kind of electrical interference—from a badly suppressed automobile ignition circuit, for example, or a faulty connection in a nearby refrigerator." Because of the source's association with passage of the constellation Vulpecula, rather than with clock time, Hewish thought perhaps some other observatory was transmitting signals when that particular spot passed overhead. He consulted a number of astronomical colleagues, but none were doing so. Still skeptical he arranged for high-precision time-signal ticks broadcast at one-second intervals to be recorded with the incoming radio pulses. "To my astonishment," he said, the pulses proved so clocklike that their time-keeping was accurate to one part in 10 million—that is, to within one second in every four months. The pulse rate later proved remarkably steady at one every 1.33730113 seconds. Furthermore, when observed in a narrow-frequency range, the pulses themselves lasted only .016 second, implying—by the same argument applied to quasar outbursts—that the source must be very small.

"Having found no satisfactory terrestrial explanation for the pulses," said Hewish, "we now began to believe they could only be generated by some source far beyond the solar system, and the short duration of each pulse suggested that the radiator could not be larger than a small planet. We had to face the possibility that the signals were, indeed, generated on a planet circling some distant star, and that they were artificial."

"Without doubt," Tony Hewish recounted later, "those weeks in December 1967 were the most exciting in my life." To announce that the beacon of another civilization had been detected would obviously create a worldwide sensation—and it might turn out, as in the case of the Soviet report on CTA 102, that a natural explanation would present itself.

"Just before Christmas," Bell recounted later, "I went to see Tony about something and walked into a high-level conference about how to present these results. We did not really believe that we had picked up signals from another civilization, but obviously the idea had crossed our minds and we had no proof that it was an entirely natural radio emission. It is an interesting problem—if one thinks one may have detected life elsewhere in the universe how does one announce the results responsibly? Who does one tell first? We did not solve the problem that afternoon, and I went home that evening very cross—here I was trying to get a Ph.D. out of a new technique and some silly lot of little green men had to choose my aerial and my frequency to communicate with us. However, fortified by some supper I returned to the lab that evening to do some more chart analysis."

Shortly before the lab closed for the night she saw, in the record from a different part of the sky, another bit of "scruff." She checked earlier recordings from the same region and found that it had occurred before. The lab would soon be locked up, but at about 1 A.M. that patch of sky would pass over her antenna and she was there to observe. "It was a very cold night," she told an interviewer, "and the telescope doesn't perform very well in cold weather. I breathed hot air on it, I kicked and swore at it, and I got it to work for just five minutes. It was the right five minutes, and at the right setting. The source gave a train of pulses, but with a different period, of about one and a quarter seconds."

Soon a total of four such sources had been found, each with its characteristic pulse rate. In keeping with astronomical tradition they were given letter-number designations. Because of their suspected artificial nature they were listed as LGM-1 (the first discovered) through LGM-4, the letters standing for "little green men."

Before they announced their discovery, however, the Cambridge group had found that LGM-1 failed one test for artificiality. This was the variation in its pulse rate to be expected if the signals came from a planet orbiting another sun. Because the pulse rate was so very uniform, even by the standards of an atomic clock, variation in the observed rate should have shown up as the planet during its orbital flight varied its speed relative to the earth—a variant of the much-used Doppler shift. This effect could not be detected, even though that of the earth's own relative motion was evident. The possibility was not ruled out that a transmitting civilization had adjusted the pulse rate to keep it steady despite orbital motion by the transmitter. This would be done, for example, if—as one radio astronomer suggested—the pulses were serving as navigational aids for long-range space travel, much as loran beacons are used by ships and planes on earth. But as it seemed less and less likely that they were artificial, they came to be known as pulsars (for pulsing radio sources), although, when Hewish presented his results at NASA's Institute for Space Studies in New York, his slides still showed the LGM designations.

Because the higher-frequency part of each radio pulse arrived ahead of the lower-frequency part, it was possible to estimate the distances to each pulsar. Low-frequency waves are slowed by electrons in space more than high-frequency waves. Therefore, calculating the distance to the point of origin was like figuring out how far two runners have traveled before the first crosses the finish line. If you know exactly how fast each of them ran and how much ahead the faster one was at the finish, it is easy to calculate how far they had to run for the winner to gain that much of a lead. While there was some uncertainty about the number of electrons adrift along the path of the radio pulses (and therefore of the slowing effect), rough distance estimates were possible, and it was clear that all the pulsars were within the Milky Way Galaxy.

But what were they? Hewish at first could think of no celestial phenomenon that might produce such rhythmic pulses. Without explaining why, he began spending a lot of time in the library of Cambridge University's optical observatory. His friends there, he said, "were surprised to see a radio astronomer taking so keen an interest in books on stellar evolution"—normally the province of optical astronomers.

He zeroed in on the manner in which stars evolve and finally collapse because, he thought, white dwarfs or neutron stars might be the answer. A 1933 proposal of Fritz Zwicky and Walter Baade of Caltech and Mount Wilson observatories that big stars, in their death throes, explosively collapse into neutron stars had almost been forgotten, but the idea that such extremely dense objects might exist had been revived, notably to explain brilliant X-ray emissions from certain points in the sky. Alastair G. W. Cameron of NASA's Institute for Space Studies in New York had proposed that superdense stars would alternately expand and contract like a beating heart. A formula defining the rate of such oscillations in terms of the density and size of the star had been derived, but when Hewish tried applying it to white dwarfs it did not work.

Such stars simply were not small enough and dense enough to oscillate at rates as fast as once a second.

When Hewish, in 1974, presented the lecture traditionally given in Stockholm by recipients of a Nobel prize, he said discovery that one of the four original pulsars had a rate of only a quarter second "made explanations involving white dwarfs increasingly difficult." Neutron stars seemed a better candidate, although whether such objects really exist remained uncertain. In 1942 Baade had pointed to a star in the center of the Crab Nebula—that spectacular residue of the supernova seen in 1054—as possibly such an object. He picked it among several stars visible in that region because, unlike the others, it seemed to be moving in company with the nebula itself against the backdrop of distant objects. His suggestion was reinforced later by his colleague Rudolph Minkowski, although at the time there was little more on which to base the proposal except the unusual nature of the star: its light seemed devoid of any spectral features at all, a peculiarity of synchrotron radiation, generated by electrons in tight orbits rather than heat.

When Hewish, with Jocelyn Bell and others of the Cambridge group, reported their discovery in the issue of *Nature* for February 24, 1968, they included the hypothetical neutron stars with white dwarfs as objects whose oscillations might produce the radio pulses. "If the suggested origin of the radiation is confirmed," they said prophetically, "further study may be expected to throw valuable light on the behaviour of compact stars and also on the properties of matter at high density."

The idea that throbbing oscillations were to blame did not seem very plausible, and it was Thomas Gold of Cornell University who realized that spin was a much more likely explanation. Gold, who with Fred Hoyle and Hermann Bondi had helped formulate the "steady state" view of the universe, was noted for his highly original—and often controversial—ideas.

If, he said, pulsars are the hypothetical neutron stars, they should be spinning very fast. As a rotating star collapses, its spin rate must increase radically, unless it sheds much of its rotational energy—its "angular momentum." This is often illustrated in terms of a figure skater whose spin rate increases when the arms, formerly spread, are brought close to the body. In the case of a neutron star the effect, Gold realized, would be extreme. For example, if a star like the sun, which spins once a month, collapsed to the assumed 10-kilometer diameter of a neutron star, it would spin 1,000 times a second! Such "millisecond" pulsars have now been observed. Also confirmed is Gold's prediction that the fastest-spinning ones should slow their pulse rates, although in a barely perceptible manner, and his explanation for what makes them pulse is still generally accepted. The star at the center of the Crab Nebula has, as Baade predicted, been shown to be a neutron star—a pulsar spinning thirty times a second—and, being visible, it is also flashing at that rate.

When Hewish won a Nobel prize for the pulsar discovery, Sir Fred Hoyle, who seems to enjoy annoying the establishment, argued that a gross injustice had been done to Bell. She herself said, "I believe it would demean Nobel prizes if they were awarded to research students, except in very exceptional cases, and I do not believe this is one of them." She pointed out that she was no longer in radio astronomy, had married (she was now Jocelyn Bell Burnell) and was engaged in other pursuits.

A more recent "false alarm" or, more properly, unexplained observation was made with the large antenna of Ohio State University, conducting perhaps the most faithful of all searches for extraterrestrial life. Its received emissions were automatically recorded, and during 1977, as a staff scientist looked over computer printouts from several days earlier, he saw a signal which, according to the project's chief scientist, Robert S. Dixon, closely matched what was expected from a transmitter on some other world. It was far narrower in bandwidth (less than 10 kilohertz) than expected from any natural source. Its reception was very brief, for as the earth rotates, the antenna can see one point in the sky for only about one minute. The direction from which the signal came was roughly the galactic center. There was no known source of man-made or astronomical emission from that direction. The observer wrote "Wow!" next to the recording and the observation has been known as "Wow!" ever since. Despite periodic observations of that spot in the sky, it has never been recorded again. Nor has it been adequately explained. It is a safe bet that there will continue to be such observations, as well as "false alarms" and even, perhaps, contrived frauds, although designers of the current SETI program, described in a subsequent chapter, believe they have made this impossible.

16

Can They Visit Us?

The discussion of communications between solar systems has dealt primarily with long-range signaling, apart from Bracewell's proposal for automated messengers. But what about travel between distant civilizations? Should we expect visitors from another world, particularly after they have detected radio emissions revealing the emergence of intelligent life in our solar system? Should we look for evidence of earlier visitations?

While early speculators thought it might be possible to fly to the moon on wings or by balloon, such thinking was terminated by the discovery that the atmosphere thins to almost nothing at a height of about 20 miles. Not until the writings of Konstantin Eduardovich Tsiolkovsky on rocket travel were thoughts of space journeys revived. His classic work, *Exploration of Space by Means of Reactive Apparatus,* was published in Russia in 1896 when Tsiolkovsky was thirty-nine years old. It sets forth many of the problems being encountered, now that space travel is a reality. By the time of his death in 1935 at the age of seventy-eight, he had written extensively on the possibility of life in other worlds, sometimes with a formidable imagination. This is reflected in his correspondence with Alexis N. Tsvetikov, then a graduate student at the Biochemical Institute of the Academy of Sciences in Kiev and later in the Department of Biophysics at Stanford University. Tsiolkovsky noted in one of his letters to Tsvetikov that in the biochemical sense life on earth "seems to be a relatively separate unit." Individuals die, he said:

However, the total amount of living matter perseveres, and even increases. We can imagine a spherical organism with the cycles of physiological processes closed completely in themselves. Such an organism will be immortal and photosynthetic, and it can develop even a higher consciousness. . . . The main activity of the highest living organisms in the Universe can be also the colonization of other worlds. Such beings, probably, could not be of spherical form, and they will not be immortal.

In the last months I have often been dangerously ill . . .

This letter was written on October 2, 1934, a few months before Tsiolkovsky died. Among his other proposals were life forms composed entirely of hydrogen (reminiscent of the living plasma cloud that, in Fred Hoyle's science-fiction novel *The Black Cloud,* invades the solar system and begins feeding on the sun). Tsiolkovsky even envisioned the ultimate in beings as a disembodied entity, "living" in space as an almost godlike island of pure consciousness. Because plants lack feeling and provide essential food for higher life forms, Tsiolkovsky saw a place for them in his superior worlds; but he regarded animals as early stages in the painful process of evolution. A superior race, he said, would painlessly eliminate them rather than see them endure the needless sufferings of evolution and the struggle for existence. He argued that, like good gardeners, such superior beings would weed out lower animal species, harmful bacteria and valueless plants, except for laboratory samples.

Tsiolkovsky ended his book *Monism of the Universe,* first published in 1925, with a number of axioms, one of which was that on at least one planet, somewhere, beings have achieved a technology permitting them "to overcome the force of gravity and to colonize the Universe." Consequently, he said, "perfection and dominance of the mind" have been spreading through the cosmos. This was initiated an infinite time in the past and colonization now is the normal manner in which life spreads. Evolution, "with all its sufferings," is rare, he said. Yet, while he believed in interstellar travel, he considered radio the chief means of communication. In a 1920 letter to the Organization of Young Technicians he predicted that, "In the near future short radio waves will penetrate our atmosphere and they will be the main means of stellar communication."

Tsiolkovsky's far-reaching speculations are somewhat reminiscent of Nikola Tesla. His mysticism was clearly distasteful to the materialistic Soviet state. It is reported that when he died, the government took all of his papers and placed them in the archives of Aeroflot, the Soviet airline. His home, in the city of Kaluga in western Russia, was converted into a museum, but not until the success of the first *Sputnik* in 1957 did he become a national hero.

The surge of excitement that accompanied the dawn of the space age stimulated new interest in the possibility of travel between solar systems. There was talk, for example, of using nuclear-powered rockets driven by the pressure of light itself; but John R. Pierce of Bell Telephone Laboratories, one of the nation's leading applied scientists, did a sober analysis of the problem with discouraging results. It was evident that to reach even the nearest sunlike stars within a human lifetime would demand a peak speed close to that of light. Yet it was inconceivable, Pierce said, that a rocket propelled by light could carry enough fuel to reach such a speed, even if there was complete conversion of the fuel into pure energy. If the energy was used, instead, to "push against" a massive object like the earth, he said, it would be somewhat easier to achieve so high a velocity, but it was hard to see how this could be applied to space travel.

Writing in the June 1959 issue of the *Proceedings of the Institute of Radio Engineers*, Pierce also discussed the possibility of scooping up hydrogen en route as fuel for such a vehicle. Throughout the galaxy there appears to be at least a sprinkling of hydrogen, and in some hydrogen clouds there are thought to be as many as 1,000 atoms per cubic centimeter (thimbleful). Pierce discussed a system using a "generous" scoop 100 square meters (more than 100 square yards) in area to snatch hydrogen en route. This collecting device, he added, would no doubt consist of force fields, such as magnetism, rather than a material substance. Taking a space ship of 17.5 tons as the smallest conceivable size for interstellar travel, he estimated that the highest speed within reach of such a system would be only 9.3 percent of the speed of light. "Clearly," he said in conclusion, "it is impossible to attain a velocity close to that of light by using interstellar matter as fuel."

Another discouraging note was sounded the next year by Edward Purcell, discoverer of the 21-centimeter line in the radio spectrum, in a lecture that he gave at Brookhaven National Laboratory on Long Island. The laboratory, where he was spending a year as research collaborator, was then the site of the world's most powerful accelerator (or atom-smasher), and his audience consisted largely of physicists. His arguments were elaborated in *Science* a year and a half later by Sebastian von Hoerner, a former associate of Frank Drake at Green Bank who was now at the Astronomisches Rechen-Institut (Astronomical Calculation Institute) in Heidelberg, Germany. Each began by calling to mind the immensity of interstellar distances. If, for example, the sun were scaled down to the size of a cherrystone, the earth would be a grain of sand 3 feet away. The nearest star would be another cherrystone 140 miles away.

Purcell pointed out that the strange "time dilation" predicted by the special theory of relativity would help passengers endure such long journeys if, in fact, they could be accelerated to a speed close to that of light. Let us assume, for example, that a vehicle accelerates at a rate equivalent to the force of gravity at the earth's surface (that is, at a rate of 1 "g"). We can be sure that human bodies could withstand such acceleration indefinitely, since the force exerted by gravity on our bodies throughout life is 1 g. Within a year such a vehicle would be moving almost as fast as light (186,000 miles a second). From then on, if the thrust of the engines remained constant, the rate of gain in speed would decrease, approaching zero but not reaching it, in the manner that mathematicians call "asymptotic." The engine thrust, no longer able to boost the speed to any degree, would produce strange effects, from the viewpoint of an observer back home. The clock rate—the inherent property "time"—on board the vehicle would slow down, approaching zero but never quite getting there. The weight of the ship and its occupants would increase; and they would be foreshortened in the direction of flight.

At the midpoint of a journey to one of the nearest sunlike stars, 12 light years away, such a vehicle would be traveling at 99 percent of light's speed. To land on a planet at its destination it would then have to decelerate at the same rate, with time still dragging its heels, until, during the final year of this outward journey, the speed dropped appreciably below that of light.

These peculiar manifestations would not be evident to those on board. Time would seem normal, as would their shape and weight. Assuming they then returned to earth in the same manner, the time for the entire journey, as measured on earth, would be 28 years (a ray of light would have made the round trip in 24 years). However, the passengers would be only 10 years older than when they started.

In a sense this is because of the peculiar relationship between time, as a dimension, and the speed of light. When we throw a

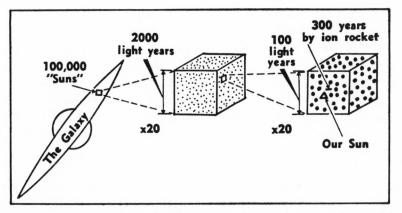

In this diagram, devised by Philip Morrison to illustrate the scale of our galaxy, a small cube of the latter's interior containing 100,000 stars, left, is expanded twentyfold to produce the center cube. A portion of the latter is again enlarged twenty times on the right. The tiny dash within that cube is the distance that would be traversed in 300 years by an ion rocket—one of the more advanced types under development.

light switch, the room seems illuminated instantaneously. In the world of our daily lives, the speed of light seems infinite and time seems invariable. But if one could accelerate almost to the speed of light, strange things would happen to time and the other dimensions. Von Hoerner pointed out that the longer such a journey was, the more extreme the effect, until finally, from the point of view of those back home, time on the space ship would virtually come to a halt. Thus, in a vehicle whose engine accelerated and decelerated it, with a force equal to gravity, a round-trip journey that seemed 20 years long to the voyagers, carrying them to a point 137 light years away, would bring them home to a world 270 years older than when they left it. A 60-year journey, by clocks on the spacecraft, would bring them back 5 million years after their departure. They would have reached out 2.5 million light years—farther than the nearest galaxy like our own.

Purcell said flatly, in his Brookhaven lecture, that the special theory of relativity, in predicting such strange effects, "is reliable." If it were not, he told his atom-smashing audience, "some expensive machines around here would be in very deep trouble." Protons in the big Brookhaven accelerator were boosted to 99.948 percent of the speed of light and their mass increased in the man-

ner predicted by relativity. Those designing the Brookhaven machine had to take this into account or it would not have worked.

Furthermore, the slowing of time, on the atomic level, is demonstrated by the extended lifetimes of the nuclear particles known
as muons, when traveling almost at the speed of light. Muons are
produced when atoms are shattered by high-energy collisions,
such as those in an atom-smasher or the impacts of the high-
energy atomic nuclei known as cosmic rays, plunging into the atmosphere from space. So brief is the lifetime of a muon that it
decays into other particles almost instantaneously, its average lifetime being 2.2 millionths of a second. Since muons produced by
cosmic rays are generated near the top of the atmosphere, they
should not survive long enough to reach the surface of the earth
despite their high speed. Yet, precisely because of this speed, their
lifetimes are prolonged in the manner predicted by Einstein, and
muons rain steadily on the earth. In other words, their time, as
we see it, is slow. If this happens on the atomic level, would it
apply to entire human beings? It is now widely believed that this
would be the case. In fact, it is hard to see why it should be
otherwise.

Having argued persuasively for the reality of time-slowing, Purcell sought to demolish the idea that anyone, on earth or anywhere else, could ever take advantage of it. He analyzed the
amount of energy required for the previously described round trip
to a star 12 light years away, in which a top velocity of 99 percent
of the speed of light would be reached at the midpoint of both
the outward and the return journeys. In terms of fuel weight, the
most efficient source of energy within our grasp is the fusion reaction of the hydrogen bomb (where the hydrogen isotopes trit-
ium and deuterium combine to form helium). Even more efficient
is the fusion that makes the sun shine. In this reaction, four hydrogen nuclei ultimately combine, under great pressure, to form
one helium nucleus. Because the binding energy required to hold
the helium nucleus together is slightly less than that in the original
hydrogen nuclei, something is left over after the reaction and
emerges as free energy. Though it is less than 1 percent of the
original mass, the energy released is formidable because of the
Einstein equation $E = mc^2$ (the energy equals the converted mass
multiplied by the speed of light squared).

To be supremely optimistic, Purcell assumed that this solar fusion process could be used with 100 percent efficiency, although

we have not yet learned to use it as a practical power source. With solar-type fusion it would require 16 billion tons of hydrogen to accelerate a 10-ton capsule to 99 percent of the speed of light—or bring it to a halt from that speed. The vehicle must be accelerated at the start, halted at its destination, accelerated for the return and halted again at the end. Since the fuel for these maneuvers must itself be accelerated, the total requirement is enormous: some 500 billion billion billion billion tons.

But, said Purcell, "this is no place for timidity, so let us take the ultimate step and switch to the perfect matter-antimatter propellant." From our present knowledge of nature, there does not seem to be any more efficient way to obtain energy than to combine matter with antimatter. When two such substances meet, they mutually annihilate each other, leaving nothing but a great deal of energy in the form of gamma rays. It is the only known process in which matter (and antimatter) can be converted entirely into energy. The gamma rays, which are at the short-wave end of the electromagnetic spectrum, could power a rocket by the equivalent of light pressure.

No one has ever really seen any antimatter. The experimenter can observe in the detector evidence that a particle of antimatter existed there briefly, and some accelerators can store particles of antimatter for limited periods, but sooner or later they meet particles of matter and vanish. If one lists all the particles that can be produced by smashing atoms, there is an "anti" particle corresponding to every one of them. The antiparticle is a mirror image of its counterpart. If the latter carries a negative charge, like the electron, then its antiparticle is positive. The antiparticle of the electron, for example, is known as the positron. It has the same mass as the electron and the strength of its electric charge is the same, but it is positive instead of negative. Discovered in 1932, it was the first hint that there is an ephemeral world of antimatter.

It seems hard to see how even the most advanced civilization could assemble a mass of antimatter and load it on a rocket, along with an equivalent mass of matter. Again Purcell set aside such problems, but he found that his hypothetical journey would still require 400,000 tons of fuel, equally divided between matter and antimatter.

Then there is the problem of interstellar gas and dust hitting the space ship like hailstones on a car's windshield. Space between the stars is not quite a vacuum. There is about one hydrogen atom per cubic centimeter and there are bits of dust and perhaps other material here and there. If you were traveling at 99 percent of the speed of light, Purcell said, this material would hit the front of your space ship in the form of radiation as intense, per square yard, as that produced by several hundred atom-smashers. "So," he said, with calculated understatement, "you have a minor shielding problem to get over before you start working on the shielding problem connected with the rocket engine." Because this engine depends on a massive output of gamma rays for its drive, the problem that it raises is formidable, Purcell said, not so much for the passengers as for the inhabitants of earth. Let us imagine that we are standing at the end of a runway watching a jet airliner take off away from us. If we were too close at the start, the blast from the engines would be scorching. But if this were a rocket powered by gamma rays and headed for a distant star, the "blast" would be devastating to any life behind it (although, as Carl Sagan later pointed out, the earth's atmosphere would act as a shield against such rays).

Purcell, in his lecture to the Brookhaven scientists, paused before ramming home his conclusion: "Well, this is preposterous, you are saying. That is exactly my point. It *is* preposterous. And remember, our conclusions are forced on us by the elementary laws of mechanics." Communication by radio waves is probably the best bet, he said, and he referred to Project Ozma, Frank Drake's attempt to intercept radio signals from other worlds, as imaginative and sound.

Nevertheless, he argued, our society is still not mature enough to engage in a large-scale search:

We haven't grown up to it. It is a project which has to be funded by the *century*, not by the fiscal year. Furthermore, it is a project which is very likely to fail *completely*. If you spend a lot of money and go around every ten years and say, "We haven't heard anything yet," you can imagine how you make out before a congressional committee. But I think it is not too soon to have the fun of thinking about it, and I think it is a much less childish subject to think about than astronautical space travel. In my view, most of the projects of the space cadets are not really imaginative. . . . All this stuff about traveling around the universe in space suits—except for *local* exploration [within the solar system] which I have not discussed—belongs back where it came from, on the cereal box.

Purcell's seeming annihilation of the idea of travel to other solar systems did not discourage everyone. In fact, it stimulated the suggestion that perhaps we may someday be able to propel the entire earth to another part of the galaxy. This was proposed, in somewhat lighthearted fashion, by Darol Froman, who had been technical associate director of the Los Alamos Scientific Laboratory in New Mexico. He was well equipped for the discussion, since the Los Alamos laboratory, operated for the Atomic Energy Commission by the University of California, was headquarters for American research on reactors for space flight. In a talk to the Division of Plasma Physics of the American Physical Society in November 1961, he noted that the sun eventually will burn out and discussed whether or not, before that dark day, it might be possible to push the earth into another solar system. The energy for this grandiose scheme would be obtained by fusion reactions, using seawater as the fuel source.

Because the oceanic supply of deuterium, the heavy form of hydrogen used in the hydrogen bomb, is insufficient to push the earth great distances, Froman proposed that it would be more reasonable to use the reaction that occurs in the sun (combining four hydrogen nuclei to form a helium nucleus), even though we are a long way from learning how to do this. The process would make it possible to use hydrogen nuclei that are abundant in the oceans. He suggested that a quarter of this fuel be allocated to escaping the sun's gravity, another quarter be held to maneuver the planet into another solar system and the remaining half be used for interstellar propulsion and for light and heat en route. The moon would be forfeited to obtain additional fuel, since, as he put it, the moon "will be no good to us anyway." In the absence of sunlight it would be virtually invisible.

Froman's earth-propulsion system could operate for as long as 8 billion years, he said, perhaps enabling a planet to outlive its parent sun and reach solar systems 1,300 light years away. It might even seem preferable, he said, to keep on traveling through the galaxy rather than go into orbit around some other sun. The oceans would then have to be replenished from time to time by gathering water from planets encountered en route. For most of us, Froman said, "the most comfortable space ship imaginable would be the earth itself. So if we don't like it here because the sun is dying or something, let's go elsewhere, earth and all. We will not have to worry about all the usual hardships of space travel. For example, the radiation problem will disappear because of shielding by the atmosphere and because we will be going at low speed. The ease and comfort of this mode of travel are shown in the next slide.''

At this point, there flashed on the screen an idyllic scene showing women golfers, pine trees and great open spaces.

Actually the problem of propelling large numbers of people to another solar system was discussed as early as 1951 by Lyman Spitzer, head of the Princeton University Observatory and noted both for his pioneering attempts to harness the power of the hydrogen bomb and for climbing the facades of the University's tallest towers. He spoke of a vehicle weighing 10,000 tons, powered by a uranium pile of perhaps 1,000 tons, generating 2 million horsepower of useful energy:

Such a ship could carry thousands of people and vast supplies anywhere in the solar system, and could even navigate to other stars, though many generations would be born, grow up and die on shipboard before such a journey were complete. However, launching such a ship from the Earth's surface to a close circular orbit would be a tremendous undertaking. With the use of chemical fuels, such a launching would require a rocket of some million tons gross weight, an achievement that would seem far, far in the future.

Among those who rebelled against the sober reasoning of Pierce, Purcell and von Hoerner was Freeman Dyson, who had suggested that a supercivilization in another solar system might redistribute the material of its planets to achieve maximum living space and maximum energy from its star. In a letter to *Scientific American* in 1964 he said the calculations of Purcell and the others were perfectly valid, insofar as they related to fuel requirements for journeys limited to human lifetimes. But what about slower trips? Engines using nuclear power could reasonably be expected to drive large space ships at a speed of a few light years per century, he said. If an intelligent race achieved a life span considerably greater than our own; or if it perfected a method of freezing its citizens, harmlessly, for prolonged hibernation, then journeys

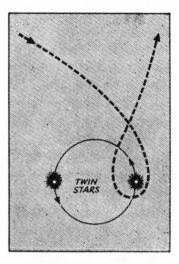

In Dyson's "gravity machine" an object is fired toward twin stars so that it circles the approaching star and is thrown back by that star's gravity, having gained much additional energy.

TWIN
STARS

reckoned in thousands of years would be conceivable. Thus, he wrote, "interstellar travel is essentially not a problem in physics or engineering but a problem in biology."

There is no reason to suppose, he asserted, that others in the galaxy have not solved this problem and that we may not ultimately do so. No doubt many would find thousand-year trips "unappealing," he said, but added: "We have no right to impose our tastes on others."

In another paper Dyson proposed a way that interstellar vehicles might pick up momentum en route. His scheme was to steal a little bit of the energy with which two very dense stars are circling one another. If, he said, a vehicle approached one such star as the star was coming toward it, the gravity of the star would whip the vehicle around in a tight orbit, sending it off into space again with far more energy than it had to begin with. It would be almost as though the vehicle had been hit by a gigantic baseball bat. The star, having transferred to the vehicle some of the energy with which it was circling its twin, would move a tiny bit closer to the star with which it was waltzing through space.

The most remarkable feature of this procedure was that, even though the vehicle underwent an explosive rate of acceleration— some 10,000 g's—no harm would come to the most delicate passenger or the most sensitive piece of equipment on board. This is because the accelerating force would be applied with almost complete uniformity to every particle of the body or instrument on board. It would not be any more uncomfortable than falling through space.

Thus, said Dyson, a vehicle could very rapidly be speeded up by more than 1,000 miles a second. The best star systems for giving vehicles such enormous accelerations, he said, would be pairs of white dwarfs, tiny "senile" stars whose density is so great that they may weigh as much as 3,000 tons per cubic inch. "It may be imagined," he wrote, "that a highly developed technological species might use white-dwarf binaries scattered around the galaxy as relay stations for heavy long-distance freight transportation." His proposal calls to mind the many occasions, since then, when the gravity of a planet has been used to boost a passing spacecraft and send it to another destination.

One of the most ardent—and controversial—champions of the feasibility of interstellar travel is Carl Sagan. His main argument was formulated while he was at the University of California at Berkeley and was presented to the American Rocket Society in 1962. He sought to show not only that such travel is possible, but that, if so, "other civilizations, aeons more advanced than ours, must today be plying the spaces between the stars."

He argued that radio waves are but a poor way to achieve a meeting of the minds between beings with utterly different histories and ways of thought. Furthermore, the radio does not permit contact between an advanced society and one that is intelligent but not yet in possession of radio technology. Nor does it allow the exchange of artifacts and biological specimens.

"Interstellar space flight sweeps away these difficulties," Sagan wrote with typical enthusiasm. "It reopens the arena of action for civilizations where local exploration has been completed; it provides access beyond the planetary frontiers, where the opportunities are limitless."

He argued that the fuel problem could be overcome by scooping up hydrogen en route and using it to power an "interstellar ramjet." The conventional ramjet was used for high-speed vehicles, such as the Bomarc missile, that scoop air into a narrowing duct, thus compressing it. The air is used to burn the fuel, and leftover gases are ejected out the rear. A characteristic feature of the ramjet is that the faster it goes, the more efficient it becomes, since the speed tends to increase the pressure differential between the scooped-up air and the surrounding atmosphere.

The interstellar ramjet was proposed in 1960 by R. W. Bussard, who, like Froman, was associated with the Los Alamos Scientific Laboratory. He explained that his engine would be a rough analogue of a conventional ramjet. It would scoop up interstellar material, which is almost entirely hydrogen, using a large part of it as fuel for a fusion reactor. The leftover matter would be squirted out the rear, providing an integral part of the propulsion system. Bussard cited the 21-centimeter observations by the Dutch at Leiden revealing the presence in various parts of the galaxy of great clouds of ionized hydrogen that could be collected magnetically. An intake area almost 80 miles in diameter would be required, he said, to achieve the needed velocity for a space ship of about 1,000 tons. "This is very large by ordinary standards," said Bussard, "but then, on any account, interstellar travel is inherently a rather grand undertaking . . ."

Sagan conceded that the problem of scooping up enough hydrogen was staggering. Amplifying Bussard's calculations, he found that in ordinary interstellar space, with only one hydrogen atom per cubic centimeter, the sweeping system would have to be 2,500 miles in diameter. However, within hydrogen clouds, where the density may be one thousand times as great, the intake, said Bussard, could be less than 80 miles wide. Perhaps, said Sagan, "starships" dart from one such cloud to another. Furthermore, he said, there may be some way to ionize material in the nonionized clouds that predominate in space, so that it can be collected magnetically. Or the starship could select paths through clouds of material that is already ionized. By magnetic techniques it may also be possible to divert particles from the passenger area, thus overcoming the radiation hazard without heavy shielding. The huge vehicle would have to begin its journey, Sagan said, with the aid of fusion-powered rocket stages. The ramjets would be used only when clear of the earth.

In his talk to the Rocket Society, Sagan said his argument was designed to "lend credence" to the possibility that interstellar vehicles may become feasible for us "within the next few centuries." We can expect, he added, "that if interstellar spaceflight is technically feasible—even though an exceedingly expensive and difficult undertaking, from our point of view—it will be developed."

He carried the possibilities of time-slowing even further than Purcell and von Hoerner. In continuous acceleration or deceleration (that is, if the engine was kept on), journeys that reached as far into the universe as galaxies millions of light years away would still be possible in the lifetime of the passengers, even though it was questionable whether any civilization would exist on their return. As noted earlier, from the earthly point of view, time on such an extended journey would virtually come to a halt. In fact, Eugen Sänger, head of the Institute of Jet Propulsion Physics at Stuttgart, Germany, had calculated that with an acceleration no greater than that of the earth's gravity, even the most distant parts of the visible universe could be reached within forty-two years, spaceship time. The incentive, said Sagan, would be greater for journeys to nearby solar systems. Even then, those on the home planet would have to wait perhaps hundreds or thousands of years for the return of their astronauts. It must, therefore, be assumed that a highly advanced society would also be stable over very long periods, preserving the records of previous expeditions and waiting patiently for the return of others. According to this hypothesis civilizations throughout the galaxy probably pool their results and avoid duplication. There may be "a central galactic information repository" where knowledge is assembled, making it far easier for those with access to such information to guess where, in the galaxy, newly intelligent life is about to appear—a problem very difficult for us, with only our own experience on one planet to go by.

Sagan likened the "spacefaring" societies to those of the European Renaissance that sent voyagers eagerly in search of new worlds on our own planet. He suggested that such societies might send out expeditions about once a year and, hence, the starships would return at about the same rate, some with negative reports on solar systems visited, some with fresh news from some already-known civilization. "The wealth, diversity and brilliance of this commerce," he said with an exuberance reminiscent of Tsiolkovsky, "the exchange of goods and information, of arguments and artifacts, of concepts and conflicts, must continuously sharpen the curiosity and enhance the vitality of the participating societies."

On the assumption that there are about a million worlds in the galaxy capable of such feats, Sagan proposed that they would visit one another about once in every thousand years and that scouts may have visited the earth from time to time in the past—perhaps a total of 10,000 times over the full span of the earth's history. One or two million years ago such visitors would have observed the emergence of primates ancestral to man and may have decided to step up the frequency of their visits to once every thousand years, Sagan said.

Is it possible, he asked, that they have visited us since the dawn of civilization? Without making direct reference to the "flying saucer" episodes of recent years, he dismissed such tales by noting that in the past few centuries, "when critical scholarship and non-superstitious reasoning have been fairly widespread," there have been no reliable reports of a visitation. However, he urged that myths and legends be reexamined for indications that such may have occurred in the distant past. He cited, for example, a suggestion in the Soviet Union that *The Book of the Secrets of Enoch* may be based on an instance in which a resident of earth was taken home by visitors and then returned to tell his bewildered countrymen about his adventures. The book, also called the *Slavonic Enoch*, is one of those, known as pseudepigrapha, which were purportedly written by biblical figures but are not accepted as such.

In it, Enoch tells that there appeared to him in a dream "two men very tall, such as I have never seen on earth. And their faces shone like the sun, and their eyes were like burning lamps; and fire came forth from their lips. Their dress had the appearance of feathers; their feet were purple, their wings were brighter than gold; their hands whiter than snow. They stood at the head of my bed and called me by my name. I awoke from my sleep and saw clearly these men standing in front of me."

They told him they had been sent by God to bring him to heaven. Then, he continued, the two beings "took me on their wings and placed me on the clouds. And lo! the clouds moved. And again (going) higher I saw the air and (going still) higher I saw the ether, and they placed me in the first heaven."

In this manner Enoch visited a succession of seven heavens, observing a multitude of wonders, including flying creatures with the feet and tails of lions and the heads of crocodiles. In the seventh heaven he met God and was instructed in the secrets of nature and of man. He was told how the earth was formed and the secrets of the planets and stars. All of this he wrote down in 366 books and, when he returned to earth, sought to impart his wisdom to his fellow men.

Somewhat similar is an account in the Old Testament book of Ezekiel. It tells how "a whirlwind came out of the north, a great cloud, and a fire infolding itself." Out of it came four creatures, each with four faces and four wings. One face was human, the others were those of an ox, a lion and an eagle. With them were great wheels, the creatures and the wheels moving in strange ways.

Another example of an ancient legend that Sagan thought should be examined critically appears in the report by Berosus of the manner in which civilization came to Sumeria—perhaps the most ancient of all civilized societies. Berosus was a Babylonian priest and historian who lived about 280 B.C. His accounts of the flood and the creation, which we have only at second hand, are very like those of the Old Testament. In the early days, he said, the people of that ancient land between the Tigris and Euphrates rivers "lived without rule and order, like the beasts of the field." Then, he said, there arose from the sea "an animal endowed with reason, who was called Oannes." Its body was like that of a fish, but with a second head beneath the fish's head and with feet on its tail like those of a man. The creature, who was able to speak, returned nightly to the ocean, being amphibious.

> This Being, in the day-time [according to the account], used to converse with men; but took no food at that season; and he gave them an insight into letters, and sciences, and every kind of art. He taught them to construct houses, to found temples, to compile laws, and explained to them the principles of geometrical knowledge. He made them distinguish the seeds of the earth, and showed them how to collect fruits. In short, he instructed them in everything which could tend to soften manners and humanise mankind.

Sagan later expressed doubt that such accounts were based on visits from another world, but said that since the legends of primitive peoples have described encounters with a superior civilization like our own, this was an incentive to search the records for evidence of a more exotic visit. He also considered it "not out of the question" that relics of such visits may be found and suggested that a hidden base may be discovered, perhaps on the far side of the moon, placed there for use by succeeding expeditions. A location such as the moon would be used, he said, lest the base be destroyed by weathering during the many centuries between visits and to avoid meddling by inhabitants of our planet. When high-resolution photographs are made of the moon, as a prelude to landing men there, he said, the possibility of such a base should be kept in mind. A monument on the moon is, in fact, featured in the movie *2001: A Space Odyssey*, produced in 1968 with Arthur C. Clarke and Stanley Kubrick as screenwriters. It was, they said, a monolith left on the moon by visitors 4 million years earlier.

Frank Drake proposed, instead, that such early visitors might have left artifacts for us to find—perhaps as a first step in establishing contact. To preserve such clues from the workings of time and from tampering by primitive inhabitants, they might have been buried in limestone caves, Drake said. Such caves would clearly attract the attention of archaeologists. The artifacts might be tagged with radioactive isotopes whose artificial origin would be evident to any sophisticated investigator. The cache, Drake said, "would then remain invisible until radiation detectors were developed."

Drake, however, has remained champion of radio communication. He has calculated that to transmit one pulse—that is, one "bit" of information—1,000 light years by radio, would cost only five cents, and few of those who have explored the problem have accepted Sagan's argument in favor of travel. They consider it a grossly uneconomical method to search for intelligent life, even though it may be the only way to discover societies before they acquire an advanced technology. If beings in other worlds have learned to live long or to hibernate, they may be able to travel between neighboring solar systems by coasting for a few centuries, instead of constantly accelerating or decelerating, although in so doing they would forfeit the time-slowing effect. But one wonders, despite Sagan's eloquent arguments, whether such trips, with all of their cost and discomfort, are necessary.

Biological specimens from another world would be of great interest, but it seems at least possible that they could be replicated by means of radio signals. The human egg cell, the size of a grain of dust, contains a mass of long-chain molecules on which are coded the information needed to construct a human being. The information is voluminous—there are presumably billions of "bits" coded into the egg—and to send it all by radio, particularly at the slow rate that may be necessary for interstellar distances, would be tedious. But by the time we can build interstellar ramjets our communications ability will certainly be much improved. We should be able to tell "them" how to build a car, a cow, a rose and a man. Nor, if radio were used, would anyone have to worry over the possibility that an arriving space ship might be carrying germs dangerous to the world being visited. Perhaps, in fact, there are galaxy-wide immigration laws forbidding travel in person.

Although even some of the more open-minded scientists shrug

off Sagan's arguments, citing the problems and limitations, the question arises: Are we arrogant to believe that we understand these limitations? Are there pertinent phenomena or peculiarities of nature unknown to us? The reasoning of men like Purcell is based on physical laws that seem immutable. We have no reason to suspect there is anything wrong with them. Yet it is well to call to mind the foresight of Benjamin Franklin, when he wrote to Joseph Priestley, the discoverer of oxygen, in 1780:

> It is impossible to imagine the height to which may be carried, in a thousand years, the power of man over matter. We may perhaps learn to deprive large masses of their gravity, and give them absolute levity, for the sake of easy transport. Agriculture may diminish its labor and double its produce; all disease may by sure means be prevented or cured, not excepting even that of old age, and our lives lengthened at pleasure even beyond the antediluvian standard. O that moral science were in as fair a way of improvement, that men would cease to be wolves to one another, and that human beings would at length learn what they now improperly call humanity!

17

UFO's

On June 24, 1947, Kenneth Arnold, a businessman from Boise, Idaho, reported that while flying a private plane near Mount Rainier, he saw a line of flying objects that he likened to "pie plates skipping over the water." Newspaper accounts called them "flying saucers." Four days later an Army Air Force's F-51 pilot said that near Lake Meade, Nevada, he saw five or six circular objects off his right wing. That night, near an Alabama air force base, several officers reported seeing in the sky a bright light that zigged and zagged at high speed. From White Sands Proving Ground in New Mexico it was reported that a pulsing light crossed from horizon to horizon in half a minute. Popular excitement soared. A month after the Arnold sighting *Life* magazine carried an article: "Flying Saucers Break Out Over the U.S."

On September 23 the Army chief of staff, General N. F. Twining, wrote to the commander of the Army Air Forces (converted later that year into a separate agency, the United States Air Force), noting the reports of "flying discs" and ordering a detailed study. He suggested various possibilities, including natural phenomena or aircraft being secretly developed by some other agency. Another alternative, he said, was "the possibility that some foreign nation has a form of propulsion possibly nuclear, which is outside our domestic knowledge."

A study of the reports had already begun at the Army's Air Technical Intelligence Center at Wright-Patterson Air Force Base near Dayton, Ohio. In response to General Twining's command, it established a special program, Project Sign. Reports continued to come in, some hoaxes or fabrications by those seeking to sell their accounts, but many by responsible citizens. Nevertheless, it became clear that many in the Air Force did not take them very

"I assure you, Madam, if any such creatures
as you describe really existed, we would be the first to know about it."

Drawing by Alan Dunn; © 1966 The New Yorker Magazine, Inc.

seriously. In February 1949, the study was renamed Project Grudge, and Sign issued its final report. Among its recommendations was that future activity on this project "should be carried on at the minimum level necessary to record, summarize and evaluate the data received on future reports and to complete the specialized investigations now in progress."

In August 1949, Grudge reported on 244 incidents, drawing heavily on the advice of Dr. J. Allen Hynek, an astronomer at Northwestern University. It estimated that 32 percent were astronomical, such as sightings of Venus, 12 percent were thought to be balloons, such as those released by the Air Weather Service or by Air Force researchers, 33 percent were classed as hoaxes or reports too fragmentary to evaluate, and 23 percent were of "unknown" origin.

"There is no evidence [said the report] that objects reported upon are the result of an advanced scientific foreign development;

and therefore they constitute no direct threat to the national security. In view of this it is recommended that the investigation and study of reports of unidentified flying objects [the term UFO had now come into vogue] be reduced in scope." The report was classified "secret" (and remained so for three years) but it recommended that two of its conclusions be made public, namely that the incidents constitute "misinterpretation of various conventional objects" or "a mild form of mass-hysteria and war nerves." The report attributed public apprehension largely to "sensationalistic reporting of many of these incidents by the press and radio." At that time the Cold War was intensifying and the Soviet Union was believed to be developing nuclear weapons and long-range missiles. Fabricating a flood of UFO reports, said the report, could produce panic that could be used "by or against an enemy."

This did not stop the UFO reports. In 1952 *Popular Science* carried three articles on the subject. There were five in *Time*, three in *Newsweek*, one each in *The New Yorker*, *Collier's*, *The New Republic* and *Reader's Digest*. An article by Donald E. Keyhoe titled "The Flying Saucers Are Real" had appeared in *True* magazine in January 1950, and he became director of NICAP, the National Investigations Committee for Aerial Phenomena, a nationwide organization that took UFO's seriously. In that same year Commander R. B. McLaughlin, USN, published in *True* "How Scientists Tracked Flying Saucers." He argued that one, sighted from White Sands Proving Ground, "came from another planet, operated by animate, intelligent beings."

A radar observation by the Army Signal Corps at Fort Monmouth, New Jersey, seemed to show an object traveling much faster than any known aircraft. This proved to be a misinterpretation of the radar, but meanwhile Project Grudge, which had been suspended, was reactivated. Its subsequent status reports, numbered 1 through 12, were classified either secret or confidential and, although later declassified, did not become generally available until published by NICAP in 1968. NICAP had joined APRO, the Aerial Phenomena Research Organization in Tucson, Arizona, as a relatively responsible nationwide association of researchers. Grudge changed its name to Project Blue Book in 1952, and by 1965 10,147 sightings had been reported of which it classed 6 percent as "unidentified." By then the project had become a low-priority effort at Wright-Patterson Air Force Base, with a three-person staff (a lieutenant colonel, sergeant and secretary). The annual number of reports received was highly variable, ranging from 169 in 1951 to 1,501 in 1952.

For a time some elements of the government took the problem more seriously, and in 1953 the Central Intelligence Agency decided to sponsor its own assessment, citing the possibility that a hostile power might swamp defense communications with a deluge of UFO reports. It convened a panel headed by H. P. Robertson, a mathematician and cosmologist at Caltech, to examine the reports. While CIA representation on the committee has never been made public, the other members were top scientists. They included Luis W. Alvarez of the University of California, a Nobelist in physics, who, at the Massachusetts Institute of Technology, had helped develop a radar that enabled planes to land in a fog. Another was Samuel A. Goudsmit, codiscoverer of electron spin. Lloyd Berkner was head of Associated Universities, which ran the Brookhaven National Laboratory and the National Radio Astronomy Observatory, where he was an enthusiastic supporter of Drake's search for extraterrestrial signals. Thornton Page was professor of astronomy at Wesleyan University in Middletown, Connecticut, and emerged as one of the few astronomers who urged that UFO's be discussed more widely by scientists. He coedited, with Carl Sagan, a 1972 book titled *UFO's, A Scientific Debate.* The panel has been criticized by UFO enthusiasts for having met for only a week, but its members had many other obligations and meager material to work on.

Its report, long secret, found no evidence of an extraterrestrial source, but it said:

> The continued emphasis on the reporting of these phenomena does, in these parlous times, result in a threat to the orderly functioning of the protective organs of the body politic. We cite as examples the clogging of channels of communication by irrelevant reports, the danger of being led by continued false alarms to ignore real indications of hostile action, and the cultivation of a morbid national psychology in which skillful hostile propaganda could induce hysterical behavior and harmful distrust of duly constituted authority.

It recommended to the CIA that "the national security agencies take immediate steps to strip the Unidentified Flying Objects of the special status they have been given and the aura of mystery they have unfortunately acquired." The CIA, with its assigned focus abroad, seems to have been reluctant to disclose its role in what was essentially a domestic problem and, for thirteen years, the report was kept secret. Rumors of such secret studies fueled the popular belief that the government knew that the earth was under surveillance, but did not wish to alarm the public.

UFO reports continued to proliferate. Some people said they had been abducted and carried off into space. The Library of Congress, having in 1968 compiled a 401-page UFO bibliography, said (without assessing reliability of the reports) that UFO's "have been tracked on military radar, and they have been photographed by astronauts, leading many to believe that they originated in outer space. They have been filmed by rocket-borne cameras and touched by farmers. They have been adored by simple people, denied by scientists, prayed to by the devout, cursed by primitives and celebrated by poets."

Finally, in 1966, the Air Force, under pressure to "do something," contracted with the University of Colorado to conduct an independent investigation. It was headed by Edward U. Condon, former president of the American Association for the Advancement of Science, head of the American Physical Society and of the National Bureau of Standards. Then in his mid-sixties, Condon had the build of a football halfback and, while a bit old for the game, was known in scientific matters to lower his head and charge the line. He could brook no nonsense, and, while trying to keep an open mind, he probably never had any serious suspicion that UFO's were extraterrestrial.

His team, including experts in radar, photo analysis, optics, astronomical phenomena and other related subjects, worked for eighteen months, making numerous field trips to UFO sites. The Stanford Research Institute in Palo Alto, California, analyzed reports of unusual radar behavior. The Ford Motor Company studied the magnetic patterns of cars that had allegedly stalled or suffered electrical failure as a UFO flew past.

In the resulting 1,485-page report fifty-nine cases were analyzed in detail from the several hundred that had been reviewed. Several seemed to defy explanation, including two involving radar. At Lakenheath Air Base in England two RAF fighters were sent aloft to investigate radar targets moving at several thousand miles an hour. A UFO got on the tail of one fighter and stayed there despite all the fighter's evasive maneuvers. In Colorado air traffic control radar detected an object following a Braniff airliner landing at Colorado Springs. As the airliner landed the object continued on, but could not be seen from the control tower.

Despite such puzzles, the study concluded that further study of UFO's had little merit, and it recommended that Project Blue Book be terminated. I had visited the Blue Book office earlier that year (1968) and been shown approximately seventy reports from a particularly interesting episode (or UFO "flap") that had occurred earlier that year:

At about 9:45 P.M. Eastern Standard Time on March 3, 1968, a woman in Indiana saw a procession of fiery objects fly overhead. "Two or three minutes later," she reported, "my cousin, my aunt and my uncle came running into the house and yelling and trying to tell me about the UFO they saw. It was at about treetop level and was seen very very clearly and was just a few yards away. All of the observers saw a long jet airplane looking vehicle without any wings. It was on fire both in front and behind. All observers also saw many windows in the UFO."

A woman in Ohio told how her dog, when the same UFO came over, lay between the garbage cans in her driveway and whimpered "like she was frightened to death." She herself had "an overpowering drive to sleep."

From Tennessee a woman reported seeing a craft with squarish windows and what seemed a riveted metal fuselage. It made no sound, which further terrified her. Other reports from motorists said the object chased them, changing course every time they turned (an illusion common to drivers who, at night, think a star or planet, which is really stationary, follows their twists and turns). One report attributed seventy-two grass fires that occurred in one area to the UFO, and a science teacher with a doctoral degree said he tried to signal to the UFO with a flashlight.

In-flight reports came from an Eastern Airlines pilot over Connecticut, a United Airlines flight over Indiana, an American Airlines flight over Pennsylvania, a Piedmont Airlines pilot over Virginia and an Air Canada pilot north of Toronto. This led air traffic controllers at centers in Indianapolis and New York to search their radars, but no unidentified objects below 60,000 feet could be detected. Most of the pilots were perplexed at what they thought was a formation of several craft only a few thousand feet above them. Two, however, guessed that what they were seeing were sections of a rocket disintegrating as it reentered the atmosphere. This was confirmed by the North American Air Defense Command, which said they were fragments of *Zond 4*, a Soviet craft launched the day before, which fell out of orbit, either by design or inadvertently.

The episode was a prime example of the frailties of human observation, so often exploited by magicians. The body of the "aircraft," the rivets on its hull, its nearness were all products of imagination and earlier conditioning. I had planned a book of my own on UFO's, assessing the limits and tricks of human perception, as well as the role of conditioning. Anyone in politics or advertising knows how one's interpretation of events can be influenced. Condon had agreed to let me use his files, as the investigation progressed, on condition that my book appear after his report. Condon's report, however, concentrated primarily on material evidence, rather than psychology. Instead I wrote the introduction to the popular edition of the report and my writing turned in other directions.

The report concluded that "nothing has come from the study of UFO's in the past twenty-one years that has added to scientific knowledge" and further extensive study "probably cannot be justified in the expectation that science will be advanced thereby." It noted that the Air Force had repeatedly concluded that UFO's did not constitute any threat to the national security and said it had found nothing to alter that view.

Early in the project the two leading civilian groups investigating UFO's, NICAP and APRO, cooperated, alerting the investigators to new incidents and helping document earlier ones. There were then rumors that Condon did not take seriously the extraterrestrial hypothesis, and relations with those groups cooled. James E. McDonald at the University of Arizona, a leading UFO protagonist, argued that such objects are one of the greatest scientific puzzles of our time and that visitations from afar are the best explanation for those that cannot otherwise be explained.

In response to such challenges to the Condon report the Air Force asked the National Academy of Sciences to review it. The findings of the resulting panel, headed by Gerald M. Clemence of Yale University, were delivered on January 8, 1969, with a covering letter by Frederick Seitz, president of the Academy. He noted that "substantial questions" had been raised as to the adequacy of the Condon research, but the panel was "unanimous in the opinion that this has been a very creditable effort to apply objectively the relevant techniques of science to the solution of the UFO problem.''

UFO reports continue to appear, and for many years Allen Hynek argued valiantly that they deserve more attention, but few scientists agreed with him. Even for the professional it is hard to envision the vastness of the distances to the nearest stars, much less those likely to shine on inhabited worlds. If those distances are ever crossed by intelligent beings, the evidence for their arrival would not be ambiguous. It would be one of the greatest events in human history and not easily hidden. There is no convincing evidence that it has occurred.

18

Where Are They?

In the 1970s and early 1980s the above question became the rallying point for SETI skeptics, leading to a revolt against the early exuberance and optimism of the search. If the galaxy is teeming with highly advanced civilizations able to cross interstellar distances, where are they? Why haven't we been visited?

The question was first asked by Enrico Fermi in about 1950 during a lunch with three fellow physicists at the Los Alamos National Laboratory. All had been associated, in various ways, with the nuclear weapons that had been developed there, and in the mid-1980s, Eric M. Jones of Los Alamos wrote to the three surviving members of that group (Fermi was deceased), asking what had led up to Fermi's question. Jones had himself proposed that a space-faring civilization, having developed vehicles that could travel at one hundredth the speed of light, could colonize the entire galaxy within 5 million years—a short time relative to the history of the earth. Fermi's three companions on that walk had been Emil Konopinski, Edward Teller and Herbert York.

In their letters to Jones all agreed that they had been en route to a Los Alamos lodge for lunch, discussing unidentified flying objects, which were then much in the news. The first "flying saucer" report had appeared about three years earlier. While the physicists dismissed the idea that these had anything to do with visitations from space, Fermi asked his companions how long they thought it might be until our technology found a way to exceed the speed of light. Konopinski said a million years, whereas Fermi thought it might come much sooner. After they had sat down to lunch, Teller said in his letter to Jones, the conversation had turned to more sober scientific subjects when Fermi suddenly said: "Where is everybody?" It was an abrupt change of subject, said

York, but "we all knew he meant extra-terrestrials." York's memory was hazy, but as he remembered it Fermi said the reason we hadn't been visited "might be that interstellar flight is impossible, or, if it is possible, always judged to be not worth the effort, or technological civilization does not last long enough for it to happen." It is said that the Hungarian physicist Leo Szilard, a participant in the bomb project and known for his impish humor, later remarked that he knew where "they" were: "They are among us, but they call themselves Hungarians."

One of the first to argue that we are, in fact, alone in the galaxy was Michael H. Hart at the National Center for Atmospheric Research in Boulder, Colorado (Hart later joined Trinity University in San Antonio, Texas). In a 1975 issue of the *Quarterly Journal of the Royal Astronomical Society* he cited the long, fruitless efforts to detect any distant signs of life. Yet he proposed that a highly advanced civilization was virtually certain to become capable of interstellar travel and colonize the galaxy. Before too long our own civilization should be similarly capable. If, then, the galaxy was full of far older civilizations, why have none visited us? This, Hart wrote, "can best be explained by the hypothesis that there are no other advanced civilizations in our Galaxy." He cited the various exotic proposals for interstellar propulsion, some of which had been described in Project Daedalus, initiated by the British Interplanetary Society in 1973 to study the feasibility of sending an unmanned probe to the nearest star, using the technology likely to be available by the year 2000. Such factors as a greatly extended life span, the use of suspended animation or freezing might make possible extremely long journeys, said Hart. Yet none have reached us. He cited various explanations, including the idea that our civilization has been quarantined by the galactic club as a "zoo," preserved as a curiosity or for fear of contamination by our perverse ways.

The zoo hypothesis had been proposed in 1973 by John A. Ball, a Harvard-Smithsonian radio astronomer, in *ICARUS, the International Journal of Solar System Studies.* After citing the puzzling absence of visitors, Ball said: "Occasionally we set aside wilderness areas, wildlife sanctuaries, or zoos in which other species (or other civilizations) are allowed to develop naturally, i.e., interacting very little with man. The perfect zoo (or wilderness area or sanctuary) would be one in which the fauna do not interact with, and are unaware of, their zoo-keepers." Advanced civilizations, he wrote later, "know that we're here but they don't care; they're ignoring us. We pose no threat and we have nothing that they want. . . . We are average, ordinary, and perhaps uninteresting. Should this answer prove true, it would deal another blow to our anthropocentric ego."

In his presentations Ball cited the explanations for no visitors proposed by radio astronomer Sebastian von Hoerner:

(a) Life and intelligence are very rare.

(b) Technology is exciting only during a short period, being surpassed by other activities.

(c) Science and technology cause severe crises (population explosion, self-destruction, genetic degeneration).

(d) Maybe there *are* many signs of life which we have not looked for or not understood.

It could be, von Hoerner wrote, that civilizations, to survive such crises, "have developed such a high degree of stabilization and regimentation that they will merge into the final crisis of stagnation (the crisis to end all crises) which even can be irreversible.

"Is this the reason," he continued, "why we do not see any sign of life and interstellar activity, among the billions of stars and galaxies we observe?" Interstellar communication, he proposed, could reawaken such a stagnating civilization.

Hart's own explanation was that since "they" are not here, "they do not exist." An extensive search for radio messages "is probably a waste of time and money," whereas "in the long run, cultures descended directly from ours will probably occupy most of the habitable planets in our Galaxy." His argument was supported by Eric Jones at Los Alamos, citing the 1974 proposal of Gerald K. O'Neill of Princeton University, an innovative developer of space technologies, that construction of space colonies was virtually inevitable. The logical place for the first one would be the L5 position in the moon's orbit. Colonies would then propagate from there to other parts of the solar system and finally to distant stars.

L5, or Lagrangian Liberation Point Number 5, is a region trailing the moon by one-third the circumference of the lunar orbit. It is one of the points in the earth-moon system in whose vicinity the gravitational attraction of both bodies is in equilibrium.

Such points had been identified in the orbit of Jupiter by the eighteenth-century French-Italian theorist Joseph-Louis, comte de Lagrange, who calculated that five of them must be produced by the combined gravitational fields of the sun and Jupiter. His prediction that small planets might have been trapped at two of them, L4 and L5, was confirmed 134 years later with the discovery of what are now known as the Trojan asteroids, each orbiting one of those points. After O'Neill's proposal a group supporting colonization formed what they called the L5 Society.

The drive for colonization, Dyson, von Hoerner and others believed, was uncontrolled population explosion. Instead of massive starvation, as predicted by Malthus, it would lead to colonization of other planets and planetary systems. Michael D. Papagiannis of Boston University, who played a key role in creation in 1984 of the special commission on SETI in the International Astronomical Union and became its first president, proposed that a "colonization wave" would sweep through the galaxy. It would, he estimated, advance 10 light years every 1,000 years. At a 1991 Russian-American conference on SETI at the University of California at Santa Cruz, William A. Newman of the University of California at Los Angeles estimated at least 1 billion years to populate the galaxy, which is probably more than 10 billion years old.

Michael A. G. Michaud, former deputy director of the Office of International Security Policy in the Department of State and a SETI enthusiast, has argued that only by colonization can we ensure that intelligence will endure. Since the lifetime of the earth and solar system are limited, he has written, "intelligence must expand its influence on the universe and thus assure its survival."

Michaud demonstrated a certain familiarity with science fiction, speculating on what might be done to avert destructive collapse or dissipation of the universe. The "organized intelligences of the universe," he said, might try to avoid this "by isolating controlled regions of the universe from the rest of space-time and universal evolution, by transferring themselves to another point in time, or by escaping this universe, perhaps to another, younger one."

Newman, on the other hand, argued that the absence of visitors has little meaning, since it depends on a succession of questionable assumptions. He summarized these as follows: "*If* extraterrestrial life is abundant and *if* space travel is relatively easy and *if* advanced civilizations will feel 'compelled' to explore the galaxy and can do so successfully, and *if* they have had enough time to do so, and *if* we have tried hard enough to find them, *then* shouldn't we see evidence of extraterrestrial life?" Absence of evidence for extraterrestrial life, he said, "should *not* be construed as evidence of absence."

Hart reaffirmed his argument at the 1979 General Assembly of the International Astronomical Union in Montreal, proposing that estimating a large number of civilizations based on the combination of probabilities devised by Frank Drake is deceptive, because so many of those factors are based almost entirely on unfounded speculations. At this time Shklovsky, the Russian who had written the book on SETI that was enlarged by Carl Sagan and played a seminal role in arousing world interest in SETI, had decided that the chances of finding and communicating with such civilizations were remote. He, too, argued that we should set colonization as our goal.

In the same year as the IAU meeting in Montreal, Hart and Ben Zuckerman of the University of Maryland, who had himself conducted searches, organized one in College Park, Maryland. Its proceedings were published in a 1982 book reviewed in *Science* by Harlan J. Smith, head of the McDonald Observatory in Texas. Smith found the book "quite one-sided," noting that several of those figuring most prominently in SETI, such as Carl Sagan and Frank Drake, had not participated. He wrote:

> Galactic civilizations, of the type that survive suicidal tendencies and develop interstellar travel, may feel no need whatever to latch onto all the real estate in the galaxy, or any more desire to interfere with indigenous life forms than we have to catch every butterfly. . . . Are we alone? We just don't know. We are unlikely to know unless we look long and hard—and maybe not even then. But either way the implications are utterly profound. If we are not alone in the galaxy, can we hope someday to qualify as members of the galactic club despite our human frailties and foolishnesses? If we are alone, we are responsible for keeping this fragile spark alive and offering what should be the blessing of sentient life to countless descendants down through the eons.

Hart was joined in his opposition to SETI by Frank J. Tipler, a mathematician at the University of California at Berkeley. In 1979 Tipler sent his paper "Extraterrestrial Intelligent Beings Do Not

Exist" to *Science*, which submitted it to at least two referees, one of whom was Carl Sagan. Tipler sent me Sagan's commentary, which, as expected, was far from enthusiastic. It began: "This is an intriguing, provocative, maddening, and deeply flawed paper." It finally appeared a year later in the *Quarterly Journal of the Royal Astronomical Society*.

A major element of Tipler's thesis was that a supercivilization would produce the "Self-Reproducing Automata" envisioned by John von Neumann, the great mathematician and pioneer in automation. "Von Neumann machines" tolerant of long space journeys and, on arrival, capable of doing even more than human beings, should long ago have populated the galaxy, Tipler said. They should have penetrated the solar system and made themselves known in unmistakable ways. One, he suggested facetiously, was constructing a "Drink Coca-Cola" sign a thousand miles wide in a conspicuous orbit. The machines would be programmed to assuredly remain under control of the civilization that made them, lest they turn against their makers.

A counter argument that civilizations, assuming they existed, should have spread through the galaxy was presented by Freeman Dyson of the Institute for Advanced Study. At the 1991 Russian-American conference on SETI in Santa Cruz he proposed "that any community of creatures adapted to living freely in the vacuum of space will spread and speciate in the Galaxy, just as life has spread and speciated on this planet. One intelligent species let loose in space may become a million intelligent species within an astronomically short time." Dyson revived his proposal of more than thirty years earlier that regions of the galaxy emitting strongly in the infrared were an especially promising place to look for such civilizations. He did not find the absence of signals surprising. "Even if the galaxy is teeming with intelligent species," he had said earlier, "I still have doubts concerning the likelihood that any of them will transmit intelligible messages for our instruction."

Tipler, in his paper, said that even though no visitors, living or mechanical, have been observed, there is a strong emotional yearning to believe in extraterrestrial life. He equated it with the belief in UFO's and the hope that they will "save us," citing a factor cited by Sagan in his contribution to a book on UFO's. "The expectation," he said, "that we are going to be saved from ourselves by some miraculous interstellar intervention works against the necessity for us to solve our own problems." After SETI, in 1984, had been given special status as Commission 51 in the International Astronomical Union, Tipler argued that SETI was not a science and should not be so dignified.

Almost coincident with Tipler's paper was a book attacking SETI by Robert T. Rood, an astronomer, and James S. Trefil, a physicist, both at the University of Virginia. They emphasized the long procession of unlikely developments in the evolution of the earth and its inhabitants, proposing that in each case, had things gone a little differently, we would not be here. Rood clung to the possibility that at least a few other civilizations might exist in our galaxy, whereas Trefil argued against even that. They did not, however, rule out the possibility of life elsewhere among the billions of other galaxies and agreed that the search for extraterrestrial intelligence should continue.

Finally, in 1985, an attack on SETI was published by Ernst Mayr of the Museum of Comparative Zoology at Harvard University, which reflected many of the arguments made two decades earlier by his predecessor George Gaylord Simpson, the authority on vertebrate evolution. Like the other doubters, including Jacques Monod, Mayr's basic argument was that so many improbable events led to the origin of life and our evolution that they had probably occurred nowhere else, making us unique: "There is, indeed, the probability that the combination and sequence of conditions that permitted the origin of life on Earth was not duplicated on a single other planet in the universe." To claim this was true, he said, was unscientific, since it could not be demonstrated, but, he added, "the claims of the proponents of extraterrestrial life and intelligence are equally outside the bounds of science."

Mayr questioned the argument that intelligence "had" to evolve because of its survival value. Vision of some sort had evolved perhaps forty times in different groups of animals, but, he argued, only a single path led from primitive nervous systems to human intelligence. "The assumption that any intelligent extraterrestrial life must have the technology and mode of thinking of late twentieth century Man is unbelievably naive," he said. The SETI program, he added, "is a deplorable waste of taxpayers' money, money that could be spent far more usefully for other purposes."

Such arguments, based on cumulative improbability, have been attacked by many authors, including Paul Horowitz of Harvard. Even without trying to assess the probabilities of planetary formation, chemical and biological evolution, and the rise of intelligence and technology, Horowitz said, "we can observe that, in all of nature's variety, there is no phenomenon that happens only once." The absence of visitors, he said, may simply mean that aliens would "rather communicate than commute."

When the coolness toward SETI reached Washington, Congress specified that none of NASA's 1982 budget be allocated for the search. Funds remaining from previous years kept the project from dying completely, but preparations had to be made for closing it down. Sagan fought back, lecturing Senator William Proxmire of the Appropriations Committee and circulating a statement that was signed by an impressive array of sixty-eight scientists and other prominent figures from a dozen nations, including seven Nobel laureates and three from the Soviet Union. As published in Science, the statement said that despite arguments on the negative side, "the only significant test . . . is an experimental one. The results, whether positive or negative, "would have profound implications for our view of the universe and ourselves." SETI also gained special international standing as the subject of a daylong session during the 1979 IAU General Assembly in Montreal, five years before it became a special commission of the union. The commission then held a SETI colloquium at Balaton, Hungary, in 1987.

The most prestigious international meeting, sponsored by the Soviet and American academies of science, was held in 1971 at the Byurakan Astrophysical Observatory in Soviet Armenia, where there had been a purely Soviet meeting seven years earlier. Those invited to the international one included such specialists as Marvin Minsky of MIT, a leading authority on artificial intelligence, and specialists in cultural evolution and biology, cryptography and electrical engineering and such well-known SETI enthusiasts as Frank Drake. Francis Crick and Charles Townes, both Nobel laureates, chaired some of the sessions. As noted by one Soviet, those seated around the U-shaped conference table were playing a "game" with hypothetical distant beings. The purpose, he said, was not to beat them but to help them win.

A Soviet radio astronomer pointed out that, to achieve maximum range, another civilization would concentrate its radio signals into an extremely narrow bandwidth and he suggested a multichannel receiver that would break up incoming emissions into increments only 1 hertz wide, anticipating the method used by American astronomers twenty years later. Minsky proposed that once a more advanced civilization had told us how to build and program a computer, it could "talk" to us in a way impossible by coded signals alone.

Ten years later, in 1981, a follow-up meeting was held at Tallinn in what was then the Estonian Soviet Socialist Republic, and there have been several conferences between Russian and American SETI participants, including the Santa Cruz one in 1991. Support by astronomical members of the National Academy of Sciences dates to the Academy's secret sponsorship of the original 1961 meeting at Green Bank. A 1972 Academy survey of the state of astronomy said, in part: "Our civilization is within reach of one of the greatest steps in its evolution: knowledge of the existence, nature, and activities of independent civilizations in space. At this instant, through this very document, are perhaps passing radio waves bearing the conversations of distant creatures. . . . In the long run, this may be one of science's most important and most profound contributions to mankind and to our civilization."

The Academy's next review, a decade later, made many of the same points. SETI "has a character different from that normally associated with astronomical research," the review said, but added that "intelligent organisms are as much a part of the universe as stars and galaxies," and were a legitimate aspect of astronomy. It is hard, the review continued, "to imagine a more exciting astronomical discovery or one that would have greater impact on human perceptions than the detection of extraterrestrial intelligence."

Finally the 1992 Academy review, chaired by John N. Bahcall of the Institute for Advanced Study in Princeton, again endorsed SETI. It said, in part: "Indeed, the discovery in the last decade of planetary disks, and the continuing discovery of highly complex organic molecules in the interstellar medium, lend even greater scientific support to this enterprise."

19

Is There Intelligent Life on Earth?

The question that forms the title of this chapter was pinned on the door of Frank Drake's office in Green Bank, West Virginia, only half in jest, for it proved the crucial unknown to emerge from the conference privately held there in November 1961.

The task of the meeting, cited briefly in the introduction to this book, was to discuss whether it might be possible to contact other worlds. The meeting was organized after the excitement generated by Drake's Project Ozma and the Cocconi-Morrison letter had penetrated the venerable halls of the National Academy of Sciences. Lloyd Berkner, acting director of the National Radio Astronomy Observatory in Green Bank, who had given the go-ahead for Ozma, was also chairman of the Academy's Space Science Board. The board had been formed in 1958, shortly after the launching of the first earth satellites, to set forth national space goals that would be scientifically sound. Its membership included two Nobel laureates, both of whom were deeply interested in problems relating to extraterrestrial life. One was Harold C. Urey, professor-at-large of chemistry at the La Jolla campus of the University of California, whose formidable intellectual energies were being directed toward meteorites, including the perplexing carbonaceous chondrites. The other, Joshua Lederberg, was professor of genetics at Stanford University, chairman of the Space Science Board's Panel on Exobiology, and destined to become a key figure in the search for life on Mars.

It was not surprising, therefore, that the board decided, in 1961, to call an informal conference at the observatory in Green Bank to assess the possibility of communication with other worlds. J.P.T. Pearman of the board staff did the organizing, and, as noted in the first chapter, it was decided not to make any public announcement. Those invited were, to a large extent, the dramatis personae of this book, though not all could come. The ones who did, in addition to Pearman, were:

Otto Struve, director of the observatory, host to the conference and its chairman. His study of slow-rotating stars had convinced him that solar systems like our own are common.

Dana W. Atchley, Jr., president of Microwave Associates, Inc., specialist in communications technology and donor of the parametric amplifier that was a key component of Project Ozma.

Melvin Calvin, who won the Nobel prize in chemistry during the conference, and who had spelled out some of the chemical processes by which life probably evolved from inanimate matter.

Giuseppe Cocconi, who, with Morrison, had proposed a search for signals on the 21-centimeter wavelength.

Frank D. Drake, who independently had initiated a search on 21 centimeters, using the 85-foot antenna at Green Bank.

Su-Shu Huang, who had calculated what types of stars would have habitable zones around them large enough and enduring enough to make them likely abodes for worlds like our own.

John C. Lilly, head of the Communication Research Institute in the Virgin Islands, where he was studying the possibility of communication between man and another rather intelligent species, the dolphin.

Philip Morrison, coauthor with Cocconi of the historic proposal for a search on 21 centimeters.

Bernard M. Oliver, vice president for research and development of the Hewlett-Packard Company and one of those who had studied in some detail the problems of interstellar communication.

Carl Sagan, with Calvin a member of the Panel on Exobiology of the Space Science Board and probably the most enthusiastic of the "exobiologists." As noted earlier, Sagan believed that travel between solar systems may be commonplace.

"The purpose of the discussions," Pearman wrote in his semi-official account, "was to examine, in the light of present knowledge, the prospects for the existence of other societies in the galaxy with whom communications might be possible; to attempt an estimate of their number; to consider some of the technical problems involved in the establishment of communication; and to examine ways in which our understanding of the problem might be improved."

On the first day, the participants were enthralled as John Lilly told of his research into interspecies communication. Only a few months earlier his book, *Man and Dolphin*, had created somewhat of a sensation, and his optimism concerning the prospects of being able to "talk" with the bottlenosed dolphins had evoked some outraged reaction in the scientific community. In short, Lilly was controversial and provocative. Indeed, he subsequently became mystical and interested in Eastern philosophies. But in 1961, to a relatively conventional audience, his credentials seemed good. He had been trained at Caltech, Dartmouth and the University of Pennsylvania, where he was made a Doctor of Medicine and later held twin associate professorships, one in medical physics and the other in experimental neurology. In 1953 he moved to the National Institute of Mental Health as a newly commissioned surgeon in the Public Health Service and became chief of the Section on Cortical Integration in the Institute's Laboratory of Neurophysiology. His scientific respectability was attested by his membership on a number of federal panels, such as the Scientific Advisory Board of the Air Force Office of Scientific Research and the Scientific Advisory Committee of the Graduate School of the National Institutes of Health. His dolphin research was supported at first by the Navy and the National Institutes of Health and later by the Air Force and NASA.

Lilly had long been interested in the brains of seagoing mammals, in part because of their immense size. The brain of a chimpanzee weighs only about one-quarter that of a human being, whereas the brain of a whale may be as much as six times heavier and that of an elephant is up to four times as heavy as a man's. In large measure this is because the brains of big animals are coarser and probably have considerably fewer nerve cells than ours. What was remarkable about the brain of the bottlenosed dolphin, according to Lilly, was not only its size, which is slightly larger than that of an average man, but the fact that, under a microscope, the density of nerve cells seems comparable to that in the human brain. In fact, he said, the cortex or outer layer of the dolphin brain is richer in folds and other structures than the equivalent part of the human brain.

In his book, Lilly sought to document the intelligence of these animals and the complexity of their "language." While they can be taught, in captivity, to vocalize outside the water for exhibition purposes, their natural mode of communication is through water and extends into a frequency range far above that of the human ear. They produce a variety of squeaks, whistles, creaks and other sounds, sometimes in very rapid succession. Lilly could easily recognize their distress call—a pair of whistles—and told of an incident which, to him, suggested the possibility that they have a complex language that can be used to make very specific requests.

The episode involved a dolphin that, during an experiment, apparently had become so chilled that it was unable to swim. Placed back in the main tank with two other dolphins, it sank to the bottom, where it was bound to suffocate unless it could reach the surface to breathe. However, it gave the distress call and the other two immediately lifted its head until the blowhole was out of the water, so that it could take a deep breath. It then sank and a great deal of whistling and twittering took place among the three animals. The two active ones then began swimming past the other so that their dorsal fins swept over its anogenital region in a manner that caused a reflex contraction of the fluke muscles, much as one can make a dog scratch itself by rubbing the right spot on its flank. The resultant action of the flukes lifted the animal to the surface, and the procedure was repeated for several hours until the ailing dolphin had recovered.

Lilly told of other instances in which these animals had come to the assistance of their fellows and of some cases in which they reportedly helped people floundering in the water.

"It is probable," he wrote, "that their intelligence is comparable to ours, though in a very strange fashion." We may be faced, he added, "with a new class of large brain so dissimilar to ours that we cannot within our lifetime possibly understand its mental processes." Instead of having speech centers as human brains do, the dolphin brain "may be doing something else entirely than what we do with our brains." In his day-to-day association with these animals he had observed sufficient humanlike behavior, he said, to encourage subtly his belief "that we shall eventually communicate with them."

Lilly then gave free rein to his imagination in discussing the implications of such an achievement. Dolphins could be used by one government to scout out the submarines of another; they could smuggle atomic bombs into enemy harbors, pick up missile nose cones at sea, help rescue the pilots of downed planes, or serve on underwater demolition teams; all of this being contingent on their loyalty to one side. He recognized the possibility of defection, and realized, too, that they might prove to be pacifists. As a form of psychological warfare, he said, they might be persuaded to sneak up on hostile submarines "and shout something into the listening gear." More valuable, he added, would be the gain in scientific knowledge, both from the experience of learning their "language" and methods of thought and from the information such animals could provide about oceanic life, their own mysterious method of navigation and so forth.

Lilly argued that to achieve a meeting of the minds with another species on earth—a species less evolved than ourselves, at least in the technological sense—was very like the problem of communication with some higher technology in another world. To prove that communication with a species, such as the bottlenosed dolphin, is not possible, he said, would take "a very long time, a lot of research, and the exploration of many possible methods." The parallel with the search for intelligent life elsewhere was obvious.

It was inevitable that Lilly's listeners should wonder: What if another world were entirely covered with water and this global ocean were dominated by a highly intelligent species, like the dolphins? Should we expect to hear signals from them? It seems unlikely. Dry land seems to be a prerequisite for a civilization that can communicate. Life that was purely aquatic could not use fire. As Morrison pointed out, it would be less likely to develop an interest in astronomy, since the creatures would not see the stars except when they poked their heads out of the water, and, without hands, they could not build telescopes. Furthermore, it can be argued that, at least on earth, intelligence appeared in response to the challenges of life on land. The only smart animals of the sea are aquatic mammals whose ancestors presumably were land mammals. The stupidity of fish of comparable size, such as the sharks, is legendary. It would appear that the whales, dolphins and other cetaceans were set on the road to intelligence while living on land.

After Lilly's presentation, Drake, like a true scientist, sought to formulate the conference's central problem as an equation, outlined in this book's introduction. He wrote it on the blackboard as follows:

$$N = R_* f_p n_e f_l f_i f_c \cdot L$$

The idea that it expresses is simple enough. The letter N, on the left, represents the number of civilizations in the galaxy that are currently capable of communicating with other solar systems.

All of the expressions on the right side of the equation are factors affecting this number. When multiplied together they give the number that we wish to know—the number of communicative societies. It was agreed that this number is crucial, for if it is large we can expect to find a civilization in our immediate part of the galaxy, whereas if there are only a handful of such civilizations they are probably separated by tens of thousands of light years; to locate them amid the host of "dead" solar systems would be extremely difficult, and the distance would be so great that only a few messages could be exchanged within a time span comparable to the entire elapsed history of the human race.

The factors appearing on the right side of the equation were as follows:

Factor One (R_*): The rate at which stars were being formed in the galaxy during the period when the solar system itself was born.

This would determine the number of stars in the galaxy near which intelligent life may recently have reached maturity—recently, that is, in the sense of the past few hundred millions of years. The astronomers at the meeting said a conservative estimate would be about one new star per year.

Factor Two (f_p): The fraction of stars with planets.

Here the experience of participants like Struve came into play. If, when stars are formed, the leftover material either coalesces into a twin star or a system of planets, then half of all stars have planets, it then being assumed that half are in two-star systems. If, however, in the formation process the leftover material is sometimes disposed of in other ways, such as forming asteroids or being blown off into space, then the fraction of stars with planets might be as low as one-fifth.

Factor Three (n_e): The number of planets, per solar system, with an environment suitable for life.

This had been investigated at some length by Su-Shu Huang. Although the uncertainties are great, it was proposed that the figure for this galaxy probably lies between one and five.

Factor Four (f_l): The fraction of suitable planets on which life actually appears.

This was Calvin's special province. He and Sagan argued that on such planets, given a time period measured in billions of years, life must sooner or later appear. As Sagan put it in one of his papers, the production of self-replicating systems is a "forced process" which inevitably occurs "because of the physics and chemistry of primitive planetary environments." Hence, the group agreed that this factor was one.

Factor Five (f_i): The fraction of life-bearing planets on which intelligence emerges.

The group was impressed by Lilly's argument that more than one intelligent species has evolved on this planet, although the example of the dolphins also raised the possibility of life that was intelligent but would not become technological.

The most telling argument here seems to be the characteristic pressure of life that drives it into every nook and cranny of the environment where, by some marvel of adaptation or ingenuity, sustenance can be found. As we look about us on the earth we see examples of this on every hand. If there is a way to live by swimming, fins will evolve. If there is a way to live by walking, legs will appear. If there is a way of life in the sky, some animals will develop wings. Mites have found a way to survive on peaks near the South Pole. Algae live in the scalding water of hot springs. In the perpetual night of the oceanic trenches, 6 miles below the waves, or adrift in the high atmosphere, one finds life. The wonderful process of evolution has, over the billions of years, pushed life into every "ecological niche" that one can imagine. As in the free enterprise system, if there is an odd way to make a living, someone will discover it sooner or later and prosper. So far as we can see, the most successful way to live on earth is to manipulate the environment by what we call intelligence. In its early stages this involves the use of tools, plants and domesticated animals to obtain clothing, shelter and food. Eventually it leads to almost complete transformation of the landscape, as is evident today to any airline passenger on our own planet.

The fork in the road that led to intelligence began very far back in the history of life on earth, for even such lowly creatures as the insects have tiny, compact computers no larger than a pinhead that are primitive brains. At least one biologist and one paleontologist—Blum at Princeton and Simpson at Harvard—doubted that intelligence is an inevitable fruit of evolution. In fact, they proposed that if there are any intelligent beings in this galaxy, they are extremely distant. But for a long succession of evolutionary "accidents," according to Simpson, we would be no smarter than a cabbage plant.

In his book *This View of Life*, he conceded that intelligence is "a marvelous adaptation" which has "survival value in a wide range of environmental conditions." It would therefore be favored by natural selection on a variety of planets, once it appeared. However, he said the factors that produced a thinking creature on earth "have been so extremely special, so very long continued, so incredibly intricate that I have been able hardly to hint at them here. Indeed, they are far from all being known, and everything we learn seems to make them even more appallingly unique."

His book was an outcry against the search for life beyond the earth, and yet he drew attention to one of the problems that those more enthusiastic about space exploration hope can be enlightened by such activity. This concerns the extent to which the paths of evolution on earth have been typical of life elsewhere. Even if life on Mars is now extinct, study of its remains, if any, should help throw light on this question.

Simpson's book had not appeared at the time of the Green Bank conference. Of those at that meeting, the one who had probably thought the most about this problem was Philip Morrison, a nuclear physicist by trade, but a voracious reader in a wide variety of fields. He argued that intelligence would always appear, sooner or later, because of "convergence." This is the tendency of species, evolving along highly diverse routes, to converge toward life forms that, because of certain basic laws, resemble one another. He cited, for example, three fundamentally different kinds of animal that evolved into the same shape: one a reptile, one a fish and the third a mammal. The reptile was the plesiosaurus, which became extinct at the end of the Mesozoic Era, some 100 million years ago. It appears to have been a terror of the deep in its heyday. The fish that he cited was the tuna, and the mammal was the dolphin. Although these creatures evolved from ancestors utterly different from one another, their final paths of evolution were determined by the laws of hydrodynamics. There is an ideal shape for bodies 6 feet in length that wish to swim. By the slow, cruel process of mutation and survival of the fittest, each of these branches of the animal kingdom produced a creature that conformed closely to this ideal. Other instances of conversion would be the appearance of eyes in vertebrates, insects and mollusks or the evolution of wings in birds, insects, mammals and fish.

So far as intelligence is concerned, Morrison saw an example of convergence in the appearance, on earth, of two rather intelligent kinds of life: men and the cetaceans (including dolphins). Being mammals, both have a common ancestor that was furry and suckled its young, but it was certainly of very low intelligence. Yet the opportunities that the environment offers to intelligence caused this characteristic to emerge at the end of two quite separate lines of evolution. An added peculiarity of man's evolution, Morrison noted, is that he has eliminated all neighboring species; he stands alone, with no close relatives in the family tree of the primates. Whether the competitors for this ecological niche eliminated their rivals by clubbing them with bats, by collecting food more efficiently or by dominance and intermarriage is unknown. This took place so long ago that there is no clear fossil evidence of what occurred. The fact, however, that this happened on earth suggests that intelligent land animals such as man do not tolerate close competitors.

The conference at Green Bank decided that a factor of one should be assigned to the emergence of intelligence—meaning that it would arise, eventually, on virtually any planet where there is life.

Factor Six (f_c): The fraction of intelligent societies that develop the ability and desire to communicate with other worlds.

Here the conferees felt the need of a sociologist, anthropologist or historian, although, as Morrison pointed out in one of his lectures, even the specialists in these fields lack the knowledge needed to be of much help. "We are trembling on the edges of speculation which our science is inadequate to handle," he said, for we have no adequate theory of the social behavior of complex societies. "Our experience, our history, is not yet rich enough to allow sound generalization." How likely is it, he asked, that populations of men, "or manlike things," would evolve a technology with explosive speed, as we have? He pleaded with the historians and sociologists to seek out guiding principles.

At Green Bank, however, Morrison did argue that the principle of convergence would apply. Specifically he sought to show that the early civilizations of China, America and the Middle East evolved separately but nevertheless along similar lines. Thus, when the Spaniards reached Central and South America, they found a civilization that was several thousand years behind Eurasia in that it had not developed an alphabet. In fact, the use of paper was unknown. Yet its monuments show that symbolic pictographs were employed to record historical events. Had the Incas and Aztecs been left to their own devices, Morrison argued, they would soon have progressed to more stylized forms, like the hieroglyphics of ancient Egypt or the early Chinese ideographs.

His point was that civilization, including primitive writing, evolved independently in Eurasia and America. The continued isolation of those regions from one another during ancient times is seemingly borne out by the fact that the contagious diseases of one were unknown in the other.

In a later discussion Morrison acknowledged that our technology could not have come about but for that remarkable event in human history, the Renaissance. In this case there was no convergence. There was only one Renaissance and it took place in only one culture. Was this, then, a freak occurrence that would not be likely to occur in other worlds? He cited, for example, the thousand-year history of classical China that evolved gunpowder, maps, printing, the compass and paper, yet leveled into a plateau of stability that endured until the New Philosophy in Europe lighted a fire that is still spreading around the world.

Just why this happened in Europe and not in China is still a favorite subject of doctoral theses. Morrison argued that had the Renaissance not come to Europe, something comparable would have been forced upon China when the status quo had run its course—when the dams had silted up, the land become impoverished, or the rice become diseased. In a different context, such a revolution has, in fact, been taking place in China over the past century.

To be weighed against this optimistic argument was the long record of civilizations that have flowered and died without ever discovering the scientific method and becoming technological. How certain can we be that this discovery is inevitable? Furthermore, it was pointed out, intelligent life may have arisen on planets deficient in the metals on which our technology is based.

As to whether or not a highly advanced society would be interested in making contact, there seemed several possibilities. If a society has conquered poverty, disease, hunger and overpopulation, greatly increasing the life span of its citizens, minimizing their labor and their worries, what would be their state of mind? Would they become bored and lazy, losing interest in science, forfeiting their intellectual curiosity? Sebastian von Hoerner, in an analysis published in *Science* shortly after the conference, pointed out that science and technology have been advanced, in large measure, though not entirely, "by the fight for supremacy and by the desire for an easy life." Both these forces, he added, "tend to destroy if they are not controlled in time: the first one leads to total destruction and the second one leads to biological or mental degeneration." Does this mean that mankind will end up vegetating in front of television, as Bracewell has suggested, or live in automated hermitages, as depicted in E. M. Forster's frightening short story "The Machine Stops"?

Morrison, in a 1960 lecture to the Philosophical Society of Washington, was more hopeful. He proposed that advanced societies throughout the galaxy are in contact with one another, such contact being one of their chief interests. They have already probed the life histories of the stars and others of nature's secrets, he said. The only novelty left would be to delve into the experience of others. "What are the novels?" he asked. "What are the art histories? What are the anthropological problems of those distant stars? That is the kind of material that these remote philosophers have been chewing over for a long time."

Of course, if such societies are numerous, they may be sated with such novelties; but he still argued that they would be interested. To discover a new realm of life would be a coup for such a civilization. If intelligent life is scarce, the motivation to make contact will be far greater—and doing so will be far more difficult. In any case the question is beset with imponderables, such as the motivation of "people" who have been operating technical institutions for long periods.

How universal is curiosity? How persistent is it? From our experience on earth, it would appear that this quality is an essential ingredient of successful intelligence. We see it, in ways that delight us, in our small mammalian friends as they go sniffing about. We know that it is the kernel of science. Morrison proposed that intelligent beings in other worlds may, over great time spans, evolve into species that lack curiosity. "They turn into some other animal and they're not interested anymore," he suggested.

However, if intellectual curiosity dies, it would seem that intelligence is likely to do so as well. The ringing words of the Norwegian explorer Fridtjof Nansen come to mind: "The history of the human race is a continuous struggle from darkness toward light. It is therefore of no purpose to discuss the use of knowledge—man wants to know and when he ceases to do so he is no longer man."

Taking all these considerations into account, the Green Bank conferees estimated that from one-tenth to one-fifth of intelligent species would engage in attempts to signal to other parts of the galaxy.

Factor Seven (L): Longevity of each technology in the communicative state.

This proved to be the factor most difficult to estimate and most critical in assessing our chances of making contact. It was here that the question arose: Is there "intelligent" life on earth? Are we smart enough to suppress our aggressions and prejudices to survive the crises that confront us? If we lack that ability, the chances are that other civilizations will lack it, too. As the conferees sat around their table in November 1961, they were very aware that their country and the Soviet Union were, in effect, holding nuclear pistols at each other's heads. It seemed conceivable that the typical lifetime of a technology sufficiently advanced to destroy itself is only a few decades. In that case, no one is calling.

We know from history that the civilizations that have arisen in the past on our own planet invariably have fallen. Is ours an exception? This time will it be a nuclear holocaust that brings down all the mighty towers of culture, instead of a new, vigorous society overrunning a senile one?

Morrison noted that on a recent visit to Cornell, Fred Hoyle, then Plumian Professor of Astronomy and Experimental Philosophy at Cambridge University, had discussed the fate of highly evolved technologies. He elaborated on the idea in a lecture at the University of Hull titled "A Contradiction in the Argument of Malthus." His reference was to the theory of Thomas Robert Malthus that population tends to increase faster than food supply, and that therefore poverty and war inevitably become the instruments for holding the population in check. Hoyle argued that as the world becomes more technological, a collapse of organization, rather than starvation, was likely to be the limiting factor. To feed and care for itself, he said, human society is bound to become more complex, and the more complex it becomes the more prone it will be to disastrous collapse. This could be the consequence of nuclear war: "Methods of warfare are certain to become more efficient." But, Hoyle said, war is not the only way in which the organization of society could collapse. Another possibility would be the appearance of a germ strain against which adequate defenses could not be mustered before it spread epidemic around an overcrowded globe. In any case he doubted that the destruction of the human race would be total. "The tattered remnants of humanity will slowly reform and re-establish themselves." The struggle for survival would bring into play laws of evolution that have been in abeyance, so far as the human species is concerned. "If I myself were a survivor," Hoyle said, peering at his audience through horn-rims, "I would live in constant terror of breaking my glasses. Such a simple event would almost certainly make further survival impossible for me." The chaos following collapse would thus eliminate evolutionary regressions, such as poor eyesight, that had appeared during the past few centuries.

The chaos would also, Hoyle argued, bring forward those with intellectual gifts. They would have to be ingenious, for they would find a world in which ready raw materials had been exhausted— quite unlike those who built the first industrial economy amid an abundance of iron, coal, oil and the like. To survive in such a situation, the world's remaining inhabitants would have to work closely together, making for the selective emergence of a more intelligent and "sociable" race. Hoyle envisaged that the cycle of catastrophe and recovery would repeat itself, speaking of as many as twenty cycles, spread over the next five thousand years. He showed the successive slow rises and steep falls as a sawtooth pattern, with the intelligence and sociability of the race increasing each time. Such improvements should ultimately enable men to do what Hoyle felt present-day man could not achieve—suppress the birth rate. In his lecture at Cornell he suggested that the break in the sawtooth pattern might finally come about through the establishment of contact with a more advanced society—one that had already achieved stability.

Likewise von Hoerner, in his *Science* article, discussed the possibility that a succession of civilizations might rise and fall on a planet during its long history. He did not, however, envisage the original species reviving each time with greater intelligence and sociability. Instead, he saw the dominant species wiped out by each cataclysm, with a new civilization emerging "out of the unaffected lower forms of life"—a far slower process than that envisaged by Hoyle. Von Hoerner calculated that only one in every 3 million stars has a technical civilization, the mean distance between them being about a thousand light years.

This was considerably more pessimistic than the estimate of Sagan, who suggested that some civilizations may achieve a global society before they discover weapons of mass destruction. In any case, he said, if they were able to live with such knowledge for a century, they would be likely to survive as long as their sun remained stable. Sagan doubted that such a civilization would succumb to a natural disaster, such as an epidemic, climate change or geologic upheaval. He estimated that one in a hundred thousand stars have advanced societies in orbit around them.

Von Hoerner's calculation of one in 3 million would make it seem impracticable to look for signals from selected stars. The entire sky would have to be scanned continuously and one would not expect other civilizations to beam signals at our sun. Rather, they would use beacons radiating in all directions. It was this idea that was behind the suggestion of Iosif Shklovsky to scan the Great Spiral Nebula in Andromeda.

In any case, von Hoerner argued, once a civilization made contact with another world, its own life expectancy would be greatly increased; for, through the knowledge that others have weathered the crisis and, perhaps, with guidance as to how this was done, the new member of the galactic community would be able to solve its problems better. On the other hand, if civilizations, once they have reached our level, survive only briefly, few, if any, will live long enough to make contact. Von Hoerner calculated that the critical longevity is in the neighborhood of forty-five hundred years. If civilizations endure a long time, compared to that figure, there is a strong chance that they will establish communications with sister societies, acquiring greatly increased wisdom and therefore an even longer life span. If the lifetimes are short, the likelihood of contact will be small.

In fact, said Morrison, if a civilization typically survives only ten years after reaching the potential for self-destruction, then, on the average, there is only one communicative site in the galaxy at any one time. There have been others in the past and there will be others in the future, but they do not last long enough to coincide. "It is a very gloomy prospect, but a possible one," he said. Indeed, he added, the discovery that this is not true would be one of the greatest rewards of detecting signals from someone else.

Von Hoerner recognized one other possibility, which was that while the average life span may be short, exceptions may alter the situation fundamentally. If two of the more stable societies have made contact "once somewhere," he said, it may have initiated a chain reaction, with new civilizations being located and "saved" before they destroyed themselves. Those who have tasted the excitement of speaking to another world might easily be inspired to long and patient efforts to broaden the network of cosmic wisdom.

At the Green Bank meeting it was concluded that the mean lifetimes must either be very long or very short. An in-between situation seemed unlikely. Thus it was proposed that the figure is either less than 1,000 years or more than 100 million years. "Fears that the value of L [lifetime] on earth may be quite short are not groundless," Pearman wrote in his account of the meeting. "However, there is at least the possibility that a resolution of national conflicts would open the way for the continued development of civilization for periods of time commensurate with stellar lifetimes."

For every light year in the distance of our observation, the number of accessible stars increases rapidly, out past 1,000 light years. Then the increase begins to drop off in some directions because the range may reach beyond the disk of the galaxy. However, this is compensated to some extent by the fact that toward the core of the galaxy the star density becomes greater than it is near our sun. In our vicinity the mean distance between stars is 9 light years, whereas toward the center of the galaxy it is only 1 light year. C. M. Cade in Britain has suggested that networks of worlds in communication with one another are more probable in this region of greater star density than in our location in the outer part of the galaxy.

The Green Bank conferees also reviewed "logical" ways to communicate. They discussed the possibility that some obvious wavelength, such as 21 centimeters, might be used only to attract attention. The signal would contain instructions as to some other, more efficient channel of communications. Morrison has drawn an analogy with the problem of establishing contact with primitive natives. One does not try calling them with television, he said. One sends a man ashore "who beats a drum." The more sophisticated channels can come later.

The participants set forth various ideas on attention-getting and eavesdropping methods. They discussed a scheme anticipating what is now under way in the SETI project, monitoring 1,000 channels in the optimum band of the radio spectrum with a radio telescope at least 300 feet in diameter, feeding data into a "very large" computer. Huang, Drake and others warned that searching for life in other worlds for hundreds or thousands of years would not hold its excitement for very long. This would apply to both individuals and governments. Drake, noting his experience with Project Ozma, pointed out that constantly acquiring nothing but negative results is most discouraging. A scientist must have some successes. Hence he proposed that a project for detection of extraterrestrial life also carry out conventional research. Perhaps time should be divided about equally between the two, he said. To keep the watchers from getting bored, Morrison proposed, it might be wise to "sneak in" false signals once in a while.

It may also be, as pointed out by Huang, that distant worlds have found it impractical to scan all candidate stars and also to beam signals toward them over enormous periods of time. If so, an eavesdropping scheme like the one proposed by Drake would be the only practicable method.

For "a high probability of success," such a system should be operated about thirty years.

Estimating the cost at $15 million, Drake concluded that "the scale of the undertaking, the shortage of people who are qualified to participate and interested in doing so, and the novel nature of the experiment all militate against early commencement of the project." Similar pessimism was expressed by Otto Struve. Despite the interest of "many distinguished scientists," he said, "those responsible for spending government money on scientific research are necessarily skeptical about financing so expensive a project that does not promise quick results." He thought a full-scale search in the near future "unlikely."

What was needed first, the Green Bank group agreed, was more knowledge concerning phenomena already within reach: the number of nearby stars that have solar systems and the possible existence of life that has evolved elsewhere. Also needed, as set forth by Pearman in his report, was "more profound analysis" of the evolutionary and sociological factors that bear on the question. If it is found that planets are abundant, then the argument in favor of a search will be enormously strengthened. Drake pointed out that the crosscorrelation method so useful in picking up weak radar signals would not only help in eavesdropping on distant radio traffic, but could also be applied to the search for planets. It had been hoped that "wiggles" in the motion of their parent stars could reveal the presence of planets, but the evidence was ambiguous, hidden amid observational errors. Drake suggested that crosscorrelation techniques be used to process large quantities of positional data and separate evidence for wiggles from the inherent errors. As in the case of his eavesdropping proposal, the method was comparable to superimposing radar echoes to raise them above the background "noise," except that, whereas the code of returning radar pulses is known, the wiggle rhythm reflecting the orbital period of a postulated planet is not. This simply meant that, by computer analysis, all possible periods had to be tested on the data. Drake later applied this method to Epsilon Eridani, one of the targets of Project Ozma, and thought he had found indications of a planet six times as large as Jupiter.

Suddenly the seemingly glamourless branch of science known as positional astronomy had taken on new meaning. For generations those who practiced it had taken photographs, made careful measurements and stored them away for comparison by future generations. This slow accumulation of data had, as Morrison put it, been viewed as old-fashioned: "Nobody helps them out; nobody gives them computing machines; nobody has lots of people working on that," but now the situation was changing. While we may learn the plenitude of solar systems and may find that life arose —or almost arose—on Mars, there is no immediate way to determine the chief unknown in the Green Bank analysis—that of the longevity of a technological civilization. The other factors in Drake's equation can be assessed by experiment or observation, but not that one. In fact, he has suggested that the only way to learn the typical longevity may be by making contact with other worlds. Yet our confidence in the future will be greatly increased if we can survive the next few decades. The life span of a society necessary for a meaningful exchange of ideas with other worlds is one for which there is no precedent on earth. Our civilization would have to endure for thousands of years. Ours is an all-or-nothing situation. Either we fail to meet the crisis or we achieve a global society, immune to most, if not all, the factors that have brought down civilizations in the past.

Meanwhile, as recounted in the next chapter, more than thirty years elapsed before a large-scale, government-funded search for other civilizations was initiated.

20

SETI

On October 12, 1992, to coincide with the 500th anniversary of Columbus's landing in the New World, two giant antennas aimed into the cosmos were turned on in the most far-reaching exploration effort ever undertaken. It was the start of NASA's ambitious Search for Extraterrestrial Intelligence, or SETI, first begun thirty-two years earlier with Drake's Project Ozma. "In the first few minutes," said John Billingham, SETI program chief at NASA's research center in Mountain View, California, "more searching will be accomplished than in all previous searches combined."

During the previous three decades, those sixty or so searches by many groups, American and foreign, had, on the one hand, produced skepticism among some scientists, leading to a "Golden Fleece" award for wasteful projects from Senator William Proxmire of Wisconsin. On the other hand, there had been impressive support from scientists, philosophers and religious leaders around the globe. The budget for fiscal year 1992, after some congressional paring, was $12.25 million, and NASA requested $13.5 million for 1993. The House Appropriations Committee proposed eliminating the entire SETI item, but in response to Senate support it ended as $12 million, with the hope that similar funding would continue annually for a total of $100 million. To avoid further congressional charges that SETI was a waste of money and highlight its potential for basic discoveries in radio astronomy, congressional supporters changed the project's name to High Resolution Microwave Survey, or HRMS, although most astronomers still refer to it as SETI.

It has had a long history. After Project Ozma, described in Chapter 15, "False Alarms," Drake participated in or advised many similar projects, most of them concentrated, as was Ozma, on or near the 21-centimeter wavelength. Soon after Ozma Nikolai S. Kardashev and G. B. Sholomitsky began their own search from a Soviet station in the Crimea. Like Drake, Kardashev has continued as a leader of Russian detection efforts, having been inspired by Shklovsky to believe that supercivilizations may be making themselves known through beacons powerful enough to be intergalactic. As noted earlier, Kardashev suggested that there may be three types of highly advanced civilization: Type One would broadcast a beacon with power equal to all of that being produced on the home planet. Type Two would somehow harness all of the power from its sun to produce an extremely strong beacon. Type Three would signal its presence with power equal to that of all the stars in its galaxy.

In 1975 and 1976 Frank Drake and Carl Sagan used the giant Arecibo antenna to search nearby galaxies for Type Two civilizations. Then Nathaniel N. Cohen of Cornell, Matthew A. Malkan of Caltech and John M. Dickey of the University of Massachusetts, searching for Type Two or Type Three civilizations, examined observations already made of twenty-five globular clusters, each a cloud of perhaps a million stars. The observations had been with dishes at Arecibo, Epping in Australia and the Haystack dish in Westford, Massachusetts.

In the Soviet Union V. S. Troitsky organized simultaneous observations from four sites as far apart as Murmansk, near Finland, and Ussuri, near Vladivostok on the Sea of Japan. The purpose was to see whether, simultaneously, they observed sporadic pulses or other signals. In another search, "Ozma II," under the direction of Patrick E. Palmer and Benjamin M. Zuckerman, both of Harvard, 674 stars were scanned at Green Bank, West Virginia, for 500 hours between 1972 and 1976. During part of that period, in the "Qui Appelle" project, 70 stars were examined by the Algonquin Radio Observatory in Ontario, Canada.

One of the longest-running searches, still under way twenty years later, was launched in 1973 by John Kraus and his colleagues at Ohio State University with a unique design that uses two widely separated reflectors facing one another. One is flat and, as the earth turns, can be tilted to receive emissions from any elevation in the sky in the wavelength region of the "water hole." The emissions are then reflected to the other fixed antenna, whose curvature reflects them back to a receiver in the middle. A number of tantalizing signals have been detected, the most famous, described in Chapter 15, "False Alarms," being the one known as "Wow!"

Another prolonged project is Serendip, which, at nominal cost, "piggybacks" its search on existing astronomical programs. It was launched in 1977 under Stuart Bowyer, professor of astronomy at the University of California at Berkeley, initially taking advantage of the university's observations with its 85-foot dish at Hat Creek, near Lassen Peak. A participant was Jill Tarter, a former graduate student in astronomy, who was destined to play a major part in SETI as its chief scientist at NASA's Ames Research Center. In 1992, Serendip 3 was added to the observing system of the giant antenna at Arecibo, and Serendip 4, with 120 million channels, was planned.

Several optical searches have been made. The *Copernicus* satellite, while in orbit, was commanded by radio to scan three stars for possible ultraviolet laser emissions. Observations of twenty-one "peculiar objects" were made with the giant Russian telescope at Zelenchukskaya in the Caucasus, looking for light pulses too short to be natural (lasting less than a very small fraction of a second), as well as narrow spectral lines indicative of a laser. The project was known as MANIA, that being the acronym of its full title in Russian, and the participants referred to themselves as "Maniacs." Optical observations from the Leuschner Observatory of the University of California at Berkeley searched the L4 and L5 libration points in the earth-moon system to see whether interstellar probes might be orbiting those points. The points, described in Chapter 18, "Where Are They?", are regions in orbit around the earth where the gravity of earth and moon compensate one another, allowing an object to orbit there indefinitely.

Searches were also conducted by the Synthesis Radio Telescope at Westerbork in the Netherlands, the Observatoire de Nançay in France, the massive dish of the Max-Planck-Institut für Radioastronomie at Effelsberg, Germany, and the Australian one at Epping. Searches from Australia were also made with NASA's Deep Space Network antenna at Tidbinbilla.

Probably the most innovative searches, before those of the SETI program that began in 1992, were led by Paul Horowitz, a physics professor at Harvard University. Using new electronics, he has been able, simultaneously, to scan large numbers of extremely narrow-band frequencies spanning a broad region close to the "magic" 21-centimeter wavelength. Since emissions from natural sources are normally spread over many frequencies, those in an extremely narrow band are apt to be artificial. Because of constantly changing relative motion between the observatory and a distant source, caused by such motions as the earth's rotation, the orbital flight of the earth around the sun, and motions of the sun relative to the center of the Milky Way Galaxy, there would be constant drift in the exact frequency of an arriving signal—the Doppler effect that alters the pitch of the horn from a passing car.

Horowitz corrects his receiving system for these motions. During each 60-second scan his receiver drifts in frequency to compensate for such motions, not listening to any one frequency long enough to record a strong signal of human origin. As a result, man-made interference has not been a problem. This, of course, assumes the sender has corrected its signal in a similar manner.

Horowitz first tried out his technique in 1978 with the dish at Arecibo. He assumed that a distant civilization had aimed radio beacons at a large number of stars, including our sun, in the hope that a civilization might be in orbit around one or more of them. The senders, knowing the motions of each target star, would have adjusted each beam so that, having been compensated for that star's motion through the galaxy, it would reach our own sun (that is, the solar system) at precisely the "magic" 21-centimeter wavelength. This meant that the only corrections needed were for the earth's own motions. The wavelength would be broadened slightly by turbulent gas en route, but not enough to mask the very-narrow-band emissions from an artificial source. On each 60-second scan 65,536 frequencies were observed at two polarizations (a total of 131,072 observations) with the receiver updated under computer control to compensate for the earth's motion. Of the 185 stars examined in Horowitz's Arecibo search, 60 were scanned more than once. All were examined for at least ten one-minute integrations. They were all within 82 light years and, broadly speaking, resembled the sun. No seemingly artificial signals were detected.

The next search by Horowitz and his colleagues, again at Arecibo but at double the 21-centimeter wavelength, was with a more elaborate system that automatically analyzed the emissions as they came in. It had been developed with the help of Stanford University and NASA's Ames Research Center, and, because it was compact and portable, was known as "Suitcase SETI." For seventy-five hours in May 1982, they scanned the same stars as previously, plus a few more, and the readings were examined for peaks of intense emission at narrow wavelengths that might indicate artificiality. A computer adjusted the receiver forty times a second to compensate for the ever-varying motions of the earth.

They did not, reported Horowitz and his colleagues, find any extraterrestrial signals. This, they said, "should not be cause for discouragement, however, since most scientists in this field believe we might have to examine a million candidate stars in order to have a reasonable chance of finding another technological civilization. . . . What is needed is a highly sensitive search, patiently carried out over several years, using a radio telescope dedicated 100% to that task."

Such a search was made possible by a new organization, the Planetary Society, with Carl Sagan and Bruce Murray, former head of the Jet Propulsion Laboratory, as prime movers. The Society provided most of the money needed to initiate such a project, using Harvard's 84-foot Agassiz telescope at the field station near the village of Harvard, west of Boston. In what was called Project Sentinel, the dish, aimed along the north-south meridian, began scanning the sky as the celestial sphere rolled past. Each day the tilt of the dish was slightly adjusted to scan a different east-west band. It took 210 days to cover the 79 percent of the heavenly sphere that passed within view.

Sentinel operated for twenty-eight months, enough time for two complete scans at 21 centimeters and one each at two of the hydroxyl wavelengths on the other side of the "water hole." The receiver was constantly corrected to match the "magic" wavelength assumed to be reaching the solar system.

The system was then converted to META (Megachannel Extraterrestrial Assay) that has since been duplicated in Argentina as META II to scan the southern heavens. They are called "megachannel" because they are simultaneously able to observe more than eight million channels, making possible tests of several "magic" frequencies and several ways in which they could be modified. These include frequency adjustments by the sender to allow for one of several forms of motion. If a signal is specifically aimed at our sun, the frequency might be adjusted to fit the sun's motion relative to the sender. Or, if all the candidate stars are in this part of the Milky Way Galaxy, the signal might be adjusted to match the motions of those stars around the galactic center. Or the compensation might be for motion of the galaxy within the entire universe—the most basic of all frames of reference. Determining this motion, in broad terms, had become possible with discovery of the faint glow left from the explosive birth of the universe— the Big Bang. The glow is, in effect, a sea of radiation through which the galaxies, with all their suns and planets, are moving. The frequency of this radiation, to an observable extent, is slightly higher in the direction of our galaxy's motion than in the opposite direction. Perhaps this provides the basic reference frame for a universal meeting ground, although for us knowledge of such motion is rudimentary. With the ability to scan millions of channels at once, the META array has been able to test all such possible modifications of the magic frequencies and, in 1993, the system was being converted to BETA with a quarter of a billion channels.

On September 29, 1985, in a ceremony at the Harvard Observatory, Steven Spielberg, director of the movie *E.T. The Extra-Terrestrial* and, with the Planetary Society, a benefactor of the project, threw a switch (with the help of his infant son, Max) and META began. Spielberg wished for success: "I just hope that there is more floating around up there than Jackie Gleason reruns."

Then, in 1992, NASA's own SETI project, dwarfing all that had come before, was launched. It had roots in the far more ambitious Cyclops proposal developed earlier at the Ames Research Center. In the early 1970s John Billingham, an Englishman, had been named office chief for SETI at Ames, having already been with NASA five years (three of them at the Johnson Space Center in Texas designing life-support systems for manned spacecraft). There followed a series of studies and workshops at Ames on techniques that might be used for an all-out SETI effort, producing reports by such SETI figures, in addition to John Billingham, as Philip Morrison and Frank Drake, Bruce Murray, Robert Dixon of Ohio State and Charles Seeger, a Cornell radio astronomer and brother of Pete Seeger, the folk singer.

From these there emerged the concept of the two-headed search now under way: a survey of the entire sky and a targeted search of about a thousand sunlike stars within a hundred light years, that is, in this corner of the galaxy. Both are heavily dependent on highly automated systems called Multi-Channel Spectrum Analyzers that can split the spectrum of incoming radio emissions into millions of different frequency channels and, with pattern-detection computers, analyze each for distant artificial sources. The prototype of the analyzer was developed at Stanford University, a short drive from Ames. The latter is responsible for the targeted search in addition to being in overall charge of SETI. The targeted search, aimed at individual stars, is currently being carried out with the world's largest dish, suspended in a bowl-shaped valley at Arecibo, Puerto Rico. Caltech's Jet Propulsion Laboratory in Pasadena is carrying out the all-sky search, using large antennas of its globe-encircling Deep Space Network and has developed its own Multi-Channel Spectrum Analyzer. In both projects antennas are employed on a "space available" basis—that is, when not allocated to a special science project or needed for some other NASA task. The one exception is a 42-meter (138-foot) antenna at Green Bank, West Virginia, which, beginning in 1995, is to work full-time on SETI's targeted search.

The all-sky search began with the new 34-meter (112-foot) dish at the Goldstone station of NASA's Deep Space Network in California's Mojave Desert. Other stations of the network are located around the world so that, as the earth turns, a distant spacecraft or other source is always in view. Besides the one at Tidbinbilla, Australia, where a 70-meter antenna may later become available, there is one at Madrid, and other giant dishes in Australia, France and Russia may also be used.

For this many-year survey the sky is divided into several hundred "frames," each a strip about 45 degrees long (one-quarter the span from horizon to horizon) and a few degrees high. In scanning one frame the antenna, for a period of about two hours, follows a "racetrack" pattern, sweeping back and forth in a series of parallel scan lines, like someone plowing a field. Coverage by each sweep overlaps the lines above and below it, but not for at least ten minutes. The antenna is thus programmed to avoid scanning neighboring lines until the earth has rotated enough so that any man-made signal will have changed its position relative to the stars. Only those sources that are still in their original celestial position will be considered candidates for the next test, which is conducted after the two-hour sweeping of the entire frame. The system compares the repeatedly recorded sources to those in a vast, computerized inventory of natural and human radio sources, such as quasars, pulsars, radio stations, spacecraft and radars—an inventory which, unfortunately, is constantly growing—thus ruling them out.

The method differs from that by which Horowitz at Harvard

discriminates between distant sources and man-made ones. Horowitz tests whether the frequency of the incoming signal drifts in response to the earth's motion, indicating that it is coming from a fixed point in space. In the all-sky survey man-made sources are supposedly eliminated if they change their celestial position during the scanning cycle. Those sources that have survived the survey's own multiple tests are finally reexamined to see whether they merit further study. The procedure is highly automated, since a sweep may produce several thousand candidates per second. The program, however, calls for a scientist always to be on hand in case something special is observed.

The full spectrum of the microwave "window" penetrating to the earth, from 1,000 to 10,000 megahertz, is examined. This includes the hydrogen and hydroxyl wavelengths of the "water hole" and all the frequencies in between. The prototype system now in use breaks up the radio waves into 2 million 20-hertz frequency channels (covering 20 or 40 megahertz, depending on whether one or two polarizations are examined). A later system will increase capacity of the frequency analyzer sixteen-fold (to 32 million channels) and for six years, beginning in 1996, it is to scan the entire sky, north and south, thirty-one times over.

At noon on Columbus Day 1992, the giant Goldstone antenna slowly tilted down from its storage position and began sweeping back and forth across a narrow band of sky. Had the stars been out, one could have seen that the band was between the constellations Hercules and Ophiuchus. Any radio emissions would, of course, have been unaffected by the sunlight. The strip of sky which had been selected for the start of this long, patient search had been designated Sky Frame No. 1. On a platform Michael J. Klein, the JPL project manager, and Samuel Gulkis, deputy project scientist, explained the search strategy. Carl Sagan pointed out that if we detect signals, "We may learn how long civilizations last" and discover an "Encyclopedia Galactica." In front of them, under a sun-shielding canopy, were several busloads of scientists, family and press. A high school band from nearby Barstow gave the proceedings a festive air.

Sky Frame No. 1 included the star GL615.1A, which was the first looked at by the targeted survey at Arecibo. The two projects were launched simultaneously with, at Arecibo, such SETI figures as Billingham, Tarter, Drake and Oliver. The range of frequencies being examined, 1,000 to 3,000 megahertz, is narrower than that

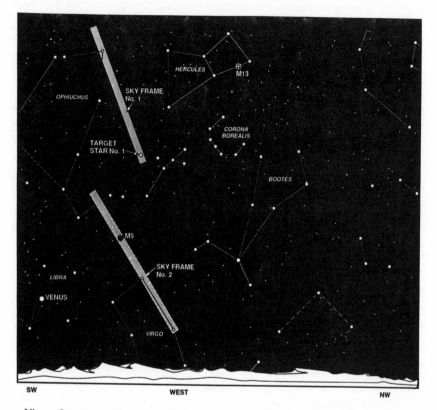

View of western sky as seen from Goldstone at 7:00 P.M. October 12, 1992.

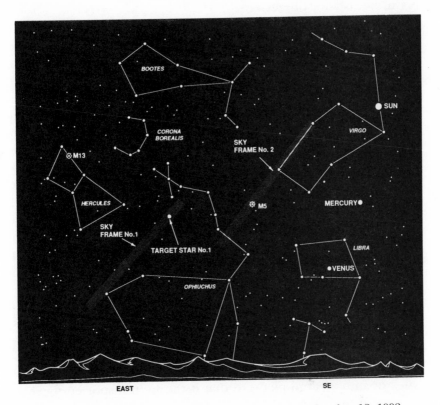

View of eastern sky as seen from Goldstone at noon October 12, 1992.

of the all-sky survey, but with far more sensitivity. In searching that spectrum of frequencies the analyzer scans six bands of different widths—from 1 hertz (in 14,370,048 channels) to 28 hertz (in 523,216 channels). The search is expected to continue into the next century, with antennas of the Deep Space Network and others in both hemispheres.

Each star is scanned from 300 to 1,000 seconds, making it possible to observe pulsed signals (the all-sky search views one spot for only a few seconds, whereas Horowitz's META does so for two minutes). Stars within 20 light years are being scanned for 1,000 seconds since, as Peter R. Backus, principal investigator of the project, put it, "planets orbiting those stars have had time to receive early television transmissions and to have responded." The very nearest stars, he said, may be observed for the longest feasible time "in order to achieve the highest possible sensitivity, perhaps even to leakage radiation." In other words, we might detect their most powerful radars or television stations.

As with the all-sky search, an archive is used to eliminate known sources. Another test, as in Ozma, compares the signal detected by the main antenna from a tiny point in the sky with that from an antenna with a broad view. If the signal is observed in the latter, the source is probably local. If the source meets this test, the observation may be repeated, but with the antenna aimed slightly to one side. If the signal disappears, the original sighting may merit further study. It was reported at the start of the searches that Caltech's antennas at Owens Valley stood ready to check on any suspicious feature. The targeted search analyzer fits onto a flatbed truck or into a cargo plane so that it can be transported to whatever site is available.

Despite extensive automation, an operator will always be on hand. During the observation the frequency of the signal may "drift" up or down, which, after correction for the earth's own motion, could mean that its home planet is rotating or in orbit. If, in any case, the source is moving with the stars, said Backus, "other observatories will be contacted for independent confirming observations. It is only after an independent site has observed the signal, that the discovery will be announced."

As described in the final chapter, a protocol has been drafted for those circumstances, but such a discovery would be very difficult to keep quiet. Strange emissions may be observed that, at first, seem unrelated to intelligent life, but are exciting and perplexing.

At the inauguration of the all-sky search Carl Sagan cited "enigmatic" emissions being recorded in the Harvard survey. During the summer of 1992, a preprint describing about sixty such events, recorded during the five years since META began operating, was circulated to SETI participants. It was followed by a revised version that cut the number to thirty-seven. While the emissions may originate among the stars, each has disappeared too soon for this to be confirmed. As of this writing none from the same celestial spot has been detected a second time.

What makes these events striking is the way in which their narrow-band "magic frequency" drifts precisely in the manner that neutralizes the effect of the earth's motions. This is what one would expect for a source among the stars. Another clue to their possible extraterrestrial origin may be the tendency of the sources to lie in the direction of the Milky Way. Paul Horowitz points out that thirty-seven events among his system's millions upon millions of observations suggest a statistical fluke, but he hopes his improved system, scanning each spot twice within seconds, may resolve the mystery.

A somewhat similar phenomenon is being observed by the targeted search at Arecibo. In this case, a signal from the target star has been detected as many as six times during one hour. Its very narrow bandwidth fluctuated slightly in frequency, as though coming from a planet in an orbit like that of the earth, but these fluctuations did not seem very uniform. When the telescope looked to one side of the star—that venerable test for an extraterrestrial source—the signal disappeared. After an hour the star set behind the horizon, and by the next day the signal had disappeared. All told, only four or five such events had been observed, said Dr. Kent Cullers, leader of the signal detecting team, in November 1992. The trouble with verification by another observatory, he pointed out, is that none on earth has comparable sensitivity and these emissions are very weak. Like Horowitz, he noted that in his system, making "billions of observations per second," such events were almost bound to occur.

Nevertheless, being unexplained, the observations call to mind the initial reports on pulsars described in Chapter 15, "False Alarms." There is, however, no indication they are pulsed or otherwise artificial. None is reobserved later and they remain, as Sagan said, an "enigma."

21

Celestial Syntax

Not long after the 1961 Green Bank conference, Frank Drake sent to all participants a strange communication. It was, he believed, of the sort that might be received as the initial message from another race of beings and it manifested the thought that he and others had given to the question: how can we exchange ideas with those who differ from us in ways we cannot guess—certainly in physiognomy, possibly in methods of thought and logic?

Drake's message consisted of a series of pulses sent in a fixed rhythm, but with many gaps. The pulses he wrote as ones; the gaps as zeros. In other words, it was a binary code, the simplest of all communication systems in that it makes use of only two symbols. It is the one commonly used in computers. His colleagues worked over it, but despite some familiarity with the workings of Drake's mind, they found its meaning difficult to extract. However, one of them, Bernard Oliver, expanded—and thus simplified—the message which appears, as received, on the next page.

A peculiar feature of this message is that it consists of 1,271 ones and zeros—in computer terminology, 1,271 "bits" of information. To anyone mathematically inclined, the number 1,271 will be recognized as the product of two prime numbers: 31 and 41 —"prime" because they can be divided only by 1 or themselves. This suggests the possibility that the message should be written out in 41 lines of 31 bits each, with the zeros, since they were pauses in the transmission, left as blank spaces. As shown on page 282, this arrangement is unenlightening.

The other possibility, 31 lines of 41 bits, is shown on page 283.

It is clear that the transmitting planet is inhabited by two-legged creatures that apparently are bisexual and mammalian. The circle

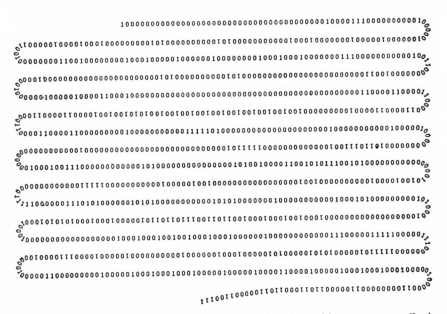

The message of 1,271 "bits" (all ones and zeros) as it would appear transcribed on continuous tape.

at the upper left is their sun, with its planets spaced below it. The manlike creature points to the fourth planet, where the civilization resides. The wavy line emanating from the third planet shows it to be covered with water, and below it is a fishlike symbol. Apparently the civilization has visited this planet and found it to have marine life. The planets are numbered, down the left side, in a binary code. Symbols across the top represent the atoms of hydrogen, carbon and oxygen, indicating that the chemical basis of life there is like our own. The scale on the right, labeled in the binary code, shows that the creatures are eleven units high, the units presumably being the wavelength of the signal—21 centimeters. This works out to 7.5 feet.

Drake's cryptogram was devised not merely for his amusement and that of his colleagues. He sought to stimulate some hard-headed thinking about the decipherment problem. What was needed, Morrison said, was a new specialty: "anti-cryptography," or the designing of codes as easy as possible to decipher. He de-

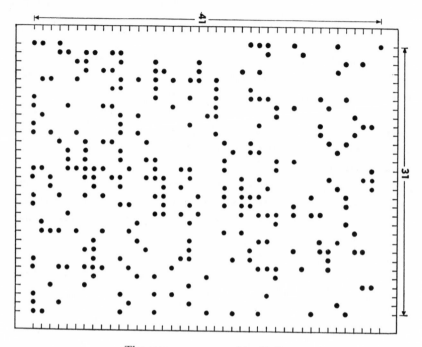

The message arranged in 41 lines.

vised a scheme for coaching distant creatures in the establishment of a television link. Such coaching, it is widely thought, would be of immense importance, for both sides would have almost nothing in common. All that we can count on is that all beings sufficiently intelligent for interstellar communication must have a mathematics based on numbers and on the simple concepts of addition, subtraction, division, equality, and so forth. They must be aware of the structure of the various atoms. Yet their higher mathematics, their logic, their way of representing atomic structure, may differ radically from our own. J. Robert Oppenheimer, the brilliant physicist who led the effort to produce the first atomic bomb, said those of a distant civilization would have great difficulty explaining their science to us, for the "language" they used to de-

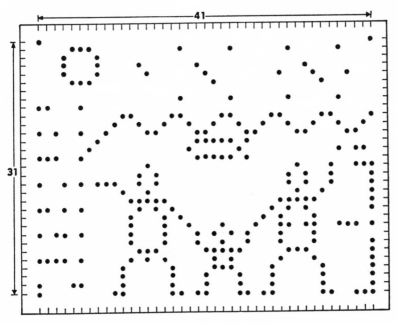

The message arranged in 31 lines.

scribe, for example, phenomena in nuclear physics could be utterly alien to our own.

A more complex message was devised by the Drake-Sagan group to be transmitted from the Arecibo dish at the 1974 ceremony marking completion of its new reflecting surface. The message consisted of 1,679 binary characters, again a multiple of two prime numbers: 23 and 73. Written out as 73 lines, each of 23 characters, it displayed only a single human figure, but representations of hydrogen, carbon, oxygen and phosphorus atoms, chemical formulas for sugars and nucleic acids, the structure of DNA and an outline of the Arecibo antenna.

The message, aimed at Messier 13, a globular cluster of 300,000 stars, took 169 seconds to send. In thirty-five minutes it had passed the orbit of Jupiter and after five hours and twenty minutes it left the solar system. It is expected to reach Messier 13 in A.D. 25,974. It will be 52,000 years from now before a reply could reach us.

When the British Broadcasting Company reported on the transmission it said the wavelength was the "magic" frequency of 21 centimeters. Sir Martin Ryle became outraged. He was not to be ignored, for in addition to being Astronomer Royal of Britain, he was head of the observatory where pulsars were discovered and developer of the multi-antenna system for which he had won a Nobel prize. In messages to Frank Drake, the International Astronomical Union and the American Astronomical Society he questioned whether letting other civilizations know of our presence was wise, lest they prove hostile—a question to be discussed further in the next chapter.

Drake replied that the BBC account was in error. The transmission was not on 21 centimeters and was essentially ceremonial. The likelihood that it would be received was virtually nil. Responding to Drake's letter, Ryle said he and his colleagues were "greatly relieved" to learn the message was not sent on 21 centimeters, but they were still concerned about any very powerful transmissions from earth. This question, he wrote, "came up at the time when the first pulsar signals looked as though they might indicate extraterrestrial life.

"If one believes that there is a finite chance of transmissions of the power we can now produce on Earth being received by another civilization, (and presumably you do), then one must also consider the consequences of revealing our existence to such a civilization. Whether they saw Earth as a useful place for colonization or mineral extraction or anything else, it seems likely to be detrimental to the present inhabitants." Ryle urged that the question be taken up at the next IAU General Assembly.

Meanwhile, largely at Sagan's initiative, "messages" were attached to four craft destined to sail beyond the solar system. The first messages were plaques carried by *Pioneer 10* and *Pioneer 11*, which, in 1972, sailed into interstellar space after flying past Jupiter and Saturn. Each, devised by Sagan, his wife, Linda Salzman, and Frank Drake, consisted of a 6-by-9-inch plate on which were inscribed sketches of unclothed male and female human figures plus a star-shaped diagram of fifteen radiating lines indicating, at the center, the location of the sender. The longest of these lines showed the direction and distance to the center of the galaxy. The other fourteen showed distances to pulsars, each identified by its pulse rate.

The public reaction included a small amount of huffing and puffing by those who objected to the display of sexual characteristics on the two figures. One letter-writer to *The Los Angeles Times* said, in part, "Isn't it enough that we must tolerate the bombardment of pornography through the media of film and smut magazines? Isn't it bad enough that our own space agency officials have found it necessary to spread this filth even beyond our own solar system?''

The more recent messages were on the two Voyager spacecraft launched in 1977 past Jupiter and the outer planets. Each carried a 12-inch copper LP disk with a player needle and instructions on how to use it—what Sagan called ''a bottle thrown into the cosmic ocean.'' If played at 16⅔ revolutions per minute, the records, during two hours, carried a wide variety of messages from earth. These included 116 photographs and diagrams and ninety minutes of music, including samples of Bach, Mozart, Mexican mariachi, Peruvian panpipes, Indian raga and Navajo night chants, as well as greetings in fifty-five languages and good wishes from Kurt Waldheim, secretary-general of the United Nations, and Jimmy Carter, president of the United States. Carter's optimistic message, written rather than spoken, said: ''We human beings are still divided into nation states, but these states are rapidly becoming a single global civilization.'' As with the earlier messages, the likelihood that they would reach anyone was negligible.

Concerning the television option, while ways to tell a recipient how to build a TV were proposed by Morrison as well as by Drake, the limit on speed of transmission is formidable. The raster forming the American television screen consists of 525 lines scanned thirty times a second, which is fast enough so that our eyes see only a smoothly changing picture. But the farther away the transmitter, the more powerful or prolonged must be each pulse, or ''bit'' of information, to stand out above the background noise. This already has presented problems in designing the television systems for exploration of the solar system—such as the one in *Mariner 4* that transmitted pictures of Mars 140 million miles to earth. Whereas commercial television stations can send a picture in one-thirtieth of a second, it took *Mariner* 8.5 hours to do so. Each image was reconstructed by computer from 240,000 bits that had been transmitted at a rate of 8.3 per second. While the transmitters used for communication between solar systems would be far more powerful than those of *Mariner 4*, the distances would be vastly greater. It therefore appears that interstellar sending speeds will have to be far slower, although this should not be too troublesome, since the messages will in any case take years to reach their destination. Also we, in our ignorance, cannot set lim-

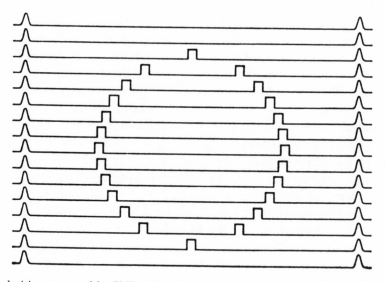

A television proposal by Philip Morrison would identify the beginning and end of each line in a TV "roster" with a sharp pulse. A square in each line would then describe a circle.

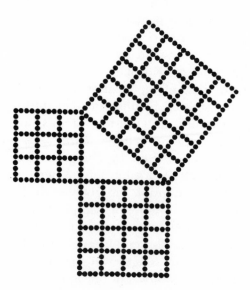

From a series of such lines, the Pythagorean Theorem could be illustrated.

its on the skill and ingenuity of our distant friends in devising ways to speed up their sending rates.

Without television it would be very difficult to arrive at a mutually understandable language. Russell F. W. Smith, linguist and associate dean of general education at New York University, has cited the difficulties in deciphering lost languages on our own planet. At the 1963 convention of the Audio Engineering Society in New York, he pointed out that, but for the discovery by Napoleon's troops of a basalt slab at Rosetta near the mouth of the Nile, Jean François Champollion could not have deciphered the hieroglyphics of ancient Egypt. The priests of Ptolemy V had placed an inscription on the stone in three alphabets: the ancient hieroglyphic, its successor, the more abbreviated demotic, and the Greek.

Despite the lack of a cosmic Rosetta Stone, some mathematicians believe that a language suitable for radio communication between different worlds can be evolved through the use of symbolic logic. The effort to develop a purely logical language, in which all mathematical reasoning could be expressed, began at the end of the nineteenth century with the work of several mathematicians, in particular Giuseppe Peano in Italy. In 1900 Alfred North Whitehead, a lecturer in mathematics at Trinity College, Cambridge, went to Paris for a mathematical congress, bringing with him one of his most brilliant students, Bertrand Russell. They were greatly excited by news of Peano's new application of logic and ultimately decided to collaborate in carrying it further. The fruit of this collaboration was their monumental three-volume work *Principia Mathematica*, published between 1910 and 1913. Thereafter, Whitehead went to Harvard and became one of the foremost philosophers of his time. His collaborator became the third Earl Russell, philosopher, winner of a Nobel prize in literature and, even in his ninetieth year, a gaunt, frail figure demonstrating in Trafalgar Square against nuclear weapons testing.

Their work on "logistic language" had been preceded by efforts to produce an international language that would bring the world closer together. The first to be widely used was Volapük, invented in 1880 by an Austrian priest. This was followed seven years later by Esperanto, but the mathematician Peano thought these had failed to escape from the arbitrary and illogical syntax of tongues that had evolved in the chance manner of nature. In 1903 he produced his Interlingua, derived from classical Latin but with a simplified syntax. It was long used in abstracting scientific articles.

These developments led Hans Freudenthal, professor of mathematics at the University of Utrecht, to attempt extending the "logistic language" of Whitehead and Russell into something intelligible to beings with whom we have nothing in common except intelligence. He called it "Lincos," as a short form of "Lingua Cosmica." The logical exposition of the language, as might take place in an extended interstellar message, was contained in his book, *Lincos: Design of a Language for Cosmic Intercourse,* published in the Netherlands in 1960. Actually, he pointed out, such a language already may be established as the vehicle for cosmic intercourse. "Messages in that language might unceasingly travel through the universe," he said.

His language was "articulated," so to speak, by unmodulated radio signals of varying duration and wavelength. These represented "phonemes" which could be combined to constitute concepts or words. The lexicon and syntax were built up gradually, much in the way that a first-grade reader uses a new word over and over in various ways, then introduces another word: "See John run. John runs to Mary . . . ," etc.

Freudenthal's book-length "language lesson" began with elementary arithmetical concepts, but soon passed on to more abstract ideas. Thus, at an early stage, the idea of right and wrong was conveyed. This was done by presenting a series of mathematical statements, all of which were obviously correct. Each of them was followed by the new word: "good." Then a mistaken statement was given, followed by another new word: "bad." Later such abstract concepts as honesty, lying, understanding and the like were presented. The teaching process used "actors" or "voices," somewhat in the Socratic method, their dialogues illustrating the meanings of new words.

To represent the radio signals in writing, Freudenthal used a mixture of mathematical, biological and linguistic symbols including some of those employed earlier by Whitehead and Russell. He pointed out that Lincos had not yet been developed to where it could explain the diversity of human individuals, but he did use it to express some of the information touched on in Drake's pictorial message, namely, the manner of human reproduction. This was set forth, in Lincos, as follows:

Ha Inq *Ha*:

$x \in$ Hom.$\rightarrow$: Ini.xExt$\cdot$ $-$:Ini$\cdot$Cor$x$.Ext $=$
   Cca.Sec 11 $\times$ 10^{10111}:

$\bigvee x \cdot x \in$ Bes.$\wedge$: Ini.xExt$\cdot$ $-$:Ini$\cdot$Cor$x$.Ext $>$Sec 0:

$x \in$ Hom.$\rightarrow$ $\bigvee$ $\ulcorner y.z\lrcorner y$ $\smile$ $z \in$ Hom.$\wedge\cdot y =$ .Mat $x\cdot\wedge\cdot z =$ .Pat$x$:

$\bigvee x \cdot x \in$ Bes.$\wedge$ $\bigvee$ $\ulcorner y.z\lrcorner y$ $\smile$ $z \in$ Bes.$\wedge\cdot y =$ .Mat $x\cdot\wedge\cdot z =$ .Pat$x$:

$x \in$ Hom.$\rightarrow$: $\wedge t$: Ini$\cdot$Cor$x$.Ext:Ant:t.Ant:Ini.x Ext
   $\rightarrow$:t Cor$x$.Par$\cdot t$ Cor.Mat x:

$\bigvee x$:$x \in$ Bes.$\wedge\cdot\wedge t$.Etc:

$x \in$ Hom.$\wedge$:$s =$ Ini$\cdot$Cor x.Ext$\cdot$
   $\rightarrow$:$\bigvee$ $\ulcorner u.v\lrcorner s$ Cor$u$.Par$\cdot s$ Cor.Mat$x$:
   $\wedge$ PauAnt.$s\cdot$Cor v.Par:PauAnt.$s\cdot$Cor.Pat$x$
   $\wedge$:s Cor x.Uni$\cdot s$ Cor$u$.s Cor$v$:

$\bigvee x$:$x \in$ Bes.$\wedge$.Etc:

Hom$=$Hom Fem.$\cup$.Hom Msc:

Hom Fem$\cap$Hom Msc$=$ $\ulcorner\ \lrcorner$:

Car:$\uparrow x\cdot$Nnc$x$Ext.$\wedge$.$x \in$ Hom Fem
   Pau$>$Car:$\uparrow x\cdot$Nnc$x$Ext.$\wedge$.$x \in$ Hom Msc:

$y =$ Mat$x$.$\wedge$.$y \in$ Hom$\cdot\rightarrow$.$y \in$ Hom Fem:$\wedge$:

$y =$ Pat$x$.$\wedge$.$y \in$ Hom$\cdot\rightarrow$.$y \in$ Hom Msc:

$x \in$ Hom$\cup$Bes:$\rightarrow$:Fin.Cor$x\cdot$Pst.Fin$x$#

The three-letter symbols were derived from Latin roots. For example, "Fem" means "female"; "Msc" means "male." In his "summary" of this passage, Freudenthal translated it, in part:

> The existence of a human body begins some time earlier than that of the human itself. The same is true for some animals. Mat, mother. Pat, father. Before the individual existence of a human, its body is part of the body of its mother. It has originated from a part of the body of its mother and a part of the body of its father.

Freudenthal commented cryptically that the last paragraph of the coded text "is somewhat premature. Its notions," he continued, "will not be used in this volume." He believed that television, including a three-dimensional form, will come only at an advanced stage of interstellar communication, since analytic geometry will be necessary to explain the method. This he considered too advanced for beginners in cosmic discourse. His approach was thus fundamentally different from that of Drake, Morrison and others, who believed instructions for television contact could be conveyed as the essential first step. In fact, the usefulness of Lincos has been seriously questioned by other mathematicians, such as Lancelot Hogben, who considered it doubtful that what seems purely logical to us would appear so to distant minds, schooled in utterly different methods of thought.

Another coding system, also based on mathematics, was devised by Lambros D. Callimahos, a cryptographer with the National Security Agency and wartime member of the U.S. Army's Signal Intelligence Service. In the March 1966 *IEEE Spectrum*, he showed how his method could be used to express two of the most famous puzzles in the mathematics of this planet: Fermat's "last theorem" and Goldbach's "conjecture." Perhaps "the others" could signal us how to solve them.

A very different approach to interspecies communication concerns attempts to "talk" with animals. The example of dolphins, discussed at the 1961 conference in Green Bank, was again on the agenda at the 1991 Russian-American conference on SETI at Santa Cruz. Recent dolphin research was described by Lori A. Marino, psychologist at the State University of New York at Albany, and Diana Reiss of the department of speech and communication at San Francisco State University. Others have compared the difficulty that might be encountered by a far more advanced culture trying to teach us their language to that encountered by those who try to teach chimpanzees.

The question of whether or not we might be able to appreciate music from another world has been analyzed by Sebastian von Hoerner, who has speculated widely on the implications of SETI. He believes that our music might be meaningful to at least some civilizations, since it is the most abstract of the arts, devoid of anything peculiar to the earth and built on universal features of acoustics. "Some other civilizations in space may have no music at all, for various biological or mental reasons," he wrote in a 1974 issue of *Psychology of Music*. "Some others may have vastly different things which they call music but what are incomprehensible to us, for similar reasons. But it seems that some of our basic musical principles are universal enough to be expected at a good fraction of other places." The tempered chromatic scale of twelve tones seems first to have been proposed by Aristoxenus in 350 B.C., he said, but was not generally adopted until the seventeenth century.

Some classical music of India uses a twenty-two-note scale. The music in some areas of Southeast Asia and Africa is based on only five notes, but the twelve-tone scale seems the most logical and universal. Will we then be greeted by new, yet comprehensible, music from other worlds? Will we learn of new composers in a class with Bach, Mozart, Beethoven and the nineteenth-century Romantics who found a way to touch our souls? It would be a rich reward, if we did.

22

What If We Succeed?

On the night of October 30, 1938, there were manifestations of panic in widely scattered parts of the United States. They began shortly after Orson Welles, the actor and producer, sat in front of a microphone in the New York studios of the Columbia Broadcasting System to introduce an adaptation of *The War of the Worlds*, written in 1898 by H. G. Wells. Orson Welles told his nationwide audience how, in recent years, people on earth had gone about their daily lives in complacence: "Yet across an immense ethereal gulf, minds that are to our minds as ours are to the beasts in the jungle, intellects vast, cool and unsympathetic regarded this earth with envious eyes and slowly and surely drew their plans against us. . . ."

The dramatization that followed, consisting of simulated news bulletins, interviews and sometimes fearful sound effects, was so realistic that thousands believed Martians had, in fact, landed in New Jersey—hideous creatures that slew all opposing them with a sinister "heat-ray." People rushed into the streets only partially clothed or struck out aimlessly across open country. Cars raced wildly through crowded cities.

Is this a hint of what might happen if we did, in fact, make contact with a superior civilization? Possibly so, according to a report submitted to the federal government in 1960. For so dramatic an effect, the encounter would presumably have to be physical, but even radio contact would lead to profound upheavals, the report said. The document was a by-product of the historic action taken by Congress on the heels of the first *Sputniks* in establishing an agency for the exploration of space. The National Aeronautics and Space Act of July 29, 1958, called for "long-range studies" of the benefits and problems to be expected from space activities.

Pursuant to this act, NASA set up a Committee on Long-Range Studies and awarded a study contract to the Brookings Institution. More than two hundred specialists were interviewed by a team led by Donald N. Michael, a social psychologist who later became director of the Peace Research Institute in Washington. Pertinent portions of the resulting report were reviewed by such figures as Lloyd V. Berkner, head of the Space Science Board, Caryl P. Haskins, president of the Carnegie Institution of Washington, James R. Killian, chairman of the Corporation of M.I.T., Oscar Schachter, director of the General Legal Division of the United Nations, and Margaret Mead, the anthropologist.

The document was submitted to NASA only a few months after Project Ozma's attempt to intercept signals from two nearby stars, and much in the minds of those who drafted it was the question of what would happen if we discovered another, far more advanced civilization. The report did not rule out the possibility of direct contact, such as the one so vividly dramatized by Orson Welles, and it suggested that artifacts left by explorers from another world "might possibly be discovered through our space activities on the Moon, Mars, or Venus." Nevertheless, it said, if intelligent life is discovered beyond the earth during the next twenty years, it most probably will be in a distant solar system and manifest itself by radio. Such circumstances, it added, would not necessarily rule out revolutionary effects:

> Anthropological files contain many examples of societies, sure of their place in the universe, which have disintegrated when they have had to associate with previously unfamiliar societies espousing different ideas and different life ways; others that survived such an experience usually did so by paying the price of changes in values and attitudes and behavior.
>
> Since intelligent life might be discovered at any time via the radio telescope research presently under way, and since the consequences of such a discovery are presently unpredictable because of our limited knowledge of behavior under even an approximation of such dramatic circumstances, two research areas can be recommended:
>
> 1. Continuing studies to determine emotional and intellectual understanding and attitudes—and successive alterations of them if any—regarding the possibility and consequences of discovering intelligent extraterrestrial life.

2. Historical and empirical studies of the behavior of peoples and their leaders when confronted with dramatic and unfamiliar events or social pressures.

Such studies, the report continued, should consider public reactions to past hoaxes, "flying saucer" episodes and incidents like the Martian invasion broadcast. They should explore how to release the news of an encounter to the public—or withhold it, if this is deemed advisable. The influence on international relations might be revolutionary, the report concluded, for the discovery of alien beings "might lead to a greater unity of men on earth, based on the 'oneness' of man or on the age-old assumption that any stranger is threatening."

Much would depend, of course, on the nature of the contact and the content of any message received. Man is so completely accustomed to regarding himself as supreme that to discover he is no more an intellectual match for beings elsewhere than our dogs are for us would be a shattering revelation. Carl Gustav Jung, the disciple of Freud who later went his own psychological way, said of a direct confrontation with such creatures: The "reins would be torn from our hands and we would, as a tearful old medicine man once said to me, find ourselves 'without dreams,' that is, we would find our intellectual and spiritual aspirations so outmoded as to leave us completely paralyzed."

The SETI project has led to a debate on whether our new acquaintances would, in fact, be "nice people" or the monstrous villains depicted by Orson Welles. Much of the population has been conditioned by science-fiction tales of evil genius at work among the stars, of death rays and battles between galaxies. This may, in part, have accounted for the popular reaction to the Welles broadcast, and the fears of such scientists as Sir Martin Ryle. Some of those concerned with the search for life in other worlds have sought to counter this attitude.

Thus Philip Morrison, in one of his lectures, questioned whether any civilization with a superior technology would wish to do harm to one that has just entered the community of intelligence. If he were looking through a microscope, he said, and saw a group of bacteria spell out, like a college band, "Please do not put iodine on this plate. We want to talk to you," his first inclination, he said, would certainly not be to rush the bacteria into a sterilizer. He doubted that advanced societies "crush out any competitive form of intelligence, especially when there is clearly no danger."

Ronald Bracewell, in a 1962 lecture at the University of Sydney, asked whether beings in another world would covet our gold or other rare substances. Do they want us as cattle or as slaves? He replied by pointing out the literally astronomical cost of transport between solar systems. Any civilization able to cover interstellar distances would hardly need us for food or raw material, which they could far more easily synthesize at home. "The most interesting item to be transferred from star to star," he said, "is information, and this can be done by radio." In 1977, at the height of the Cold War, he concluded that, since technology on earth had been honed by warfare, with a premium placed on cunning and weaponry, much the same would prevail elsewhere. "That is not to say that some gentle Buddha could not have influenced whole populations to follow quite different, less competitive paths," he said. "But if our own history is any guide, such a population would have been overrun by those who value technical mastery of nature. Technological life, rather than merely intelligent life, is what will determine membership in the galactic communication network."

Bracewell cited a 1975 report prepared by the Library of Congress for the House Committee on Science and Technology which pointed out the danger of a hasty reply: "Since we have no knowledge of their nature," it said, "we may be aiding in our own doom. Although it is tempting to assume that any civilization advanced enough to travel over interstellar space would have overcome the petty differences that cause wars, they may not be sure if we have. They may have encountered other warlike peoples and learned . . . to arrive . . . ready for combat."

Bracewell said he himself found it hard to imagine much of any threat, particularly of a familiar kind. And there is, he said, no point in remaining silent if a probe has already reached the solar system and reported our presence.

Another among those who envisioned worlds corrupted by technology was Freeman Dyson, the imaginative physicist at the Institute for Advanced Study in Princeton. In a 1964 letter to *Scientific American* he, too, questioned whether one was justified in assuming that the distant creatures with whom we may converse are moral by our standards.

"Intelligence may indeed be a benign influence," he said, "creating isolated groups of philosopher-kings far apart in the heavens and enabling them to share at leisure their accumulated wisdom.'' On the other hand, he added, "intelligence may be a cancer of purposeless technological exploitation, sweeping across a galaxy as irresistibly as it has swept across our own planet." Assuming interstellar travel at moderate speeds, he said, "the technological cancer could spread over a whole galaxy in a few million years, a time very short compared with the life of a planet."

What our detectors will pick up is a technological civilization, he argued, but it will not necessarily be intelligent, in the pure sense of the word. In fact, he continued, it may more likely be "a technology run wild, insane or cancerously spreading than a technology firmly under control and supporting the rational needs of a superior intelligence." On the other hand, he said, there is also the possibility that a "truly intelligent" society might no longer feel the need of, or be interested in, technology. "Our business as scientists is to search the universe and find out what is there. What is there may conform to our moral sense or it may not. . . . It is just as unscientific to impute to remote intelligences wisdom and serenity as it is to impute to them irrational and murderous impulses. We must be prepared for either possibility and conduct our searches accordingly."

Nevertheless, it seems hard to believe that a race of villains or a civilization run amok would not have blown itself to smithereens long before it reached into other solar systems. As noted in Chapter 19, "Is There Intelligent Life on Earth?", an achievement of lasting peace and stability seems to be an important element in qualifying for membership in the interstellar community. While Dyson is understandably dismayed at what technology and population growth are doing to the face of our planet, a society that could not bring such trends under control would seem destined to disintegrate. The achievement of great stability and serenity would seem a prerequisite for a society willing to make the expensive and enormously prolonged effort required to contact another world.

What, indeed, do we know about the roots of evil—of greed, aggression and treachery? On earth they are clearly manifestations of a complex society that has not reached stability. Animals, as a rule, are aggressive only when necessary. The well-fed lion lies near the water hole while his traditional victims, sensing his satiety, quench their thirst in peace. Man's inherent aggression has made wars possible—though it does not necessarily initiate wars. But it would appear that, if man is to survive, this aspect of his personality will have to be controlled—an achievement that in the past few years has seemed increasingly possible.

There is, however, no agreement on whether or not we should reply. The advice of the Czech astronomer Zdenek Kopal is: "For God's sake let us not answer." George Wald, professor of biochemistry at Harvard, winner of a Nobel prize and a firm believer that life has evolved beyond earth, has said: "I can conceive of no nightmare so terrifying as establishing such communication with a so-called superior (or, if you wish, advanced) technology in outer space." To be attached, "as by an umbilical cord," to a more advanced civilization would be a "degradation of the human spirit," Wald told a 1972 symposium at Boston University. The possibly devastating effects of such a contact were cited in the Brookings Institution study commissioned by NASA soon after it was formed. The fate of the American Indian would be a case in point.

Another analysis was made by Michael Michaud, a foreign service officer who, in addition to handling such problems as European Security, NATO and arms control, had developed a special interest in SETI. While on the State Department's Iran Desk in 1972 he published an article in the *Foreign Service Journal* titled "Interstellar Negotiation," in which he listed a variety of possible dangers following contact. He rejected the argument that interstellar travel is impractical: Distance does not guarantee security "any more than an advanced civilization guarantees peaceful intentions." Regrettably, he said, "the stereotype of the benevolent, super-intelligent alien may be as unrealistic as the stereotype of the bug-eyed monsters carrying off shapely human females. . . . A species which had experienced nothing but hostile contacts with others, perhaps resulting in military conflict, would be predisposed to be hostile and to make military preparations." A great galactic empire "might regard us as nothing more than a troublesome infestation to be circumscribed or eliminated."

One of the primary roles of government is to protect its people from external threats. The dangers from beyond the earth could take a number of forms, Michaud said. "In a standard science-fiction scenario, the aliens might use super-weapons to blast us from orbiting vehicles. They might use chemical or biological fumigation. Or they might have the power to trigger an explosion in our sun, turning it into a nova which would fry the earth."

There was no certainty, he said in a later article, that the aliens would have any more concern for human beings than we do for whales and dolphins.

Since the aliens might have competing states, he wrote (showing his foreign affairs background), it would be important to learn whether the message came from only one of them. "An expansionist species might have colonized many planets, creating a far-flung empire or federation. We would need to know if we were communicating with a strong central authority, a weak one losing its grip, or a rebellious colony. We could unwittingly become involved in an interstellar civil war."

Our reaction to a signal, he said, could be either silence or a reassuring response that was prelude to "interstellar negotiation." The latter, he said, would be "the most difficult diplomacy Earthmen have ever attempted." It could last centuries, "it could risk human survival; or it could bring an incalculable richness of knowledge, physical instrumentalities, and cultural growth, and open a door to a Galactic society."

Arthur C. Clarke, one of the most knowledgeable of science-fiction writers, has pointed out, however, the difficulties of administering a galactic empire, be it benign or tyrannical, because the distances would make communication so slow. Radio messages to our nearest neighbors would take decades and those across the galaxy would require tens of thousands of years.

In a 1977–1978 journal of the American Institute of Aeronautics and Astronautics, Michaud warned that the first detection of an alien message might not be from a scientific search but by such a source as military surveillance, and might be kept secret. There would be a strong temptation, he said, "for individual nations or groups to conduct separate dialogues with the aliens to exploit contact for their own purposes." When contact became public knowledge, that "almost certainly would cause many more humans to attribute events on Earth to aliens; there would be an upsurge in witch-hunting and UFO sightings."

On the other hand, he said, "many of us would simply be excited by this outside stimulus, with its suggestion of a break with conventionality and of new prospects for the future; the world needs a shared adventure." He cited Philip Morrison's argument that contact with alien worlds could bring us a richer store of information than that inherited by medieval Europe from ancient Greece. It could lead to a new and greater Renaissance. What an enriching experience it would be to learn of the histories, political and economic organizations and cultural achievements of entirely different civilizations! It is sometimes argued that we might learn a cure for cancer or how best to harness the fusion reaction that powers the sun, but what we learn medically from beings that evolved via a different biochemical route might be meaningless and we might long since have conquered the fusion reaction ourselves.

In time, said Michaud, communication with a civilization more advanced than ours could be "the beginning of the end of Man as we have known him." Our culture "might fade and vanish, becoming a quaint historical memory, as we merged with a superior culture."

In the early 1980s a committee of the International Academy of Astronautics began discussing what to do if signals were detected, and a number of papers on the subject were presented at subsequent annual congresses of the International Astronautical Federation. Allan E. Goodman of the School of Foreign Service at Georgetown University proposed a code of conduct at the federation's 1986 congress in Innsbruck, Austria. Finally a draft prepared by Michaud and two key SETI figures, John Billingham and Jill Tarter, was presented and approved at the federation's 1990 congress in Dresden, Germany. Titled "Principles for Activities Following the Detection of Extraterrestrial Intelligence," it has subsequently been approved by the Committee on Space Research (COSPAR), a surviving element of the International Geophysical Year of 1957–58, as well as by other international organizations. It has been put before the International Astronomical Union and is designed for ultimate endorsement by the United Nations, possibly after the first evidence of extraterrestrial intelligence is observed.

The parties to the declaration are "We, the institutions and individuals participating in the search for extraterrestrial intelligence." One of its key principles is that no detection be announced until all parties to the declaration have been informed, so that they can try to confirm the observation. If the evidence is credible, the IAU's Central Bureau for Astronomical Telegrams in Cambridge, Massachusetts, is to be informed, as well as the secretary-general of the United Nations, as provided by the UN Outer Space Treaty. An intergovernmental treaty on the subject is not recommended at this stage.

"The discoverer should have the privilege of making the first public announcement," the declaration says, providing also that: "No response to a signal or other evidence of extraterrestrial intelligence should be sent until appropriate international consultations have taken place," as specified in a separate declaration yet to come.

Possible provisions of the latter were discussed by John Billingham at the 1991 Russian-American conference in Santa Cruz. Any message, he said, should be on behalf of all humanity, speaking "with one voice." The decision to act should be by an appropriate, all-encompassing international body. Silence is to be maintained until it is finally decided whether or not to respond and what to say if we do.

Still mysterious is the nature of those we might be dealing with. Because we have information on only one form of intelligent, technological life, we tend to think of such life elsewhere as resembling ourselves far more closely than is justified. The creatures in which we are interested, besides having minds, must be able to move about and to build things. That is, they must have something comparable to hands and feet. They must have senses, such as sight, touch and hearing, although the senses that evolve on any given planet will be determined by the environment. For example, it may be that, for various reasons, vision in the infrared part of the spectrum will be more useful than sight in the wavelengths visible to human eyes. Creatures fulfilling such requirements might bear little resemblance to man. As Philip Morrison has put it, they may be "blue spheres with twelve tentacles." They may be as big as a mountain or as small as a mouse, although the amount of food available would set a limit on largeness and the fixed sizes of molecules must limit the extent to which the size of a complex brain can be compressed. Life spans in some worlds might be extremely great. The cells of our bodies (with a few exceptions, such as brain cells) are constantly replenishing themselves. It would seem that barring accident or disease this should continue indefinitely, but because of some subtle influence the replacement process is imperfect. This, the essence of aging, is now under intensive study. It is not inconceivable that it can be controlled. Progress in the transplantation of human organs and the manufacture of other body components (such as heart valves) has led some of the most sober medical men to believe we may ultimately be able to extend lifetimes considerably. Lives measured in many

centuries instead of barely one would make the slowness of interstellar signaling (or travel) far more acceptable.

There is also the possibility that other civilizations will learn the
secret of aging and completely neutralize it. On earth mortality is
characteristic of multicelled, sexually reproducing creatures, but
not of single-celled ones that reproduce by cell division. It is fac-
tored into the individual cells of our bodies. While a cell from an
infant, when cultured in the laboratory, subdivides many times, it
eventually dies. A cell from an older person subdivides many fewer
times before doing so, as though there is a mortality factor hidden
in its genetic material. The nature of this factor remains elusive.
It may have evolved because it had survival value for a species. If
a species lived only long enough to produce a viable new generation, then died to make way for its offspring, the species could
evolve to meet a challenge far more rapidly than one in which
there were long intervals between generations.

Among those who believe that most, if not all, inhabitants of
other worlds are immortal is the SETI pioneer, Frank Drake. "I
fear we have been making a dreadful mistake," he wrote during
1976 in *Technology Review*, "by not focusing all searches—including
those to be accomplished by a system such as Cyclops [the proposed "orchard" of radio antennas]—on the detection of the signals of the immortals. For it is the immortals we will most likely
discover. . . . It has been said that when we first discover other
civilizations in space we will be the dumbest of them all. This is
true, but more than that, we will probably be the only mortal
civilization." Carl Sagan has discussed the possibility that the inhabitants of other worlds might not only have become immortal
but lost all motivation for "interstellar gallivanting."

Speculation on the existence and morality of those on other
worlds has a history that long predates the current surge of interest in the possibility of other inhabited earths. In the fifth century
B.C. Democritus postulated an infinite number of worlds, and his
pupil, Metrodorus of Chios, argued that they must be inhabited.
"To consider the earth the only populated world in infinite
space," he wrote, "is as absurd as to assert that in an entire field
sown with seed only one grain will grow."

Four centuries later Lucretius likewise assumed those worlds
were inhabited. The Copernican revolution, displacing the earth
from the center of the universe, set in motion a religious and
philosophical upheaval now further intensified by the reawakened

possibility that beings of uncertain morality may inhabit other worlds.

Despite the trial of Galileo, the burning of Giordano Bruno in the sixteenth century, and other suppressions of heresy, the Christian church has tended to accept new scientific ideas sooner than the public, according to Krister Stendahl, dean of the Harvard School of Theology. At a Boston University symposium in about 1965 he said: "I have studied very carefully the way in which the Christian church has lived with changed world views, from a near-Eastern view to the Ptolomeic view to the Copernican view, et cetera." The resistance to change, he said, came not from the theologians but from society as a whole. "And you would never hear a Jesus or a Buddha or a Mohammed criticizing the scientific world-view of their time," he continued. "They take for granted the world-view of their time, usually on a very popular level."

At the close of the eighteenth century the Reverend Timothy Dwight, president of Yale University, saw a moral challenge in the populations of other worlds. Dwight was a grandson of Jonathan Edwards, the chief puritanical theologian of his period, and colleges at Yale are named for them both. God, said Dwight in a sermon, had created a "countless multitude of worlds, with all their various furniture. With his own hand he lighted up at once innumerable suns, and rolled around them innumerable worlds. All these . . . he stored, and adorned, with a rich and unceasing variety of beauty and magnificence; and with the most suitable means of virtue and happiness."

By the mid-nineteenth century, while Father Angelo Secchi, the great Jesuit astronomer, was helping lay the foundations of modern astrophysics, he was also worrying about the religious implications of the vast universe opening before him. Could it be, he asked, that God populated only one tiny speck in this cosmos with spiritual beings: "It would seem absurd to find nothing but uninhabited deserts in these limitless regions," he wrote. "No! These worlds are bound to be populated by creatures capable of recognizing, honoring and loving their Creator."

Both Protestant and Catholic theologians pursued this subject during the early decades of the twentieth century, many of them viewing it in terms of man's spiritual history as set forth in their dogma. It was proposed by Catholic authorities that beings in other worlds could be in a variety of "states."

For example, they might be in a state of grace, such as that enjoyed by Adam and Eve before they succumbed to the serpent's temptation. They might be in the fallen state of mankind after the expulsion from Eden. They might have undergone these stages and been redeemed by some action of God. They might be in a "state of integral nature," midway between man and angel and immune to death; or they might have proved so evil that redemption was denied them. C. S. Lewis, the Anglican lay theologian, proposed that the vast distances between solar systems may be a form of divine quarantine: "They prevent the spiritual infection of a fallen species from spreading"; they block it from playing the role of the serpent in the Garden of Eden. In 1960 Father Daniel C. Raible of Brunnendale Seminary of the Society of the Precious Blood in Canton, Ohio, said in the Catholic weekly *America* that intelligent beings elsewhere "could be as different from us, physically, as an elephant is from a gnat." To be "human" in the theological sense they need only be "composites of spirit and matter." As God could create billions of galaxies, Raible reasoned, "so He could create billions of human races each unique in itself." To redeem such races, he said, God could take on any bodily form. "There is nothing at all repugnant in the idea of the same Divine Person taking on the nature of many human races. Conceivably, we may learn in heaven that there has been not one incarnation of God's son but many."

The idea of God incarnate in some strange, if not grotesque, creature is to some Catholics unacceptable. Such a dissenting point of view was expressed by Joseph A. Breig, a Catholic journalist, in a dialogue published in a 1960 issue of *America.* There can have been only one incarnation, one mother of God, one race "into which God has poured His image and likeness," he said.

In reply Father L. C. McHugh, an associate editor of the magazine, wrote: "Does it not seem strange to say that His power, immensity, beauty and eternity are displayed with lavish generosity through unimaginable reaches of space and time, but that the knowledge and love which alone give meaning to all this splendor are confined to this tiny globe where self-conscious life began to flourish a few millennia ago?"

Soon after his appointment as president of the University of Notre Dame, Theodore M. Hesburgh, an early and ardent supporter of SETI, visited Green Bank to witness the original Ozma search. He found the thought of life in other worlds entirely compatible with his Catholic faith. The search for it, he said, deserves "the serious and prolonged attention of many professionals from a wide range of disciplines—anthropologists, artists, lawyers, politicians, philosophers, theologians—even more than that, the concern of all thoughtful persons, whether specialists or not."

Like the Catholics, the Protestants have been concerned with whether or not God could have taken on bodily form elsewhere. Paul Tillich, one of the foremost Protestant theologians, argued that there is no real reason why such incarnations could not have taken place: "Incarnation is unique for the special group in which it happens, but it is not unique in the sense that other singular incarnations for other unique worlds are excluded. . . . Man cannot claim to occupy the only possible place for Incarnation. . . . The manifestation of saving power in one place implies that saving power is operating in all places."

Astronomy tells us that new worlds are constantly evolving and old ones are becoming uninhabitable. The lifetime of our own world is thus limited, Tillich said, but this "leaves open other ways of divine self-manifestation before and after our historical continuum." Elsewhere he wrote, with characteristic boldness: "Our ignorance and our prejudice should not inhibit our thoughts from transcending our earth and our history and even our Christianity."

The hint of Tillich that, though worlds come and go, spiritual life somewhere in the universe goes on forever has been carried further by John Macquarrie of the University of Glasgow, in lectures that he gave in 1957 at Union Theological Seminary in New York. He took his cue from the "steady state" theory proposed by Fred Hoyle, Hermann Bondi and Thomas Gold, in which, although the universe is constantly expanding, there is continuous creation of new matter to fill the resulting voids. "It might well be the case," Macquarrie said, "that the universe has produced and will continue to produce countless millions of . . . histories analogous to human history." The creation would be cyclical, in that it produces "the same kind of thing over and over again in endless variations.

"To attempt to draw ultimate conclusions about God and the universe from a few episodes of the history which has been enacted on this planet," he continued, "would seem to be a most hazardous if not impossible proceeding."

Long before Project Ozma the British mathematician and physicist Edward A. Milne proposed that only one incarnation—that of Jesus—was necessary, because when news of it reached other worlds via radio, they too would be saved. This was challenged by E. L. Mascall, university lecturer in the philosophy of religion at Oxford, who said redemption was, in effect, not exportable. Like a number of other theologians who have taken part in this dialogue, Mascall cited the argument of Saint Thomas Aquinas in the thirteenth century that the Incarnation could have taken one of several forms: instead of God the Son appearing on earth, he said, it might have been God the Father or God the Holy Ghost. He reasoned that none of these alternatives was ruled out on theological grounds. If so, today's scholars argue, why could there not also be incarnations elsewhere?

At the conclusion of his analysis, Mascall was almost apologetic for discussing such hypothetical problems: "Theological principles tend to become torpid for lack of exercise," he said, "and there is much to be said for giving them now and then a scamper in a field where the paths are few and the boundaries undefined; they do their day-to-day work all the better for an occasional outing in the country."

Actually, such "outings" have been taken by churchmen as sober as William Ralph Inge, dean of St. Paul's Cathedral in London from 1911 to 1934 and known as "the gloomy dean" for his criticism of modern life. Like Father Secchi, he considered it intolerably presumptuous to believe that spiritual life exists only on this one planet. In a series of lectures at Lincoln's Inn Chapel between 1931 and 1933 he discussed Plato's concept of a "Universal Soul," and its parallels with Christian views on the Holy Spirit. This, he said,

raises the question whether there is soul-life in all parts of the universe. I do not think we need follow Fechner in believing that every heavenly body has a soul of its own. [Gustav Theodor Fechner, the German experimental psychologist, preached a highly animistic philosophy.] But I would rather be a star-worshipper than believe with Hegel (if he was not merely speaking impatiently, which is likely enough) that the starry heavens have no more significance than a rash on the sky, or a swarm of flies. There may be, and no doubt are, an immense number of souls in the universe, and some of them may be nearer to the divine mind than we are.

It is, perhaps, a sign of the times and of the convergence of Catholicism and Protestantism that theologians of both faiths have been moved by the discussion of the Incarnation in a poem, "Christ in the Universe," by the English turn-of-the-century poet Alice Meynell:

> . . . No planet knows that this
> Our wayside planet, carrying land and wave,
> Love and life multiplied, and pain and bliss,
> Bears, as chief treasure, one forsaken grave. . . .
>
> But, in the eternities,
> Doubtless we shall compare together, hear
> A million alien Gospels, in what guise
> He trod the Pleiades, the Lyre, the Bear.
>
> O, be prepared, my soul!
> To read the inconceivable, to scan
> The million forms of God those stars unroll
> When, in our turn, we show to them a Man.

There are some faiths that have long preached the existence of many worlds, and so for them the shock of discovery would be lessened. Notable among these are the Buddhists and Mormons. The doctrine of the latter, the Church of Jesus Christ of Latter-Day Saints, is set forth in a series of revelations including one titled the "Visions of Moses, as revealed to Joseph Smith the Prophet, in June, 1830." This tells how Moses "beheld many lands; and each land was called earth, and there were inhabitants on the face thereof." God told Moses:

"And worlds without number have I created. . . . But only an account of this earth, and the inhabitants thereof, give I unto you. For behold, there are many worlds that have passed away by the word of my power. And there are many that now stand, and innumerable are they unto man; but all things are numbered unto me, for they are mine and I know them. . . . And as one earth shall pass away, and the heavens thereof even so shall another come; and there is no end to my works, neither to my words."

This is strikingly similar to the "steady state" concept advanced by such men as Tillich and Macquarrie.

The holy books of Buddhism, in their own glittering way, speak of many worlds. Such a book is the *Saddharma-Pundarika* or *Lotus of the True Law*. The Lord appears before an assembly of Bodhisattvas, or wise men, gathered from countless worlds and numbering eight times the grains of sand in the river Ganges. He tells them of many worlds, of golden people, jeweled trees, perfumed winds, wafting showers of petals.

While the Hindus believe in the transmigration of souls into other bodies or spiritual states, their various cosmologies do not envisage other inhabited worlds. As for the Confucians, their attention has been centered on earthly matters, in particular the behavior of man, even though some early Chinese scholars spoke of other earths with their own heavens shining upon them. The thirteenth-century philosopher Teng Mu wrote (as translated by Joseph Needham): "It is as if the whole of empty space were a tree, and heaven and earth were one of its fruits. Empty space is like a kingdom, and heaven and earth no more than a single individual person in that kingdom. Upon one tree there are many fruits, and in one kingdom many people. How unreasonable it would be to suppose that beside the heaven and earth which we can see there are no other heavens and no other earths."

About a dozen centuries ago Jewish thinkers also conceived of the cosmos in somewhat a steady-state manner. This is reflected in one of the commentaries on the Bible, prepared long ago by scholarly rabbis and entitled the *Midrash Rabba*. It states that "The Holy One, blessed be he, builds worlds and destroys them." However, according to contemporary Jewish theologians there has not been much speculation on the religious implications of extraterrestrial life.

Some religious thinkers have been troubled at the attempts of biologists in the laboratory to produce something that could be considered a primitive form of life, regarding this as usurping the role of God. George Wald of Harvard has scoffed at this, arguing that life is built into the chemistry of the universe, poised to emerge wherever conditions are right. Such experimenters are not "creating" life, he said. They are simply trying to establish the conditions whereby life can emerge. Life, he wrote, is "part of the order of nature. It has a high place in that order, since it probably represents the most complex state of organization that matter has achieved in our universe."

It is true that even as our personal lives are numbered, so is that of the earth itself. It is destined to be incinerated when the sun becomes a red giant or frozen when the sun burns out many billions of years hence. As Bertrand Russell pointed out, products of human genius—all the labors of the ages, all the artistic, literary and architectural achievements of past millenia—are doomed to extinction when the life of this planet ends.

Will this be the end of all life? asked Russell's contemporary, Sir James Jeans. "Is this, then, all that life amounts to—to stumble, almost by mistake," into existence. Is it our destiny "to stay clinging on to a fragment of a grain of sand until we are frozen off, to strut our tiny hour on our tiny stage with the knowledge that our aspirations are all doomed to final frustration, and that our achievements must perish with our race, leaving the universe as though we had never been?"

The answer is either communication or colonization. Once the Genome Project has provided a full genetic blueprint of a human being, we will be able, given a time frame of millions, or even billions, of years, to transmit that information to any society that wants to build replicas. Assuming we do not suffer from contact with a superior civilization, the positive effect could be an enrichment beyond anything the human race has hitherto experienced.

The Jesuit scholar and philosopher Teilhard de Chardin, in his book *The Phenomenon of Man* (not published, because of its unorthodoxy, until after his death in 1955), cited the "mutual fecundation" that would occur from the encounter between two civilizations that had evolved independently. Likewise the astrophysicist Alastair Cameron, in the introduction to his anthology *Interstellar Communication*, said "there may be millions of societies more advanced than ourselves in our galaxy alone. If we can now take the next step and communicate with some of these societies, then we can expect to obtain an enormous enrichment of all phases of our sciences and arts. Perhaps we shall also receive valuable lessons in the techniques of stable world government."

One can hope that knowledge of such a civilization, and how it overcame prejudice, hatred and war, might enable us to leapfrog centuries ahead. Until we make contact we will never know how much we can learn from those whose biology, way of life, medical history, and internal problems are very different from our own. Nevertheless true wisdom may be a torch—one that we have not yet received, but that can be handed down to us from a civilization late in its life and passed on by our own world as its time of extinction draws near. Thus, as our children and grandchildren offer some continuity to our personal lives, so communion with cosmic manifestations of life would join us with a far more magnificent form of continuity.

Our world is undergoing revolutionary changes. As we devastate our planet with industrialization, pollution, highways, housing and haste, the restoration of the soul that comes from contemplating nature unmarred by human activity becomes more and more inaccessible. Through our material achievements, we face the ultimate prospect of a society dominated by mediocrity, conformity and comfort-seeking. The world desperately needs a global adventure to rekindle the flame that burned so intensely during the Renaissance, when new worlds were being discovered on our own planet and in the realms of science. Some of our fellows have trod the moon and others may tread Mars, but the possibility of ultimately "seeing" worlds in other solar systems, however remote, is an awesome prospect. "The soul of man was made to walk the skies," the English poet Edward Young wrote in the eighteenth century:

> how great,
> How glorious, then, appears the mind of man,
> When in it all the stars and planets roll!
> And what it seems, it is. Great objects make
> Great minds, enlarging as their views enlarge;
> Those still more godlike, as these more divine.

A similar view was presented by the Spanish-American philosopher George Santayana. Those who commune with the eternal, he said, come close to immortality. "He who, while he lives, lives in the eternal, does not live longer for that reason. Duration has merely dropped from his view; he is not aware of or anxious about it; and death, without losing its reality, has lost its sting. The sublimation of his interest rescued him, so far as it goes, from the mortality which he accepts and surveys."

The universe that lies about us, visible only in the privacy, the intimacy of a clear night, is incomprehensibly vast, and the conclusion that life exists across the vastness seems inescapable. We cannot yet be sure whether it lies within reach, but common sense says we are not alone. According to Lee DuBridge, former head of Caltech and science adviser to President Nixon, while it remains uncertain whether there are other inhabited worlds or we are alone in this vast universe, either alternative is "mind-boggling."

When one looks out among the stars and contemplates the possibility that the miracle of life may have occurred only on this one planet, the thought is frightening, for our stewardship has not been notable. In numerous ways we have transformed the earth and its atmosphere to suit us, but not the millions of other species. In 1992 the National Academy of Sciences issued a report saying that if current trends continue, a quarter of the world's species of plant and animal may have vanished within fifty years. In the past, evolution has not allowed one species to dominate the planet for long. There always emerged a predator, a disease, a competitor, or exhaustion of its sustenance that brought it in check or to extinction. So far the human race, through exercise of its "intelligence," has avoided this, but can it do so indefinitely? An awesome responsibility falls on our shoulders. Our heritage becomes cosmic, and measures to preserve the species and the environment cannot easily be dismissed as too costly. To meet the challenge is the supreme test of our intelligence.

Bibliography

General Sources (with pertinent citations)

Billingham, John, ed., *Life in the Universe*, MIT Press (Cambridge, 1981).

Bracewell, Ronald, *The Galactic Club—Intelligent Life in Outer Space*, W. H. Freeman (San Francisco, 1974–75).

Cameron, A. G. W., ed., *Interstellar Communication*, W. A. Benjamin, Inc. (New York, 1963).

Chipman, Ralph, ed., *The World in Space*, A United Nations survey of space activities and issues, chapter 1.8, Prentice-Hall (Englewood Cliffs, N.J., 1982).

Drake, Frank, "On Hands and Knees in Search of Elysium," *Technology Review*, vol. 78, no. 7 (June 1976), pp. 22–29.

Feinberg, Gerald, and Robert Shapiro, *Life Beyond the Earth, the Intelligent Earthling's Guide to Life in the Universe*, William Morrow (New York, 1980).

Kellerman, K., and B. Sheets, eds., *Serendipitous Discoveries in Radio Astronomy*, National Radio Astronomy Observatory (1984).

Morrison, Philip, John Billingham and John Wolfe, eds., *The Search for Extraterrestrial Intelligence—SETI*, NASA (1977).

Murray, Bruce, Samuel Gulkis and Robert E. Edelson, "Extraterrestrial Intelligence: An Observational Approach," *Science*, vol. 199 (Feb. 3, 1978), pp. 485–92.

Oliver, Bernard M., "The Search for Extraterrestrial Intelligence," *Engineering and Science*, Dec. 1974–Jan. 1975, pp. 7–11 and p. 30 passim.

Papagiannis, Michael D., "Recent progress and future plans on the search for extraterrestrial intelligence," *Nature*, vol. 318 (Nov. 14, 1985), pp. 135–40.

Ponnamperuma, Cyril, and A. G. W. Cameron, *Interstellar Communication: Scientific Perspectives*, Houghton Mifflin Co. (Boston, 1974).

Sagan, Carl, and Frank Drake, "The Search for Extraterrestrial Intelligence," *Scientific American*, May 1975, pp. 80–89.

Shklovsky, I. S., and Carl Sagan, *Intelligent Life in the Universe*, Holden-Day, Inc. (San Francisco, 1966).

Smith, Marcia S., "Possibility of Intelligent Life Elsewhere in the Universe," Report prepared for the Committee on Science and Technology, U.S. House of Representatives, 94th Congress, Nov. 1975, revised for the 95th Congress, 1977, Serial O.

Swift, David W., *SETI Pioneers. Scientists Talk about Their Search for Extraterrestrial Intelligence*, University of Arizona Press (Tucson, 1990).

1. Spheres within Spheres

Flammarion, Camille, *La Pluralité des Mondes Habités, Etude où l'on expose les conditions d'habitabilité des terres célestes discutées au point de vue de l'astronomie, de la physiologie et de la philosophie naturelle*, 2nd ed. (Paris, 1864). Book One deals with early speculations.

Heath, Sir Thomas, *Aristarchus of Samos, the Ancient Copernicus*, Clarendon Press (Oxford, 1913).

Laertius, Diogenes, *Lives of Eminent Philosophers*, with an English translation by R. D. Hicks, Harvard University Press (Cambridge, 1931–1950). See sections on Anaxagoras, Zeno and other early thinkers.

Lucretius, *The Nature of the Universe*, translated with an introduction by Ronald Latham, Penguin Books (Baltimore, 1951), pp. 29, 91, 92.

Munitz, M. K., ed., *Theories of the Universe from Babylonian Myth to Modern Science*, The Free Press (Glencoe, Ill., 1957). See texts of relevant documents from ancient, Renaissance and post-Renaissance periods.

Plato, *Timaeus*, from *The Collected Dialogues of Plato*, Edith Hamilton and H. Cairns, eds., Bollingen Series 71, Pantheon (New York, 1961), pp. 1164–65.

Robertson, H. P., "The Universe," *Scientific American*, Sept. 1956, p. 73.

Teng Mu, quoted by Philip Morrison, *Bulletin of the Philosophical Society of Washington*, vol. 16 (1962), p. 81.

2. Science Reborn

Borel, Pierre, quoted by Flammarion, *op. cit.*, pp. 30–31.

Bruno, Giordano: see *Giordano Bruno, His Life and Thought, With Annotated Translations of His Work, On the Infinite Universe and Worlds*, by Dorothea Waley Singer, Henry Schuman (New York, 1950), pp. 257, 304; also *Historical Trials* by Sir John Macdonnell (Oxford, 1927), p. 83; and *Giordano Bruno* by J. L. McIntyre (London, 1903), p. 99.

Digges, Thomas, quoted by Munitz, *op. cit.*, p. 188.

Dingle, Herbert, "Cosmology and Science," *Scientific American*, Sept. 1956, p. 228.

de Bergerac, Cyrano, *Voyage à la Lune*, extract in Flammarion, *op. cit.*, pp. 509–14.

de Fontenelle, Bernard Le Bovier, *Conversations on the Plurality of Worlds* (London, 1809), with a "Memoir of the Life and Writings of the late Monsieur de Fontenelle," by M. de Voltaire, pp. 5, 59, 78, 98. (Originally published in 1686 as *Entretiens sur la pluralité des mondes.*)

Huygens, Christiaan, *Cosmotheros*, quoted by Munitz, *op. cit.*, pp. 221–22.

Milton, John, *Paradise Lost*, Book Eighth, lines 145 ff.

Newton, Isaac, letter to Richard Bentley, Dec. 10, 1692, quoted by Munitz, *op. cit.*, p. 211.

Paine, Tom, "The Age of Reason," *The Selected Works of Tom Paine and Citizen Tom Paine*, The Modern Library of the World's Best Books, 1946, pp. 303, 305, 312, 316, 321, passim.

Pope, Alexander, *An Essay on Man*, Epistle I, lines 19–28.

Wilkins, Bishop John, *Mathematical Magick* (1648), quoted by Vilhjalmur Stefansson in *Northwest to Fortune*, Duell, Sloan & Pearce (New York, 1958), p. 333.

3. Is Our Universe Unique?

Herschel, William, "On the Construction of the Heavens," see excerpts in Munitz, *op. cit.*, pp. 264–68.

Kant, Immanuel, *Universal Natural History and Theory of the Heavens*, see excerpts in *ibid.*, pp. 231–49.

Leavitt, H. S., "Discovery of the Period-Magnitude Relation," in *Source Book in Astronomy 1900–1950*, H. Shapley, ed., Harvard University Press (Cambridge, 1960), pp. 186–89, excerpted from "Periods of 25 Variable Stars in the Small Magellanic Cloud," *Harvard Circular*, No. 173 (1912).

———, "1777 Variables in the Magellanic Clouds," *Annals of Harvard College Observatory*, vol. 60, no. 4 (1908), pp. 87–108.

Shapley, Harlow, "From Heliocentric to Galactocentric," in *Source Book in Astronomy, op. cit.*, pp. 319–24, excerpted from *Star Clusters*, Harvard Observatory Monographs, No. 2, McGraw-Hill (New York, 1930), pp. 171–78.

———, *Of Stars and Men, The Human Response to an Expanding Universe*, Beacon Press (Boston, 1958), pp. 9, 53, 55.

Struve, Otto, and Velta Zebergs, *Astronomy of the 20th Century*, Macmillan (New York, 1962), chapters 15 ("Pulsating Variable Stars"), 19 ("The Milky Way") and 20 ("Galaxies").

de Vaucouleurs, Gérard, "The Supergalaxy," *Scientific American*, July 1954, pp. 30–35.

Wright, Thomas, *An Original Theory or New Hypothesis of the Universe* (1750), quoted by Munitz, *op. cit.*, p. 226, and by Munitz in *Space, Time and Creation, Philosophical Aspects of Scientific Cosmology*, The Free Press (Glencoe, Ill., 1957), p. 22.

4. Seeking Other Suns

Abt, Helmut A., "The Rotation of Stars," *Scientific American*, April 1977, pp. 46–53.

Huang, S. S., "A Nuclear-Accretion Theory of Star Formation," *Publications of the Astronomical Society of the Pacific*, vol. 69 (Oct. 1957), pp. 427–30.

———, "The Problem of Life in the Universe and the Mode of Star Formation," *ibid.*, vol. 71 (Oct. 1959), pp. 421–24, reprinted in Cameron, *op. cit.*, as Item 7.

———, "Life-Supporting Regions in the Vicinity of Binary Systems," *ibid.*, vol. 72 (April 1960), pp. 106–14, reprinted in Cameron, *op. cit.*, as Item 8.

———, "The Limiting Sizes of the Habitable Planets," *ibid.*, vol. 72 (Dec. 1960), pp. 489–93, reprinted in Cameron, *op. cit.*, as Item 9.

———, "Occurrence of Life in the Universe," *American Scientist*, vol. 47 (Sept. 1959), pp. 397–402, reprinted in Cameron, *op. cit.*, as Item 6.

———, "Life Outside the Solar System," *Scientific American*, April 1960, pp. 55–63.

Russell, H. N., "The Spectrum-Luminosity Diagram," *Source Book in Astronomy, op. cit.*, pp. 253–62, excerpted from *Popular Astronomy*, vol. 22 (1914), pp. 275–94, 331–51.

Shajn, G. A., and Otto Struve, "On the Rotation of the Stars," *Source Book in Astronomy, op. cit.*, pp. 116–23, originally in *Monthly Notices of the Royal Astronomical Society*, vol. 89 (1929), pp. 222–39.

Shapley, Harlow, "Crusted Stars and Self-Heating Planets," *Matematica y Fisica Teorica, Serie A*, vol. 14 (1962), Tucumán (Argentina) National University.

Spitzer, Lyman, Jr., "The Beginnings and Future of Space Astronomy," *American Scientist*, vol. 50 (Sept. 1962), pp. 473–84.

Struve, Otto, *Stellar Evolution*, Princeton University Press (1950), p. 130.

———, "Proposal for a Project of High-Precision Stellar Radial Velocity Work," *The Observatory*, vol. 72 (Oct. 1952), pp. 199–200.

———, *The Astronomical Universe*, Condon Lectures, Oregon State System of Higher Education (Eugene, 1958), p. 26.

Struve, Otto, and Zebergs, *op. cit.*, p. 112, chapters 10 ("Spectral Classification"), 11 ("Stellar Atmospheres and Spectroscopy") and 14 ("Double Stars").

5. The Solar System: Exception or Rule?

Alfvén, Hannes, and Gustaf Arrhenius, *Evolution of the Solar System*, NASA (1976).

Cameron, A. G. W., "The History of Our Galaxy" and "The Origin of the Solar System," in Cameron, *op. cit.*, Items 2 and 3.

———, "The Formation of the Sun and Planets," *Icarus*, vol. 1 (May 1962), pp. 13–69.

Eucken, Arnold, "Über den Zustand des Erdinnern," *Die Naturwissenschaften* (Berlin, April–June 1944), pp. 112–21.

Flammarion, *op. cit.*, pp. 48–49, 266–69 (on the views of Bode and Kant).

Fowler, W. A., "Nuclear Clues to the Early History of the Solar System," *Science*, vol. 135 (March 23, 1962), pp. 1037–45.

Schmidt, O. J.: see B. J. Levin, "The Origin of the Solar System," *New Scientist*, vol. 13 (Feb. 8, 1962), p. 323.

Swedenborg, Emanuel, "Earths in Our Solar System Which Are Called Planets and Earths in the Starry Heaven Their Inhabitants, and the Spirits and Angels There from Things Heard and Seen," from the Latin of Emanuel Swedenborg, Swedenborg Society (London, 1962).

Ter Haar, D., and A. G. W. Cameron, "Historical Review of Theories of the Origin of the Solar System," in *Origin of the Solar System, Proceedings of a conference held at the Goddard Institute for Space Studies, New York, January 23–24, 1962*, Robert Jastrow and A. G. W. Cameron, eds., Academic Press (New York, 1963).

———, "Origin of the Solar System," *Annual Reviews of Astronomy and Astrophysics*, vol. 26 (1988), pp. 441–72.

Urey, H. C., *The Planets, Their Origin and Development*, Yale University Press (New Haven, 1952).

Urey, H. C., "Diamonds, Meteorites, and The Origin of the Solar System," *Astrophysical Journal*, vol. 124 (Nov. 1956), pp. 623–37.

Whipple, F. L., "History of the Solar System," paper presented at the Centennial Meeting, National Academy of Sciences (Washington, D.C., Oct. 21, 1963).

6. Seeing Infrared

Dietrich, Jane, "Discoveries from IRAS," (unsigned), *Engineering and Science*, vol. 47, no. 3 (1984), pp. 7–11.

Dyson, Freeman J., "Search for artificial sources of infrared radiation," *Science*, vol. 131 (June 3, 1960), p. 1667.

Gatley, Ian, D. L. DePoy, and A. M. Fowler, "Astronomical Imaging with Infrared Array Detectors," *Science*, vol. 242 (Dec. 2, 1988), pp. 1264–69.

Henbest, Nigel, "First Light on Starbirth," *New Scientist*, Aug. 27, 1987, pp. 46–49.

Levy, E. H., "Protostars and Planets: Overview from a planetary perspective," *Protostars and Planets II*, University of Arizona Press (Tucson, 1984), pp. 3–16.

Levy, E. H., J. I. Lunine, and M. S. Matthews, eds., *Protostars and Planets*

III, University of Arizona Press (Tucson, 1993 [in press]). This volume has several contributions on planet formation.

Overbye, Dennis, "Mapping the heat of heaven," *Discover*, April 1983, pp. 82–85.

7. Planet-Hunting

Abt, Helmut A., "The Companions of Sunlike Stars," *Scientific American*, April 1977, pp. 96–104.

Angel, J. R. P., A. Y. S. Cheng and N. J. Woolf, "A space telescope for infrared spectroscopy of Earth-like planets," *Nature*, vol. 322 (July 24, 1986), pp. 341–43. This was followed on December 11 by an exchange of commentaries by these authors and Bernard Burke.

Bailes, M., A. G. Lyne and S. L. Shemar, "A planet orbiting the neutron star PSR1829-10," *Nature*, vol. 352 (July 25, 1991), pp. 311–13.

Black, David, "It's all in the timing," *Nature*, vol. 352 (July 25, 1991), pp. 278–79.

———, "Finding and Studying Other Planetary Systems," *Planetary Report*, Nov.–Dec. 1987, pp. 20–21.

Black, David C., and Graham C. J. Suffolk, "Concerning the Planetary System of Barnard's Star," *Icarus*, vol. 19, (1973), pp. 353–57.

Burke, Bernard F., "Detection of planetary systems and the search for evidence of life," *Nature*, vol. 322 (July 24, 1986), pp. 340–41.

Diner, David J., and John F. Appleby, "Prospects for planets in circumstellar dust: sifting the evidence from Beta Pictoris." *Nature*, vol. 322, (July 31, 1986), pp. 426–38.

Esposito, Larry W., chairman, *Strategy for the Detection and Study of Other Planetary Systems and Extrasolar Planetary Materials*, National Research Council, 1990.

Fabian, A. C., and Ph. Podsiadlowski, "Binary precursors for planet?" *Nature*, vol. 353 (Oct. 31, 1991), p. 801.

Field, George, and Donald Goldsmith, *The Space Telescope, Eyes Above the Atmosphere*, Contemporary Books, Inc. (Chicago, 1989).

Gehrels, T., and R. S. McMillan, eds., *The Spacewatch Report*, No. 5, Tucson, Arizona, Feb. 13, 1991.

Hanson, Robert B., B. F. Jones and D. N. C. Lin, "The astrometric position of T Tauri and the nature of its companion," *Astrophysical Journal*, vol. 270 (July 1, 1983), L-27-L30.

Harrington, Robert S., and Betty J. Harrington, "Van Biesbroeck 8 and van Biesbroeck 10: Dark Companions of Nearby Star," *Mercury*, Jan.–Feb. 1985, pp. 14–15.

Lindley, David, "Solar systems beyond our own," *Nature*, vol. 334 (Aug. 11, 1988), p. 467.

O'Leary, Brian, "Searching for Other Planetary Systems," *Sky and Telescope*, Aug. 1980, pp. 111–13.

Podsiadlowski, Ph., J. E. Pringle and M. J. Rees, "The origin of the planet orbiting PSR1829-10," *Nature*, vol. 352 (Aug. 29, 1991), pp. 783–84.

van de Kamp, Peter, "Planetary Companions of Stars," *Vistas in Astronomy*, A. Beer, ed., vol. 2, Pergamon Press (London, 1956), pp. 1040–48.

————, "Barnard's Star as an Astrometric Binary," *Sky and Telescope*, vol. 26 (July 1963), pp. 8–9.

————, "Astrometric Study of Barnard's Star from Plates Taken with the 24-inch Sproul Refractor," *Astronomical Journal*, vol. 68, (Sept. 1963), pp. 515–21.

Walgate, Robert, "Emerging solar systems in view," *Nature*, vol. 304 (Aug. 21, 1983), p. 681.

Wolszczan, A. and D. A. Frail, "A planetary system around the millisecond pulsar PSR1257+12," *Nature*, vol. 355 (Jan. 9, 1992), pp. 145–47.

Wuers, R. A. M. J., E. P. J. van den Heuvel, M. H. van Kerkwuk, and D. Bhattacharya, "Genesis of a pulsar's planets," *Nature*, vol. 355 (Feb. 13, 1992), p. 593.

8. Creation or Evolution?

Akabori, Shiro, "On the Origin of the Fore-protein," *Proceedings of the First International Symposium on the Origin of Life on the Earth*, Aug. 1957, Pergamon Press (London, 1959), pp. 189–96.

Barghoorn, E. S., see "Woodring Conference on Major Biologic Innovations and the Geologic Record," by P. E. Cloud, Jr., and P. H. Abelson, *Proceedings of the National Academy of Sciences*, vol. 47 (Nov. 15, 1961), pp. 1705–12.

Bernal, J. D., "The Problem of Stages in Biopoesis," *Proceedings of the First International Symposium . . . , op. cit.*, pp. 38–53.

Darwin, Charles, letter to G. C. Wallich, March 28, 1882, and an earlier letter, from *Notes and Records of the Royal Society of London*, vol. 14, no. 1 (1959), quoted by M. Calvin in *Chemical Evolution*, Condon Lectures, Oregon State System of Higher Education (Eugene, 1961), p. 2.

Dubos, R. J., *Louis Pasteur, Free Lance of Science*, Little, Brown (Boston, 1950), Chapter 6.

Haldane, J. B. S., "The Origin of Life" (1928), published in *The Rationalist Annual* (1929), pp. 3–10, and in *The Inequality of Man and Other Essays* (London, 1932), pp. 148–60.

————, "Radioactivity and the origin of life in Milne's cosmology," *Nature*, vol. 153 (May 6, 1944), p. 555.

————, "Origin of Life," *New Biology*, no. 16, Penguin (1954).

van Helmont, J.-B., quoted by Louis Pasteur, *Oeuvres* (Paris, 1922), vol. 2, p. 329.

Michelet, Jules, *La Mer* (Paris, 1861), pp. 116–17.

Oparin, A. I., *Origins of Life*, translated with annotations by Sergius Morgulis, Dover Publications (New York, 1953). See particularly historical sections in Chapters 1 and 2 and p. 246.

Pasteur, Louis, *Oeuvres, op. cit.*, vol. 2, pp. 202–5, 328–46.

————, *Comptes Rendus hebdomadaires des Séances de l'Académie des Sciences* (Paris), vol. 50 (1860), pp. 303–7, 849–54; vol. 51 (1860), pp. 348–52; vol. 56 (1863), pp. 734–40.

Pouchet, F. A., *ibid.*, vol. 47 (1858), pp. 979–84; vol. 48 (1859), pp. 148–58, 546–51; vol. 57 (1863), pp. 765–66.

Vallery-Radot, René, *The Life of Pasteur* (London, 1911), vol. 1, pp. 92–112.

9. Panspermia

Anders, Edward, "Pre-biotic organic matter from comets and asteroids," *Nature*, vol. 342 (Nov. 16, 1989), pp. 255–57.

Anders, Edward, Ryoichi Hayatsu and Martin H. Studier, "Interstellar Molecules: Origin by Catalytic Reactions on Grain Surfaces?", *Astrophysical Journal*, vol. 192 (Sept. 1, 1974), L101–L105.

Bally, John, "Interstellar Molecular Clouds," *Science*, vol. 232 (April 11, 1986), pp. 183–92.

Belton, Michael J. S., "The Wobbling Nucleus of Halley's Comet," *The Planetary Report*, vol. VII, no. 2 (March–April 1987), pp. 8–10.

Bonnet, Roger-M, Roald Z. Sagdeev, Minoru Oda and Burton I. Edelson, *Encounter '86, An International Rendezvous with Halley's Comet*, The European Space Agency BR-27 (Nov. 1986).

Buhl, David, "Molecules and Evolution in the Galaxy," *Sky and Telescope*, March 1973, pp. 156–58.

Chown, Marcus, "Organics or organisms in Halley's nucleus," *New Scientist*, April 17, 1986, p. 23.

Chyba, Christopher, and Carl Sagan, "Endogenous production, exogenous delivery and impact-shock synthesis of organic molecules: an inventory for the origins of life," *Nature*, vol. 355 (Jan. 9, 1992), pp. 125–131.

Chyba, Christopher F., Paul J. Thomas, Leigh Brookshaw and Carl Sagan, "Cometary Delivery of Organic Molecules to the Early Earth," *Science*, vol. 249 (July 27, 1990), pp. 366–73.

Crick, F. H. C., and L. E. Orgel, "Directed Panspermia," *Icarus*, vol. 19 (1973), p. 341.

Flam, Faye, "Seeing Stars in a Handful of Dust," *Science*, vol. 253 (July 26, 1991), pp. 380–81.

Goldsmith, Donald, and Nathan Cohen, "The Great Molecule Factory in Orion," *Mercury*, Sept.–Oct. 1991, pp. 148–152.

Hoyle, Sir Fred, and Chandra Wickramasinghe, *Evolution from Space, a Theory of Cosmic Creationism*, Simon and Schuster (New York, 1982), Dent edition 1981.

———, "The case for life as a cosmic phenomenon," *Nature*, vol. 322 (Aug. 7, 1986), pp. 509–10.

Irvine, William M., "Chemistry Between the Stars," *The Planetary Report*, Nov.–Dec. 1987, pp. 6–9.

Keller, H. U., R. Kramm and N. Thomas, "Surface features on the nucleus of comet Halley," *Nature*, vol. 331 (Jan. 21, 1988), pp. 227–40.

Kissel, J., and F. R. Krueger, "The organic component in dust from comet Halley as measured by the PUMA spectrometer on board Vega 1, *Nature*, vol. 326 (April 23, 1987), pp. 755–60.

Ponnamperuma, Cyril, ed., *Comets and the Origin of Life*, Proceedings of the Fifth College Park Colloquium on Chemical Evolution, University of Maryland, D. Reidel Publishing Co. (1981).

Thompson, W. Reid, and Carl Sagan, "Organic Chemistry on Titan— Surface Interactions," *Proceedings of the Symposium on Titan*, Toulouse, France, Sept. 1991 European Space Agency publication SP338, Noordwijk, Netherlands, 1992.

Verschuur, Gerrit L., "Molecules Between the Stars," *Mercury*, May–June 1987, pp. 66–76.

Wickramasinghe, D. T., and D. A. Allen, "Discovery of organic grains in comet Halley," *Nature*, vol. 323 (Sept. 4, 1986), pp. 44–46.

Wilkening, Laurel L., ed., *Comets*, with 48 collaborating authors. University of Arizona Press (Tucson, 1982).

Yokoo, Hiromitsu, and Tairo Oshima, "Is Bacteriophage ThetaX174 DNA a message from an Extraterrestrial Intelligence?" *Icarus*, vol. 38 (1949), pp. 148–53.

10. Fossils from Space

Anders, Edward, and G. G. Goles, "Theories on the Origin of Meteorites," *Journal of Chemical Education*, vol. 38 (Feb. 1961), pp. 58–66.

———, "Meteorite Ages," *Reviews of Modern Physics*, vol. 34 (April 1962), pp. 287–325.

———, "Age and Origin of Meteorites," paper presented at annual meeting, National Academy of Sciences (Washington, D.C., April 29, 1964).

———, "Meteoritic Hydrocarbons and Extraterrestrial Life," *Annals of the New York Academy of Sciences*, vol. 93 (Aug. 29, 1962), pp. 651–64.

Anders, Edward, and F. W. Fitch, "Search for Organized Elements in

Carbonaceous Chondrites," *Science*, vol. 138 (Dec. 28, 1962), pp. 1392–99.

Anders, Edward, "On the Origin of Carbonaceous Chondrites," *Annals of the New York Academy of Sciences*, vol. 108, art. 2, *op. cit.*, pp. 514–33; see also "Panel Discussion," pp. 611–12.

Anders, Edward, Eugene R. DuFresne, Ryoichi Hayatsu, Albert Cavaillé, Ann DuFresne and Frank W. Fitch, "Contaminated Meteorite," *Science*, vol. 146 (Nov. 27, 1964), pp. 1157–61.

Bernal, J. D., "The Problem of the Carbonaceous Meteorites," *Times Science Review* (Summer 1961), pp. 3–4.

———, "Comments," in symposium on "Life-forms in meteorites," *Nature*, vol. 193 (March 24, 1962), pp. 1127–29.

———, "Is There Life Elsewhere in the Universe?" *The Listener*, April 26, 1962, pp. 723–24.

———, see "Panel Discussion," *Annals of the New York Academy of Sciences*, vol. 108, art. 2, *op. cit.*, pp. 606–15.

Berthelot, M., *Comptes Rendus, op. cit.*, vol. 67 (1868), p. 849.

Berzelius, J. J., *Kongl. Vetenskapsacademiens Handlingar* (Stockholm, 1834), p. 148.

Bourrelly, Pierre, "Panel Discussion," *Annals of the New York Academy of Sciences, op. cit.*, p. 614.

Briggs, Michael H., "Organic constituents of meteorites," *Nature*, vol. 191 (Sept. 16, 1961), pp. 1137–40.

Calvin, M., and S. K. Vaughn, "Extraterrestrial Life: Some Organic Constituents of Meteorites and Their Significance for Possible Extraterrestrial Biological Evolution," *Space Research: Proceedings of the First International Space Science Symposium*, Hilde Kallmann, ed., North-Holland Publishing Co. (Amsterdam, 1960), pp. 1171–91.

Chao, E. C. T., E. M. Shoemaker and B. M. Madsen, "First Natural Occurrence of Coesite," *Science*, (July 22, 1960), pp. 220–22.

Claus, George, and B. Nagy, "A microbiological examination of some carbonaceous chondrites," *Nature*, vol. 192 (Nov. 18, 1961), pp. 594–96.

Cloëz, S., Daubrée, et al., *Comptes Rendus, op. cit.*, vol. 58 (1864), pp. 984–90; vol. 59 (1864), pp. 37–40.

Farrell, M. A., "Living Bacteria in Ancient Rocks and Meteorites," *American Museum Novitates*, no. 645, American Museum of Natural History (New York, July 18, 1933).

Fitch, F. W., and E. Anders, "Organized element: possible identification in Orgueil Meteorite," *Science*, vol. 140 (June 7, 1963), pp. 1097–1100.

———, "Observations on the Nature of the 'Organized Elements' and Carbonaceous Chondrites," *Annals of the New York Academy of Sciences*, vol. 108, art. 2, *op. cit.*, pp. 495–513.

Fitch, F. W., H. P. Schwarcz and E. Anders, " 'Organized elements' in carbonaceous chondrites," *Nature,* vol. 193 (March 24, 1962), pp. 1123–25.

Gilbert, L. W., ed., (description of the Alais fall), *Annalen der Physik,* vol. 24 (1806), pp. 189–94.

Gilvarry, J. J., "Origin and nature of lunar surface features," *Nature,* vol. 188 (Dec. 10, 1960), pp. 886–91.

Hahn, Otto, *Die Meteorite (Chondrite) und ihre Organismen* (Tübingen, 1880).

Haldane, J. B. S., "Origin of Life," *op. cit.,* pp. 18–25.

Kulik, L. A., "The Tunguska Meteorite," *Source Book in Astronomy 1900–1950, op. cit.,* pp. 75–79.

Lipman, C. B., "Are There Living Bacteria in Stony Meteorites?" *American Museum Novitates, op. cit.,* no. 588 (Dec. 31, 1932).

Mason, Brian, "Origin of chondrules and chondritic meteorites," *Nature,* vol. 186 (April 16, 1960), pp. 230–31.

———, *Meteorites,* John Wiley & Sons (New York, 1962).

Meinschein, W. G., "Hydrocarbons in the Orgueil Meteorite," *Proceedings of Lunar and Planetary Exploration Colloquium,* North American Aviation, Inc. (Downey, Calif., Nov. 15, 1961), vol. 2, no. 4, pp. 65–67.

Mueller, G., "The Properties and Theory of Genesis of the Carbonaceous Complex within the Cold Bokkeveld Meteorite," *Geochimica et Cosmochimica Acta,* vol. 4 (Aug. 1953), pp. 1–10.

Mueller, G., "Interpretation of micro-structures in carbonaceous meteorites," *Nature,* vol. 196 (Dec. 8, 1962), pp. 929–32.

Nagy, Bartholomew, W. G. Meinschein and D. J. Hennessy, "Mass Spectroscopic Analysis of the Orgueil Meteorite: Evidence for Biogenic Hydrocarbons," *Annals of the New York Academy of Sciences,* vol. 93 (June 5, 1961), pp. 25–35.

———, Discussion of the paper by Anders (cited above), *Annals of the New York Academy of Sciences,* vol. 93 (Aug. 29, 1962) pp. 658–64.

———, "Aqueous, Low Temperature Environment of the Orgueil Meteorite Parent Body," *Annals of the New York Academy of Sciences,* vol. 108, art. 2, *op. cit.,* pp. 534–52.

Nagy, Bartholomew, K. Fredriksson, H. C. Urey, G. Claus, C. A. Andersen and J. Percy, "Electron Probe Microanalysis of Organized Elements in the Orgueil Meteorite," *Nature,* vol. 198 (April 13, 1963), pp. 121–25.

Nagy, Bartholomew, K. Fredriksson, J. Kudynowski and L. Carlson, "Ultra-violet spectra of organized elements," *Nature,* vol. 200 (Nov. 9, 1963), pp. 565–66.

Nagy, Bartholomew, M. T. J. Murphy, V. E. Modzeleski, G. Rouser, G. Claus, D. J. Hennessy, U. Colombo and F. Gazzarrini, "Optical activity in saponified organic matter isolated from the interior of the orgueil meteorite," *Nature,* vol. 202 (April 18, 1964), pp. 228–33.

Palik, P., "Further life-forms in the orgueil meteorite," *Nature*, vol. 194 (June 16, 1962), p. 1065.

Papp, A., "Fossil Protobionta and Their Occurrence," *Annals of the New York Academy of Sciences*, vol. 108, art. 2, *op. cit.*, p. 462. See also "Panel Discussion," *ibid.*, p. 613.

Sagan, Carl, "Indigenous Organic Matter on the Moon," and "Biological Contamination of the Moon," *Proceedings of the National Academy of Sciences*, vol. 46 (April 15, 1960), pp. 393–402.

Sisler, F. D., *Proceedings of Lunar and Planetary Exploration Colloquium, op. cit.*, pp. 67–73.

Spratt, Christopher, and Sally Stephens, "Meteorites That Have Struck Home," *Mercury*, March–April 1992, pp. 50–56.

Staplin, F. L., "Microfossils from the Orgueil Meteorite," *Micropaleontology*, vol. 8 (July 1, 1962), pp. 343–47.

Sylvester-Bradley, P. C., and R. J. King, "Evidence for Abiogenic Hydrocarbons," *Nature*, vol. 198 (May 25, 1963), pp. 728–31.

Timofeyev, B. V., "On the Occurrence of Organic Remains in Chondritic Meteorites," abstracts of papers presented at the 4th Astrogeological Meeting Geological Society of the USSR, May 1962. See also *Pravda* (Moscow), May 11, 1962, p. 6.

————, see *Ogonyok*, no. 42 (Moscow, Oct. 1962), pp. 20–21.

Urey, H. C., "Origin of life-like forms in carbonaceous chondrites," *Nature*, vol. 193 (March 24, 1962), pp. 1119–23.

————, "Lifelike Forms in Meteorites," *Science*, vol. 137 (Aug. 24, 1962), pp. 623–27.

————, "Panel Discussion," *Annals of the New York Academy of Sciences*, vol. 108, art. 2, *op. cit.*, pp. 606–15.

————, *The Planets, op. cit.*, pp. 193–210.

Urey, H. C., "Diamonds, Meteorites, and the Origin of the Solar System," *op. cit.*

————, "Primary and Secondary Objects," *Journal of Geophysical Research*, vol. 64 (Nov. 1959), pp. 1721–37.

————, "The Chemical History of Meteorites," paper presented at the annual meeting, National Academy of Sciences (Washington, D.C., April 29, 1964).

Urey, H. C., and H. Craig, "The Composition of the Stone Meteorites and the Origin of the Meteorites," *Geochimica et Cosmochimica Acta*, vol. 4 (Aug. 1953), pp. 36–82.

Wöhler, F., "Neuere Untersuchungen über die Bestandtheile des Meteorsteines vom Capland," *Sitzungsberichte, Mathematisch-Naturwissenschaftlichen Classe der Kaiserlichen Akademie der Wissenschaften* (Vienna, 1860), vol. 41, pp. 565–67.

Wood, J. A., "Chondrules and the Origin of the Terrestrial Planets," *Nature*, vol. 194 (April 14, 1962), pp. 127–30.

11. Life's Origin—Freak or Inevitable?

Bridgwater, D., J. H. Allaart, J. W. Schopf, C. Klein, M. R. Walter, E. S. Barghoorn, P. Strother, A. H. Knoll and B. E. Gorman, "Microfossil-like objects from the Archaean of Greenland: a cautionary note," *Nature*, vol. 28, (Jan. 8, 1981), pp. 51–53.

Büeler, Hansruedi, Marek Fischer, Yolande Lang, Horst Bluethmann, Hans-Peter Lipp, Stephen J. DeArmond, Stanley B. Prusiner, Michel Aguet and Charles Weissmann, "Normal development and behaviour of mice lacking the cell-surface PrP protein," *Nature*, vol. 356 (April 16, 1992), pp. 577–82.

Cairns-Smith, A. G., *Genetic Takeover and the Mineral Origins of Life*, Cambridge University Press (1982).

Calvin, Melvil, "Chemical Evolution and the Origin of Life," *American Scientist*, vol. 44 (July 1956), pp. 248–63.

———, "Origin of Life on Earth and Elsewhere," *University of California Radiation Laboratory Reports*, UCRL 9005 and 9440 (1959, 1960).

———, *Chemical Evolution*, Condon Lectures, *op. cit.*

Chesebro, Bruce, "PrP and the scrapie agent," *Nature*, vol. 356 (April 16, 1992), p. 560. See also report in same issue by You Geng Xi et al.

Chyba, Christopher, and Carl Sagan, "Endogenous production, exogenous delivery and impact-shock synthesis of organic molecules: an inventory for the origin of life," *Nature*, vol. 355 (Jan. 9, 1992), pp. 125–32.

Cloud, Preston, "How Life Began," *Nature*, vol. 296 (March 18, 1982), p. 198.

De Duve, Christian, *Blueprint for a Cell: The Nature and Origin of Life*, Neil Patterson Publishers (Burlington, N.C., 1991).

Eigen, Manfred, William Gardiner, Peter Schuster and Ruthild Winkler-Oswatitsch, "The Origin of Genetic Information," *Scientific American*, April 1981, pp. 88–118.

Fox, S. W., and Kaoru Harada, "The Thermal Copolymerization of Amino Acids Common to Proteins," *Journal of the American Chemical Society*, vol. 82 (July 20, 1960), pp. 3745–51.

———, "How Did Life Begin?" *Science*, vol. 132 (July 22, 1960), pp. 200–8.

———, "Synthesis of Uracil under Conditions of a Thermal Model of Prebiological Chemistry," *Science*, vol. 133 (June 16, 1961), pp. 1923–24.

Fox, S. W., and Shuhei Yuyama, "Abiotic Production of Primitive Protein and Formed Microparticles," *Annals of the New York Academy of Sciences*, vol. 108, art. 2 (June 29, 1963), pp. 487–94.

Garay, A. S., and J. A. Ahlgren-Beckendorf, "Differential interaction of chiral Beta-Particles with enantiomers," *Nature*, vol. 346 (Aug. 2, 1990), pp. 451–53.

Gee, Henry, "Something completely different," (on Burgess shale and other similar fossils), *Nature*, vol. 358 (Aug. 6, 1992).

Klein, Harold P., chairman, et al., Committee on Planetary Biology and Chemical Evolution, *The Search for Life's Origins*, National Research Council, National Academy of Sciences, 1990.

Lewin, Roger, "RNA Catalysis Gives Fresh Perspective on the Origin of Life," *Science*, vol. 231 (Feb. 7, 1986), pp. 545–46.

Maher, Kevin A., and David J. Stevenson, "Impact frustration of the origin of life," *Nature*, vol. 331 (Feb. 18, 1988), pp. 612–14.

Miller, S. L., "A Production of Amino Acids under Possible Primitive Earth Conditions," *Science*, vol. 117 (May 15, 1953), pp. 528–29.

———, "Production of Some Organic Compounds under Possible Primitive Earth Conditions," *Journal of the American Chemical Society*, vol. 77 (May 12, 1955), pp. 2351–61.

Miller, S. L., and H. C. Urey, "Organic Compound Synthesis on the Primitive Earth," *Science*, vol. 130 (July 31, 1959), pp. 245–51.

Moore, Linda, Brenton Knott and Neville Stanley, "The Stromatolites of Lake Clifton, Western Australia—Living Structures Representing the Origins of Life," *Search*, vol. 14 (Dec. 1983–Jan. 1984), pp. 309–14.

Nisbet, E. G., "RNA and hot-water springs," *Nature*, vol. 322 (July 17, 1986), p. 206.

Oró, Joan, "Comets and the Formation of Biochemical Compounds on the Primitive Earth," *ibid.*, vol. 190 (April 29, 1961), p. 389.

Oró, Joan, A. P. Kimball, R. Fritz and F. Master, "Amino Acid Synthesis from Formaldehyde and Hydroxylamine," *Archives of Biochemistry & Biophysics*, vol. 85 (Nov. 1959), pp. 115–30.

Oró, Joan, and A. P. Kimball, "Synthesis of Purines under Possible Primitive Earth Conditions—I. Adenine from Hydrogen Cyanide," *ibid.*, vol. 94 (Aug. 1961), p. 217.

Oró, Joan, and C. L. Guidry, "A novel synthesis of polypeptides," *Nature*, vol. 186 (April 9, 1960), pp. 156–57.

Oró, Joan, and S. S. Kamat, "Amino-acid synthesis from hydrogen cyanide under possible primitive earth conditions," *ibid.*, pp. 442–43.

Patrusky, Ben, "Before There Was Biology," *Mosaic*, vol. 15, no. 6 (1984).

Ponnamperuma, Cyril, R. M. Lemmon, R. Mariner and M. Calvin, "Formation of Adenine by Electron Irradiation of Methane, Ammonia, and Water," *Proceedings of the National Academy of Sciences*, vol. 49 (May 1963), pp. 737–40.

Ponnamperuma, Cyril, R. Mariner and C. Sagan, "Formation of adenosine by ultraviolet irradiation of a solution of adenine and ribose," *Nature*, vol. 198 (June 22, 1963), pp. 1199–1200.

Ponnamperuma, Cyril, C. Sagan and R. Mariner, "Ultraviolet Synthesis of Adenosine Triphosphate Under Possible Primitive Earth Conditions," *Research in Space Science, Special Report No. 128*, Smithsonian Institution Astrophysical Observatory (Cambridge, Mass., July 10, 1963).

Russell, M. J., A. J. Hall, A. G. Cairns-Smith and P. S. Braterman, "Submarine hot springs and the origin of life," *Nature*, vol. 336 (Nov. 10, 1988), p. 117.

Urey, H. C., "Primitive Planetary Atmospheres and the Origin of Life," *Proceedings of the First International Symposium . . . , op. cit.*, pp. 16–22.

Wald, George, "The Origins of Life," *Proceedings of the National Academy of Sciences* (Aug. 1964).

Vidal, Gonzalo, "The Oldest Eukaryotic Cells," *Scientific American*, Feb. 1984, pp. 48–57.

Waldrop, M. Mitchell, "Goodbye to the Warm Little Pond?", *Science*, vol. 250 (Nov. 23, 1990), pp. 1078–80.

Zielinski, Wojciech S. and Leslie E. Orgel, "Autocatalytic synthesis of a tetranucleotide analogue," *Nature*, vol. 327 (May 28, 1987), pp. 346–47.

12. Mars

Baker, V. R., R. G. Strom, V. C. Gulick, J. S. Kargel, G. Komatsu and V. S. Kale, "Ancient oceans, ice sheets and the hydrological cycle on Mars," *Nature*, vol. 352 (Aug. 15, 1991), pp. 589–94, with erratum on pp. 86–87 of Nov. 7 issue.

Carr, Michael H., "The Surface of Mars: A Post-Viking View," *Mercury*, Jan.–Feb. 1983, pp. 2–15.

Cooper, Henry S. F., Jr., *The Search for Life on Mars, Evolution of an Idea*, Holt, Rinehart and Winston (New York, 1980).

Flammarion, Camille, *La Planète Mars et ses Conditions d'Habitabilité* (Paris, 1892).

Horowitz, N. H., "The Design of Martian Biological Experiments," *Life Sciences and Space Research*, vol. 2 of *Proceedings of the Symposia on Extraterrestrial Biology and Organic Chemistry, Methods for the Detection of Extraterrestrial Life and Terrestrial Life in Space*—Warsaw, 1963, M. Florkin, ed., North-Holland Publishing Co. (Amsterdam, 1964).

———, *To Utopia and Back, the Search for Life in the Solar System*, W. H. Freeman and Company (San Francisco, 1986).

Kargel, Jeffrey S., and Robert O. Strom, "The Ice Ages of Mars," *Astronomy*, vol. 20, no. 12 (Dec. 1992), pp. 40–45.

Karlsson, Haraldur R., Robert N. Clayton, Everett K. Gibson, Jr., and Toshiko K. Mayeda, "Water in SNC Meteorites: Evidence for a Martian Hydrosphere," *Science*, vol. 255 (March 13, 1992), pp. 1409–11.

Kuprevich, V. F., President, Academy of Sciences of the Byelorussian SSR, on enriching earth with Martian flora and fauna: "Earth, Life, Space," *Tekhnika-Molodezhi*, no. 9 (Moscow, 1961), p. 11.

Lederberg, J., and C. Sagan, "Microenvironments for Life on Mars," *Proceedings of the National Academy of Sciences*, vol. 48 (Sept. 1962), pp. 1473–75.

Levin, G. V., A. H. Heim, J. R. Clendenning and M.-F. Thompson, " 'Gulliver'—A Quest for Life on Mars," *Science*, vol. 138 (Oct. 12, 1962), pp. 114–21.

Lowell, Percival, *Mars*, Houghton Mifflin (Boston, 1895), pp. 201 and 208.

Masursky, Harold, "Mars," Chapter 8 in *The New Solar System*, J. Kelly Beatty et al., eds., Sky Publishing Company (Cambridge, Mass., 1981) (new edition, 1990).

McEwen, Alfred S., "New Martian Paradigms," *Reviews of Geophysics, Supplement*, pp. 290–96, April 1991, U.S. Annual Report to the International Union of Geodesy and Geophysics 1987–1990.

NASA, *Viking 1—Early Results*, NASA SP-408 (Washington, D.C., 1976).

Sagan, Carl, "On the Origin and Planetary Distribution of Life," *Radiation Research*, vol. 15 (Aug. 1961), pp. 174–92.

Saheki, Tsuneo, "Martian Phenomena Suggesting Volcanic Activity," *Sky and Telescope*, vol. 14 (Feb. 1955), pp. 144–46.

Salisbury, F. B., "Martian Biology," *Science*, vol. 136 (April 6, 1962), pp. 17–26.

Shklovsky, I. S., *Vselennaya Zhizn Razum* (*Universe, Life, Intelligence*), Press of the Academy of Sciences of the USSR (Moscow, 1962). See especially Chapter 17, "Sputniki Marsa—iskusstbennuiye?" ("The Satellites of Mars—Artificial?") English edition, *Intelligent Life in the Universe*, Carl Sagan, ed., Holden-Day (San Francisco, 1966).

Sinton, W. M., "Spectroscopic Evidence for Vegetation on Mars," *Astrophysical Journal.*, vol. 126 (Sept. 1957), pp. 231–39.

———, "Further Evidence of Vegetation on Mars," *Science*, vol. 130 (Nov. 6, 1959), pp. 1234–37.

———, "Evidence of the Existence of Life on Mars," *Advances in the Astronautical Sciences*, *op. cit.*, pp. 543–51.

Tikhov, G. A., *Journal of the British Astronomical Association*, vol. 65 (1955), pp. 193–204.

Todd, David, proposal for Mars telescope in Chile: see *New York Times*, 1921, Sept. 7, p. 2; Sept. 8, p. 17. Gets government cooperation on radio silence: *ibid.*, 1924, Aug. 21, p. 11; Aug. 28, p. 6. Flammarion, aged 82, reaffirms his belief in a superior civilization on Mars: *ibid.*, Aug. 22, p. 13; Aug. 24, p. 30.

13. The Uniquely Rational Way

Burke, B. F., "Contributions of Overseas Observatories," *Physics Today*, vol. 14 (April 1961), pp. 26–29.

Cocconi, Giuseppe, letter to Lovell, June 29, 1959, published as an appendix to *The Exploration of Outer Space*, by Sir Bernard Lovell (Oxford, 1962). In referring to ten planets in the solar system, Cocconi apparently includes the planet that may have disintegrated to form the asteroid belt.

Cocconi, Giuseppe, and Philip Morrison, "Searching for interstellar communications," *Nature*, vol. 184 (Sept. 19, 1959), pp. 844–46, reprinted in Cameron, *Interstellar Communication, op. cit.*, as Item 15.

Jansky, K. G., "Directional Studies of Atmospherics at High Frequencies," *Proceedings of the IRE* (Institute of Radio Engineers), vol. 20 (1932), pp. 1920–32.

———, "Electrical Disturbances Apparently of Extraterrestrial Origin," *ibid.*, vol. 21 (Oct. 1933), pp. 1387–98.

———, see *New York Times*, May 5, 1933, p. 1, and June 28, 1933, p. 1.

Lovell, Sir Bernard, "Search for Voices from Other Worlds," *New York Times Magazine*, Dec. 24, 1961, pp. 18 ff.

Marconi, G. M., reports strange interference with transoceanic radio traffic. See *New York Times*, 1920, Jan. 27, p. 7; Jan. 28, p. 5; Jan. 29, p. 1; Jan. 30, p. 18; Jan. 31, p. 24. Reportedly hears signals from Mars. *Ibid.*, 1921, Sept. 2, p. 1; Sept. 3, p. 4.

Pettengill, G. H., H. W. Briscoe, J. V. Evans, E. Gehrels, G. M. Hyde, L. G. Kraft, R. Price and W. B. Smith, "A Radar Investigation of Venus," *Astronomical Journal*, vol. 67 (May 1962), pp. 181–90 (on the Millstone Hill observations).

Price, R., P. E. Green, Jr., T. J. Goblick, Jr., R. H. Kingston, L. G. Kraft, Jr., G. H. Pettengill, R. Silver and W. B. Smith, "Radar Echoes from Venus," *Science*, vol. 129 (March 20, 1959), pp. 751–53.

Tesla, Nikola: see "Nikola Tesla," A Commemorative Lecture by A. P. M. Fleming, given to The Institute of Electrical Engineers, Nov. 25, 1943; also *Prodigal Genius, the Life of Nikola Tesla*, by John J. O'Neill, Ives Washburn (New York, 1944); also *Return of the Dove*, by Margaret Storm, A Margaret Storm Publication (Baltimore, 1959); also *Electrical Genius—Nikola Tesla*, by A. J. Beckhard, Julian Messner, Inc. (New York, 1959).

———, quoted by L. I. Anderson in "Extraterrestrial radio transmissions," letter to *Nature*, vol. 190 (April 2, 1961), p. 374.

Victor, W. K., and R. Stevens, "Exploration of Venus by Radar," *Science*, vol. 134 (July 7, 1961), pp. 46–48 (on Goldstone observations).

Victor, W. K., R. Stevens and S. W. Golomb, *Radar Exploration of Venus: Goldstone Observatory Report for March–May 1961*, Technical Report No.

32–132, Jet Propulsion Laboratory, California Institute of Technology (Pasadena, Aug. 1, 1961).

14. Other Channels

Basler, Roy P., George L. Johnson and Richard R. Vondruke, "Antenna concepts for interstellar search systems," *Radio Science*, vol. 12 (Sept.–Oct. 1977), pp. 845–58.

Bracewell, R. N., "Communications from superior galactic communities," *Nature*, vol. 186 (May 28, 1960), pp. 670–71, reprinted in Cameron, *Interstellar Communication, op. cit.*, as Item 25.

————, "Life in the Galaxy," lecture at Summer Science School, University of Sydney, Jan. 8–19, 1962, published in *A Journey Through Space and the Atom*, ed. by S. T. Butler and H. Messel, Nuclear Research Foundation (Sydney, 1962); also reprinted in Cameron, *op. cit.*, as Item 24.

Buyakas, V. I., et al., "Infinitely Built-up Space Radio Telescope," Space Research Institute, Academy of Sciences of the USSR (Moscow, 1977 [mimeo].)

Conway, R. G., K. I. Kellermann and R. J. Long, "The Radio Frequency Spectra of Discrete Radio Sources," *Monthly Notices of the Royal Astronomical Society*, vol. 125 (1963), p. 261.

Cordes, James M., "Astrophysical masers as amplifiers of ETI signals," *Proceedings of the Third Decentennial US-USSR Conference on SETI*, Astronomical Society of the Pacific (1992–93, in press).

Dyson, F. J., "Search for Artificial Stellar Sources of Infrared Radiation," *Science*, June 3, 1960, p. 1667, reprinted in Cameron, *op. cit.*, as Item 11. See also letters in *Science*, July 22, 1960, pp. 250–53.

Edie, L. C., "Messages from Other Worlds," letter in *Science*, vol. 136 (April 13, 1962), p. 184.

Eshleman, Von R., "Gravitational Lens of the Sun: Its Potential for Observations and Communications over Interstellar Distances," *Science*, vol. 205 (Sept. 14, 1979), pp. 1133–35.

Golay, M. J. E., "Note on the Probable Character of Intelligent Radio Signals from Other Planetary Systems," *Proceedings of the IRE*, vol. 49 (May, 1961), p. 959, reprinted in Cameron, *op. cit.*, as part of Item 19.

Kardashev, N. S., "The Transmission of Information by Extraterrestrial Civilizations," *Astronomical Journal* (*Astronomichesky Zhurnal*), vol. 41 (March–April 1964), pp. 282–87. Translation in *Soviet Astronomy-AJ*, vol. 8, no. 2 (Sept.–Oct., 1964), pp. 217–26.

Oliver, B. M., "Some Potentialities of Optical Masers," *Proceedings of the IRE*, vol. 50 (Feb. 1962), pp. 135–41, reprinted in Cameron, *op. cit.*, as Item 22.

————, "Radio Search for Distant Races," *International Science and Technology*, no. 10 (Oct. 1962), pp. 55–60.

————, "Interstellar Communication," published in Cameron, *op. cit.*, as Item 29.

————, "The Search for Extraterrestrial Intelligence," *Mercury*, March 1973, pp. 11–12.

Oliver, B. M., and John Billingham, co-directors, *Project Cyclops—A Design Study of a System for Detecting Extraterrestrial Intelligent Life*, revised edition 1973, published jointly by Stanford University and NASA Ames Research Center.

Schwartz, R. N., and C. H. Townes, "Interstellar and interplanetary communication by optical masers," *Nature*, vol. 190 (April 15, 1961), pp. 205–8, reprinted in Cameron, *op. cit.*, as Item 23.

Sullivan, W. T. III, S. Brown and C. Wetherill, "Eavesdropping: The Radio Signature of the Earth," *Science*, vol. 199 (Jan. 27, 1978), pp. 377–87.

Webb, J. A., "Detection of Intelligent Signals from Space," Institute of Radio Engineers Seventh National Communications Symposium Record: *Communications–Bridge or Barrier* (1961), p. 10, reprinted in Cameron, *op. cit.*, as Item 18.

Wetherill, C., and W. T. Sullivan, "Eavesdropping on the Earth," *Mercury*, March–April 1979, pp. 23–28.

15. False Alarms

Burnell, S. Jocelyn Bell, "Little Green Men, White Dwarfs or Pulsars?" *Cosmic Search*, Jan. 1979, pp. 17–21.

Drake, F. D., "How Can We Detect Radio Transmissions from Distant Planetary Systems?" *Sky and Telescope*, vol. 19 (Jan. 1960), pp. 140–43, reprinted in Cameron, *Interstellar Communication, op. cit.*, as Item 16.

————, "Project Ozma," *Physics Today*, vol. 14 (April 1961), pp. 40–46.

————, "Project Ozma," *McGraw-Hill Yearbook of Science and Technology, 1962*, reprinted in Cameron, *op. cit.*, as Item 17.

Drake, F. D., "A Reminiscence of Project Ozma," *Cosmic Search*, Jan. 1979, pp. 11–14.

Drake, F. D., and Dava Sobel, *Is Anyone Out There?*, Delacorte Press (1992). Pp. 21–43 are on Project Ozma.

Kardashev, N. S., "The Transmission of Information by Extraterrestrial Civilizations," *Electronics Express*, vol. 6, no. 10 (1964), from *Astronomicheskii Zhurnal*, vol. 41 (March–April 1964), pp. 282–87.

Lunan, Duncan A., "Space Probe from Epsilon Bootis" *Spaceflight*, vol. 15, no. 4 (April 1973), pp. 122–31.

Sholomitsky, G. B., "Variability of the Radio Source CTA-102," *Infor-*

mation Bulletin on Variable Stars, Commission 27 of the I.A.U., No. 83 (Konkoly Observatory, Budapest, Feb. 27, 1965).

Slish, V. I., "Angular size of radio stars," *Nature,* vol. 199 (Aug. 17, 1963), p. 682.

Struve, Otto, "Astronomers in Turmoil," *Physics Today,* vol. 13 (Sept. 1960), pp. 22–23.

Thomas, Shirley, "Frank D. Drake," *Men of Space, Profiles of Scientists Who Probe for Life in Space,* vol. 6, Chilton Books (Philadelphia, 1963), pp. 62–89.

16. Can They Visit Us?

Berosus: see *Cory's Ancient Fragments of the Phoenician, Carthaginian, Babylonian, Egyptian and Other Authors,* by E. Richmond Hodges (London, 1876), p. 57.

Bussard, R. W., "Galactic Matter and Interstellar Flight," *Astronautica Acta,* vol. 6 (1960), pp. 179–94.

Charles, R. H., ed., *The Book of the Secrets of Enoch,* translated from the Slavonic by W. R. Morfill, Clarendon Press (Oxford, 1896), Chapters 1, 3.

Drake, F. D., "The Radio Search for Intelligent Extraterrestrial Life," *op. cit.*

Dyson, F. J., "Gravitational Machines," published in Cameron, *Interstellar Communication, op. cit.,* as Item 12.

Forward, R. L. "Roundtrip Interstellar Travel Using Laser-pushed Light Sails," *Journal of Spacecraft and Rockets,* vol. 21, no. 2 (March–April 1984), pp. 187–95.

————, "A National Space Program for Interstellar Exploration," *Hughes Research Laboratories Research Report 492,* Submitted to the Subcommittee on Space Science and Applications of the House Committee on Science and Technology, July 1975.

Froman, Darol, "The Earth as a Man-Controlled Space Ship," *Physics Today,* vol. 15 (July 1962), pp. 19–23.

von Hoerner, Sebastian, "The General Limits of Space Travel," *Science,* vol. 137 (July 6, 1962), pp. 18–23, reprinted in Cameron, *op. cit.,* as Item 14.

Mallove, Eugene F., Robert L. Forward, Zbigniew Paprotny and Jurgen Lehmann, "Interstellar Travel and Communication: A Bibliography," *Journal of the British Interplanetary Society,* vol. 33, no. 6 (June 1980), pp. 201–48.

Pierce, J. R., "Relativity and Space Travel," *Proceedings of the IRE,* vol. 47 (June 1959), pp. 1053–61.

Purcell, Edward, "Radioastronomy and Communication Through

Space," *Brookhaven Lecture Series,* no. 1, Brookhaven National Laboratory (Nov. 16, 1960), reprinted in Cameron, *op. cit.,* as Item 13.

Sagan, Carl, "Direct Contact Among Galactic Civilizations by Relativistic Interstellar Spaceflight," *Planetary and Space Science,* vol. 11 (1963), pp. 485–98.

Sparks, Jared, *The Works of Benjamin Franklin,* Whittemore, Niles and Hall (Boston, 1856), vol. 8, p. 418.

Spitzer, Lyman, Jr., "Interplanetary Travel Between Satellite Orbits," *Journal of the American Rocket Society,* vol. 22 (March–April, 1952), p. 93. (Originally published in *Journal of the British Interplanetary Society*).

Tsiolkovsky, K. E.: see letter by A. N. Tsvetikov in *Science,* vol. 131 (March 18, 1960), pp. 872, 874. Excerpts from Tsiolkovsky letters were also provided to the author by Dr. Tsvetikov.

17. UFO's

Condon, Edward U., scientific director, and Daniel S. Gillmore, ed., *Scientific Study of Unidentified Flying Objects,* E. P. Dutton & Co. (New York, 1968).

Sagan, Carl, and Thornton Page, eds., *UFO's—A Scientific Debate,* Cornell University Press (1972).

18. Where Are They?

Bahcall, John N., chairman, Astronomy and Astrophysics Survey Committee, *The Decade of Discovery in Astronomy and Astrophysics,* National Research Council/National Academy of Sciences (1991).

Ball, John A., "The Zoo Hypothesis," *Icarus,* vol. 19 (1973), pp. 347–49.

Fang, T. C., and N. N. Kiang, "On the Properties of Life," *Scientia Sinica,* vol. XX, no. 3 (May–June 1977), pp. 280–86.

Field, George B., chairman, *Astronomy and Astrophysics for the 1980's,* National Academy of Sciences (1982).

Finney, Ben. R., and Eric M. Jones, *Interstellar Migration and the Human Experience,* University of California Press, 1985.

Hart, Michael H., "An Explanation for the Absence of Extraterrestrials on Earth," *Quarterly Journal of the Royal Astronomical Society,* vol. 16 (1975), pp. 128–35.

von Hoerner, Sebastian, "Population Explosion and Interstellar Expansion," *Journal of the British Interplanetary Society,* vol. 28 (1975), pp. 691–712. An earlier version (in German) appeared in *Einheit und Vielheit, Festschrift für Carl Friedrich v. Weizsäcker zum 60. Geburtstag,* Vandenhoeck & Ruprecht (Göttingen, 1972).

———, "Astronomical Aspects of Interstellar Communication," *Astronautica Acta,* vol. 18 (1973), pp. 421–30.

Jones, Eric M., "Colonization of the Galaxy," *Icarus*, vol. 28 (1976), pp. 421–22.

———, " 'Where Is Everybody?' An Account of Fermi's Question," *Los Alamos Report LA-10311-MS*, issued March 1985.

Kuiper, T. B. H., and M. Morris, "Searching for Extraterrestrial Civilizations," *Science*, vol. 196 (May 6, 1977), pp. 616–21.

Michaud, Michael A. G., "The Extraterrestrial Paradigm—Improving the Prospects for Life in the Universe," *Interdisciplinary Science Reviews*, vol. 4, no. 3 (1979), pp. 177–92.

Newman, William I., and Carl Sagan, "Galactic Civilizations: Population Dynamics and Interstellar Diffusion," *Icarus*, vol. 46 (June 3, 1981), pp. 293–327.

Oliver, B. M., "Proximity of Galactic Civilizations," *Icarus*, vol. 25 (1975), pp. 360–67.

O'Neill, Gerard K., "The colonization of space," *Physics Today*, Sept. 1974, pp. 32–40.

———, *The High Frontier*, William Morrow and Company (New York, 1977).

Papagiannis, Michael D., "Bioastronomy: The Search for Extraterrestrial Life," *Sky and Telescope*, June 1984, pp. 508–11.

Rood, Robert T., and James S. Trefil, *Are We Alone—The Possibility of Extraterrestrial Civilizations*, Charles Scribner's Sons (New York, 1981).

Sagan, Carl, "Extraterrestrial Intelligence: An International Petition" (with names), *Science*, vol. 218 (Oct. 29, 1982), p. 426.

Tipler, Frank J., "Extraterrestrial Beings Do Not Exist," *Physics Today*, April 1981, pp. 9, 70–71. See also responses in *Physics Today*, March 1982.

———, "Extraterrestrial Intelligence: A Skeptical View of Radio Searches," *Science*, vol. 219 (Jan. 14, 1983), pp. 110–12.

———, "The Most Advanced Civilization in the Galaxy Is Ours," *Mercury*, Jan.–Feb. 1982, pp. 5–16, 25 ff.

19. Is There Intelligent Life on Earth?

Blum, H. F., "Perspectives in Evolution," *American Scientist*, vol. 43 (Oct. 1955), pp. 595–610.

———, "On the Origin and Evolution of Living Machines," *ibid.*, vol. 49 (Dec. 1961), pp. 474–501.

———, "On the Origin and Evolution of Human Culture," *ibid.*, vol. 51 (March 1963), pp. 32–47.

Cade, C. M., "Communicating with Life in Space," *Discovery* (May 1963) pp. 36–41.

von Hoerner, Sebastian, "The Search for Signals from Other Civilizations," *Science*, vol. 134 (Dec. 8, 1961), pp. 1839–43, reprinted in Cameron, *Interstellar Communication, op. cit.*, as Item 27.

Hoyle, Fred, *A Contradiction in the Argument of Malthus, The St. John's College Cambridge Lecture 1962–63 delivered at the University of Hull 17 May 1963*, University of Hull Publications (1963).

Lilly, J. C., *Man and Dolphin*, Doubleday (New York, 1961).

Marino, Lori A., and Diana Reiss, "Self-recognition in the Bottlenose Dolphin: a Methodological Test Case for the Study of Extraterrestrial Intelligence," *Proceedings of the Third Decennial US-USSR Conference on SETI* Astronomical Society of the Pacific (1992–93 [in press]).

Morrison, Philip, "Interstellar Communication," a paper read before the Philosophical Society of Washington, Oct. 7, 1960, *Bulletin of the Philosophical Society of Washington*, vol. 16, no. 1 (1962), pp. 59–81, reprinted in Cameron, *Interstellar Communication, op. cit.*, as Item 26.

Pearman, J. P. T., "Extraterrestrial Intelligent Life and Interstellar Communication: An Informal Discussion," Item 28 in Cameron, *Interstellar Communication, op. cit.*

Simpson, G. G., "The Nonprevalence of Humanoids," *Science*, vol. 143 (Feb. 21, 1964), pp. 769–75.

United States Congress, House of Representatives, Committee on Science and Astronautics, *Panel on Science and Technology*, 87th Congress, 2nd Session Hearings, 4th Meeting, March 22, 1962 (Washington, D.C.), pp. 73–74.

20. SETI

Backus, Peter R., "The NASA SETI Microwave Observing Project Targeted Search," *Proceedings of the Third Decennial . . . , op. cit.* SETI.

Bowyer, Stuart, et al, "The Berkeley Parasitic Program," *Icarus*, vol. 53 (1985), p. 147.

Colomb, F. R., E. Hurrell, J. C. Olalde, G. E. Lemarchand, "SETI Activities in Argentina," International Astronomical Union 21st General Assembly, July 23–Aug. 1, 1991.

Horowitz, Paul, "A Search for Ultra-Narrowband Signals of Extraterrestrial Origin," *Science*, vol. 201 (Aug. 25, 1978), pp. 733–35.

Horowitz, Paul, Brian S. Matthews, John Forster, Ivan Linscott, Calvin C. Teague, Kok Chen and Peter Backus, "Ultranarrowband Searches for Extraterrestrial Intelligence with Dedicated Signal-Processing Hardware," *Icarus*, vol. 67 (1986), pp. 525–39.

Morrison, Philip, John Billingham and John Wolfe, eds., *SETI: The Search for Extraterrestrial Intelligence*, NASA SP-419, 1977.

Olsen, E. T., E. B. Jackson and S. Gulkis, "SETI Site Surveys of RFI in the Band 1.0 GHz–10.4 GHz," *Proceedings of the Third Decennial . . . , op. cit.*

Peterson, A. M., and K. S. Chen, "The Multi-Channel Spectrum Analyzer," Appendix II to *Proceedings of the First I.A.U. Symposium on the Search for Extraterrestrial Life*, pp. II-1 to II-15.

Klein, M. J., D. J. Burns, C. F. Foster, M. F. Garyantes, S. Gulkis, S. M. Levin, E. T. Olsen, H. C. Wilck, and G. A. Zimmerman, "The NASA SETI Microwave Observing Project Sky Survey," *Proceedings of the Third Decennial . . .* , *op. cit.*

Kraus, John, *Big Ear*, Cygnus-Quasar Books (Powell, Ohio, 1976).

Shapley, Deborah, "Astronomers' Search for Other Worlds Brings Contact with Politics," *Science*, vol. 200 (June 30, 1978), pp. 1467–68.

Tarter, Jill C., "SETI and Serendipity," *Acta Astronautica*, vol. 11, no. 7–8 (1984), pp. 387–91.

21. Celestial Syntax

Bracewell, R. N., "Life in the Galaxy," *op. cit.*

Freudenthal, Hans, *Lincos, Design of a Language for Cosmic Intercourse*, North-Holland Publishing Co. (Amsterdam, 1960), p. 14 and item 4-18-8.

von Hoerner, Sebastian, "Universal Music?" in *Psychology of Music*, vol. 2, no. 2 (1974), pp. 18–28.

Hogben, Lancelot, "Astraglossa, or First Steps in Celestial Syntax," an address to the British Interplanetary Society, Nov. 6, 1952, reprinted in Hogben, *Science in Authority*, W. W. Norton (New York, 1963).

———, review of *Lincos, Nature*, vol. 192 (Dec. 2, 1961), pp. 826–27.

Morrison, Philip, "Interstellar Communication," *op. cit.*, pp. 78 ff.

———, see illustrations to "Are We Being Hailed from Interstellar Space?" by G. A. W. Boehm, *op. cit.*

Sagan, Carl, Linda Salzman Sagan and Frank Drake, "A Message from Earth," *Science*, vol. 175 (Feb. 25, 1972), pp. 881–84.

Smith, R. F. W., see "Communication with Extraterrestrial Beings Called Improbable Unless Man Can Signal in Two Systems of Thought," *Science Fortnightly* (Oct. 30, 1963), p. 8.

Wooster, Harold, moderator, Session 1–5, 1965 IEEE Military Electronics Conference, Washington, D.C., "Communication with extraterrestrial intelligence," *IEEE Spectrum*, March 1966, pp. 153–63.

22. What If We Succeed?

Billingham, John, "SETI Post-Detection Protocols: What Do You Do After Detecting a Signal?", *Proceedings of the Third Decentennial . . .* , *op. cit.*

Billingham, John, Michael A. G., Michaud and Jill Tarter, "The Declaration of Principles for Activities Following the Detection of Extraterrestrial Intelligence," with text, *Proceedings of the Third Decennial . . .* , *op. cit.*

Bracewell, "Life in the Galaxy," *op. cit.*

Breig, J. A., and Fr. L. C. McHugh, "Other Worlds—for Man," *America,* vol. 104 (Nov. 26, 1960), pp. 294–96.

Cameron, *Interstellar Communication, op. cit.,* p. 1.

Cantril, H., *The Invasion from Mars, A Study in the Psychology of Panic, with the complete script of the famous Orson Welles Broadcast,* Princeton University Press (1940).

Chardin, Pierre Teilhard de, *The Phenomenon of Man,* Harper (New York, 1959), p. 286.

Dyson, F. J., letter in *Scientific American,* April 1964, pp. 8–10.

Goodman, Allan E., "The Diplomatic Implications of Discovering Extraterrestrial Intelligence," *Mercury,* March–April 1987, pp. 56–57.

Hamilton, William, "The Discovery of Extraterrestrial Intelligence, A religious response," *The Humanist,* vol. XXXVI, no. 3 (May–June 1976), pp. 24–26.

Harford, James, "Rational Beings in Other Worlds," *Jubilee* (May 1962), pp. 17–21.

Inge, W. R., *God and the Astronomers,* Longmans, Green & Co. (London, 1933), p. 268.

Jung, C. G., quoted by Harford, *op. cit.,* p. 21.

Macquarrie, J., *The Scope of Demythologizing—Bultmann and his Critics,* SCM Press (London, 1960), pp. 60–61, 179.

Mascall, E. L., *Christian Theology and Natural Science—Some Questions on Their Relations,* Longmans, Green & Co. (London, 1956), pp. 36–45.

Michael, D. N., "Proposed Studies on the Implications of Peaceful Space Activities for Human Affairs," Brookings Institution report to NASA (Washington, D.C., Dec., 1960), pp. 182–84.

Michaud, Michael A. G., "Interstellar Negotiation," *Foreign Service Journal,* Dec. 1972, pp. 10–14.

Michaud, Michael A. G., John Billingham and Jill Tarter, "A Reply from Earth? A Proposed Approach to Developing a Message from Humankind to Extraterrestrial Intelligence After We Detect Them," Presented at the IAF Congress, Oct. 1990 (Dresden).

Midrash Rabba, commentary on Genesis, chapter 3, verse 7, Dr. H. Freedman, ed. and transl., Soncino Press (London, 1939), vol. 1, pp. 23–24.

Oppenheimer, J. R., "On Science and Culture," *Encounter* (Oct. 1962), pp. 6–7.

Raible, Fr. Daniel C., "Rational Life in Outer Space?" *America,* vol. 103 (Aug. 13, 1960), pp. 532–35.

Russell, Bertrand, *Mysticism and Logic,* quoted in *The Exploration of the Universe,* Part One of *The Citizen and the New Age of Science,* L. B. Young, ed., American Foundation for Continuing Education (Chicago, 1961), vol. 2, p. 368.

The Saddharma-Pundarika or The Lotus of the True Law, transl. by H. Kern, *Sacred Books of the East*, Clarendon Press (Oxford, 1884), vol. 21, chapters 14 and 15.

Secchi, Fr. Angelo, *et al.*, see W. D. Muller, *Man Among the Stars*, Criterion Books (New York, 1957), chapter 13 ("Religion in Space").

Shapley, H., *Of Stars and Men., op. cit.*

Smethurst, A. F., *Modern Science and Christian Beliefs*, Abingdon Press (New York, 1955), pp. 96–97.

Smith, Joseph, *The Pearl of Great Price*, The Church of Jesus Christ of Latter-Day Saints (Salt Lake City, 1952), chapter 1 ("The Book of Moses").

Tillich, Paul, *Systematic Theology*, vol. 2 (Chicago, 1957), pp. 95–96, 100–101.

——, see "Paul Tillich—Mystic, Rationalist, Universalist," by A. P. Stiernotte, *Crane Review*, vol. 4 (Spring 1962), pp. 175–76.

——, quoted in *Newsweek*, Oct. 8, 1962, p. 115.

Wald, George, "Theories of the Origin of Life," *The Voice of America Forum Lectures, Biology Series*, no. 20, 1960–61. U.S. Information Agency (Washington, D.C.), p. 7.

Zubek, T. J., "Theological Questions on Space Creatures," *The American Ecclesiastical Review*, vol. 145 (Dec. 1961), pp. 393–99.

Acknowledgments

Many specialists helped to bring this 1993 edition up to date. Particularly helpful were Eugene H. Levy and his colleagues at the University of Arizona, who provided material from the 1991 Protostars and Planets III meeting in Tucson. Also of special importance was the role of Jill Tarter, John Billingham, Seth Shostak, and their SETI colleagues at the NASA Ames Research Center and the nearby SETI Institute. They provided especially useful material from the Third Decennial US-USSR Conference on SETI, which had taken place in Santa Cruz a short time earlier. Essential, as well, was the further help of Frank D. Drake in telling of his seminal and long-term role in the search.

The book also owes much to specialists in many fields who reviewed the original edition and made numerous suggestions. The author has tried to incorporate as many of their proposals as possible, but there are still uncertainties and controversies affecting various aspects of the search. Since this book is intended chiefly for laymen, no attempt has been made to set forth all arguments on each side. Rather, to communicate the flavor of the debates, only those aspects most readily understood have been included.

The following, in some cases identified by their earlier titles, have generously reviewed portions of the original manuscript within their special areas of competence: Edward Anders, Associate Professor of Chemistry, Enrico Fermi Institute for Nuclear Studies, University of Chicago; Melvin Calvin, Professor of Chemistry, University of California at Berkeley; A. G. W. Cameron, Institute for Space Studies, Goddard Space Flight Center, National Aeronautics and Space Administration; Frank D. Drake, Jet Propulsion Laboratory, California Institute of Technology; Christian T. Elvey, University Research Professor, University of Alaska; Su-

Shu Huang, Goddard Space Flight Center, NASA; Peter van de Kamp, director, Sproul Observatory, Swarthmore College; Brian H. Mason, Curator of Mineralogy, American Museum of Natural History; Stanley L. Miller, Associate Professor of Chemistry, University of California, San Diego (La Jolla); Philip Morrison, Visiting Professor of Physics, Massachusetts Institute of Technology; Bartholomew Nagy, Professor of Chemistry, Fordham University; Edward M. Purcell, Professor of Physics, Harvard University; Freeman H. Quimby, Chief, Exobiology Programs, Bioscience Programs Division, NASA; Carl Sagan, Smithsonian Astrophysical Observatory; Harlow Shapley, emeritus Director, Harvard College Observatory; Lyman Spitzer, Jr., Director, Princeton University Observatory; Kai Aa. Strand, Scientific Director, U.S. Naval Observatory; Charles H. Townes, Provost, Massachusetts Institute of Technology; Harold C. Urey, Professor-at-Large of Chemistry, University of California, San Diego (La Jolla); and Gerard H. de Vaucouleurs, Associate Professor of Astronomy, University of Texas.

It is not possible to cite all who have assisted in furnishing material and in other ways. Among those who were especially helpful in this regard were: Leland I. Anderson of Minneapolis; Giuseppe Cocconi of CERN, the European Organization for Nuclear Research near Geneva; Sidney W. Fox, Director, Institute for Space Biosciences, Florida State University; Lancelot Hogben, Vice-Chancellor and Principal, University of Guyana, British Guiana; Norman H. Horowitz, Professor of Biology, California Institute of Technology; John A. Kessler, Lincoln Laboratory, Massachusetts Institute of Technology; Gerard P. Kuiper, Director, Lunar and Planetary Laboratory, University of Arizona; Fr. Clement J. McNaspy, S.J., Associate Editor, *America*; Alfred P. N. F. Stiernotte, Professor of Theology, Theological School, St. Lawrence University; Alexis N. Tsvetikov, Department of Biophysics, Stanford University; and Richard S. Young, Chief, Exobiology Division, Ames Research Center, NASA.

My foremost debt is to Frank D. Drake, for advice and assistance, and to Philip Morrison, whose lectures on this subject were the initial inspiration for this book.

Superb editorial assistance for this revised edition was rendered by Hugh Rawson, my editor at NAL Penguin, and my wife, Mary. Of inestimable help were my agent, Fifi Oscard, and the research facilities of the New York Public Library, particularly its Science and Technology Division, and those of *The New York Times*. A considerable portion of the material in this book was collected in the service of *The Times* and my thanks are due that newspaper for permission to draw upon this material and on art work used to illustrate it. I am also grateful to the following for permission to use previously published material:

Drawings: *Fortune*, for several drawings from the March 1961 issue, reprinted by courtesy of *Fortune Magazine*; Dr. Su-Shu Huang, for the drawing used in his article in the June 1961 issue of *Sky and Telescope*, and for two drawings used with his April 1960 article in *Scientific American; International Science and Technology* for two drawings from their October 1962 issue, pp. 55–60; Lincoln Laboratory, M.I.T., for the diagram of astronomical distances; Milton K. Munitz for two drawings from *Theories of the Universe*, Free Press, © 1957; *Scientific American* for the two drawings from the April 1960 issue, reprinted with permission; *Sky and Telescope* for the drawing in the June 1961 issue and for the diagram of Angular Momentum in the solar system, published in the January 1960 issue; George Philip & Son, Ltd., The London Geographical Institute, for star finder used as basis of a drawing; *Physics Today* for drawings on the origin of the solar system with an article by T. L. Page, vol. 1 (1948), no. 6, pp. 14–15. A number of the drawings were prepared by Andrew Sabbatini of *The New York Times*.

Text: *Of Stars and Men* by Harlow Shapley, Beacon Press, © 1958; *Interstellar Communication*, by Cameron, W. A. Benjamin, Inc., ©1963; *The Origin of Life* by Alexandr Ivanovich Oparin, translated by S. Margulis, published by Dover Publications, Inc., New York 14, N.Y. at $1.75 and reprinted through permission of the publisher. *Mars* by Percival Lowell, Houghton Mifflin Co., 1895; *God and the Astronomers* by W. R. Inge, Longmans, Green, London, 1933; *Lincos* by Hans Freudenthal, North-Holland Publishing Co., Amsterdam, 1960; *On the Nature of the Universe* by Lucretius, translated by Ronald Latham, Penguin Books, ©1951; *Origin of Life, New Biology No. 16* by J. B. S. Haldane, Penguin Books, © 1954; *Poems*, complete ed., by Alice Meynell, Chas. Scribner's Sons, 1923.

Photographs and diagrams: The California Association for Research in Astronomy (Keck Observatory); the National Radio Astronomy Observatory, Charlottesville, Virginia; The Arecibo Observatory (part of the National Astronomy and Ionosphere Center, operated by Cornell University under a cooperative agreement with the National Science Foundation); the Jet Propulsion Laboratory of the California Institute of Technology; the NASA Ames

Research Center, Moffett Field, California; the European Southern Observatory, Garching, Germany; the Martin Marietta Astronautics Group, Denver, Colorado; and Richard P. Binzel of the Massachusetts Institute of Technology for use of figures 1-1 (a, b and c) in The Spaceguard Survey, Report of the NASA International Near-Earth-Object Detection Workshop, January 25, 1992, published by the Jet Propulsion Laboratory.

Index

Page numbers in bold refer to captions.